62 Springer Series in Solid-State Sciences

Edited by Peter Fulde

Springer Series in Solid-State Sciences

Editors: M. Cardona P. Fulde H.-J. Queisser

Volume 1-39 are listed on the back inside cover

Theory of Heavy Fermions and Valence Fluctuations

Proceedings of the Eighth Taniguchi Symposium,
Shima Kanko, Japan, April 10–13, 1985

Editors: T. Kasuya and T. Saso

With 106 Figures

Springer-Verlag
Berlin Heidelberg New York Tokyo

Professor Tadao Kasuya
Dr. Tetsuro Saso

Department of Physics, Tohoku University, Aramaki Aoba,
Sendai, 980, Japan

Series Editors:

Professor Dr. Dr. h. c. Manuel Cardona
Professor Dr. Peter Fulde
Professor Dr. Hans-Joachim Queisser

Max-Planck-Institut für Festkörperforschung, Heisenbergstrasse 1
D-7000 Stuttgart 80, Fed. Rep. of Germany

ISBN-13:978-3-642-82620-7 e-ISBN-13:978-3-642-82618-4
DOI: 10.1007/978-3-642-82618-4

Softcover reprint of the hardcover 1st edition 1985

2153/3130-543210

Preface

This volume contains the proceedings of the Eighth Taniguchi International Symposium on the *Theory of Condensed Matter*, which was held at Shima Kanko Hotel in Shima, Japan, 10–13 April 1985. The topic of the Symposium was *Valence Fluctuation and Heavy Fermion Systems*, one of the most fundamental problems in present-day condensed matter physics.

The dilute Kondo problem, which is one of the most typical and unique many-body problems in condensed matter physics, developed recently into the dense Kondo and the coherent Kondo lattice problems in the 4f electron systems. It is accepted now that a large degeneracy in f-electron systems makes this latter situation possible by enhancing the single-site Kondo state relative to the inter-site magnetic interactions. Now, anomalous behavior in f-electron systems show rich variety and are called valence fluctuation phenomena as a whole. They have, however, a common feature. In the lowest temperature region, they show either heavy Fermion-like character or a narrow gap formation at the Fermi energy. Discovery of superconductivity in the heavy Fermion systems is attracting more interest. Anyway, the valence fluctuating states are thought to be of fundamental importance to bridge the gap between the localized magnetic states and the delocalized nonmagnetic states.

The present volume contains the following contributions: First there is a brief review of the experimental results and interpretations for valence fluctuation phenomena, which is followed by fundamental considerations on the various aspects of the most important interactions, the $d-f$ Coulomb interaction and the $c-f$ mixing interaction. The second contribution reviews the experimental situation for the heavy Fermion superconductivity as well as the possibility of new kinds of exotic superconductors from the experimental point of view. Then the results of the exact solution of the single-site Kondo problem are reviewed in the next contributions. The exact solution is limited to the static properties and thus various studies on dynamical properties such as transport, optical and neutron scattering are shown in the next chapters in some approximation schemes. Some preliminary work on the Kondo lattice state are shown in the following contributions and finally the possibility of new exotic superconductivity in the heavy Fermion systems is treated from the theoretical standpoint.

Following the philosophy of Mr. Taniguchi, a limited number of active researchers were invited to participate in the symposium to encourage intense discussions. In order to understand the philosophy of the Taniguchi Foundation, Mr. Taniguchi's speech delivered at the reception is given in the Foreword. We would like to express our cordial thanks to Mr. Taniguchi and the Taniguchi Foundation for giving us an opportunity to organize this successful symposium.

July 1985 *Tadao Kasuya*

Foreword

Speech by Toyosaburo Taniguchi

Dr. Kubo, Distinguished Guests, and Friends:

I am very happy, pleased and honored to be here this evening with so many distinguished guests, friends, and scholars from different parts of the world.

The Taniguchi Foundation wishes to extend a warm and sincere welcome to the many participants of the Eighth International Symposium on the Theory of Condensed Matter; this was inaugurated seven years ago through the dedicated efforts of Dr. Ryogo Kubo, who is gracing us today with his presence.

We are deeply indebted to Dr. Kubo, Dr. Kasuya, and their associates for their strenuous efforts which have been so crucial in rendering this particular symposium possible.

The success of this particular symposium shall, however, not be considered as due to the Foundation, but entirely due, I feel, to the unremitting efforts of the Organizing Committee and of all those who have lent their support to this program.

In this sense, your words of praise about the symposium, if any, should be directed to all of them. We know that it takes an enormous amount of preparation to put together the program, but their efforts pay off when they see the faces of satisfied participants, and I am also deeply appreciative of their efforts.

To date, we are proud to note that there has been a total of 53 participants in this Division from 9 countries:

Canada, Denmark, France, Ireland, Israel, Switzerland, Great Britain, the United States of America, and the Federal Republic of Germany,

with 116 participants from Japan.

And, we are all friends.

Whenever I meet with the participants in our symposia, both young and old, I am impressed by their honest dedication to their mission of pursuing the vocation they are in.

To those younger people who were made welcome by world-famous scholars, we offer our hopes that even after steadily rising to the position of respect and fame, which they will hold throughout the world in this particular field of knowledge within, say, the next 10 years from now, or 20 or 30, you will join hand in hand and help to forge closer bands of friendship and cooperation in a manner that will make a major contribution not only to academia but also to permanent world peace.

And, it is our hope that our symposia be continued as long as the fund permits, say, for several years or more.

In closing, we sincerely trust that all the participants, both Japanese and foreign, will return home, bearing pleasant memories of the Symposium and of our enjoyable time together.

Thank you.

Participants in the Taniguchi International Symposium, April 10–13, 1985 (at Shima, Japan)

Seated (*left to right*)

F. Steglich P. Schlottmann O. Gunnarsson T. M. Rice M. Tachiki T. Kasuya
K. Yosida P. Fulde P. Coleman A. C. Hewson

Second row (*left to right*)

T. Saso N. Kawakami K. Ueda K. Yamada A. Okiji Y. Kuramoto
H. Fukuyama C. M. Varma J. Kanamori

Third row (*left to right*)

K. Miyake O. Sakai S. Maekawa F. J. Ohkawa Mrs. Lotsch H. Lotsch J. Wilkins

Table of Contents

Part V Theories for Periodic Systems and Heavy Fermions

Part VI Superconductivity in Heavy Fermion Systems

Part I

Introductory Survey on Theories and Experiments

The c-f Mixing Interaction and Valence Fluctuation: Fundamental Considerations

T. Kasuya

Department of Physics, Tohoku University, Sendai 980, Japan

A brief review on the experimental studies and their interpretations for the valence fluctuation phenomena in f-electron systems, in particular for Ce-compounds, is given firstly. Fundamental characters of the intra-atomic f-f and d-f Coulomb-exchange interactions and the inter-atomic c-f mixing interactions are studied in particular in relation to the c-f mixing and the valence fluctuation. As a typical example, detailed quantitative studies for the various anomalous properties in Ce-monopnictides are explored. Finally a new type of valence fluctuation in $CeRh_3B_2$ is studied briefly.

1. Introduction

The study of rare earth compounds started practically in 1950s with the development of new technology to separate various rare earth elements. At first they were treated as a typical example of atomic-like localized magnetism due to well-localized 4f states, well screened by the 5s and 5p filled orbitals. Then the mutual magnetic interaction cannot be a direct one such as in 3d metal magnetism but should be of indirect type such as now called RKKY interaction, in which the c-f exchange interaction (c means the usual conduction band electron) originated typically by the intra-atomic d-f exchange interaction, is responsible. Detailed studies for heavy rare earth metals are typical example [1]. Discovery of Kondo state in dilute alloys changed the situation drastically, and by the end of 1960 we had the new idea that if the c-f exchange interaction is of negative sign, or antiferromagnetic, the ground-state is singlet because of the logarithmic instability at the Fermi energy, which is due to the two dimentional character of the Fermi liquid at the Fermi energy in the many-body type interaction. This means that the on-site negative c-f exchange interaction makes the localized moment unstable causing it to a singlet state [2].

In 1960s, most typical examples for dilute Kondo system were of 3d atoms such as Mn and Fe, in which we were forced to study ppm range of dilute alloys to avoid the mutual interaction effect [2]. This means that in the 3d impurity usually the mutual interaction is much stronger than the on-site Kondo binding energy, and thus the latter is easily destroyed by the former. In 1970s, however, we discovered suddenly various anomalous rare earth compounds in which the Kondo-like behavior is observed even in the dense system [3]. In these materials, different from the 3d case, the Kondo effect seems to overcome the mutual interaction. In the purely periodic systems, the ground state can not be the scattering state but should be a coherent Kondo lattice state and some of them are transformed into superconductive state. Some materials show other kinds of anomaly substantially different from the Kondo behavior in the sense that the valence is really intermediate. They may be called dynamical mixed valent system. Now we call these anomalous states as the valence fluctuating (VF) state as a whole, which is attracting a lot of theoretical and experimental physicists. In this article, we shall give a brief survey for the study of the VF state.

The actinide compounds with 5f electrons are now forming a rapidly growing field. These compounds are particularly interesting in the sense that they bridge the 3d and the 4f compounds. Some of them show the VF like behavior as in the 4f compounds and extend it to the region of more extended f character. In this sense, the 5f and 3d compounds are also treated here briefly to compare with and to study an extended view of the 4f character.

2. Brief Review of Experiments and Interpretations

Most extensively studied materials are Ce-compounds. The best way to determine the valence of rare earth (RE) atom is to analyse the double structure of L_{III} absorption spectra, that is the absorption edge structure of the 2p core electron in which each peak corresponds to the tri- or tetra-valent configuration of Ce atom [4]. The observed ratio of valency is, however, not for the initial state but for the final state. To obtain the information for the initial state, some many body effects so called shake up and down processes should be taken into account. It is now recognized that the ambiguity due to the above many body effect is within ±0.1 for the valence and the valence of any Ce-compound thus obtained does not exceed 3.3. Even for such a small region of valence change, 3 ~ 3.3, however, observed character changes very much as shown later. Note that anomaly of lattice parameter is the easiest and most common measure for the anomaly in valence but the quantitative estimation of valence based on the lattice constant gives serious error in some case [5].

In crystals, therefore, Ce atom is kept in principle in the trivalent state, that is $4f^1$. When this $4f^1$ level is well below the Fermi energy E_F, the normal magnetic behavior is expected. Note that the $4f^2$ level is always well above E_F in Ce-compounds. In reality, however, it is rather difficult to find Ce-compounds which are really normal. As the $4f^1$ level approaches the Fermi energy anomalous properties are enhanced. As a dilute alloy it shows a typical Kondo behavior and further even for a pure Ce compound it shows the so called dense Kondo behavior very often. A typical example is shown in Fig. 1 for $La_{1-x}Ce_xB_6$ series [6]. Usually the Kondo temperature T_K, which corresponds to the binding energy of the Kondo singlet state, depends sensitively on the

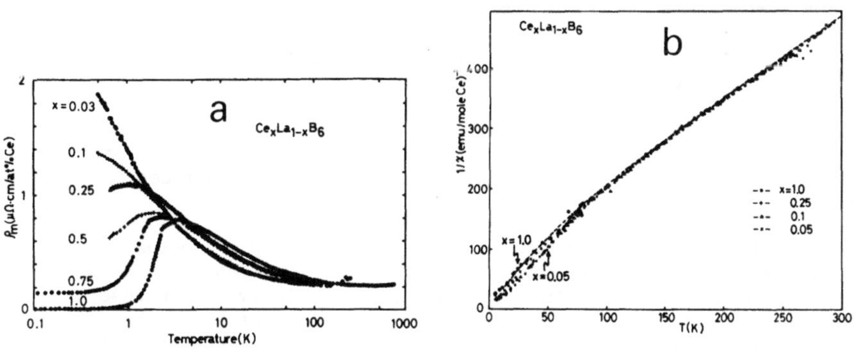

Fig.1 Magnetic resistivity $\rho_m = (\rho - \rho(LaB_6))/x$, (a), and inverse susceptibility $\chi^{-1} = x(\chi - \chi(LaB_6))^{-1}$, (b), of $Ce_xLa_{1-x}B_6$ per Ce-atom are plotted as a function of log(T) and T, respectively, for several values of x. The value of θ_p obtained from the high temperature region is 60 K, common to all the samples, indicating the Kondo effect. The Kondo temperature T_K is estimated to be 1 K for x = 0.03 sample while to be 2 K for x = 1, which is evaluated by scaling ρ vs. log(T) data at 10 to 20 K. (ref.6)

lattice constant because of the c-f mixing interaction. In the present system, the lattice parameter is primarily determined by the boron network and thus does not depend so much on RE. As the result, T_K does not change so much through the series, and thus the anomalous behavior per Ce is nearly x-independent at higher temperature than T_K indicating that each Kondo state on each Ce atom behaves nearly independently at higher temperature even for dense system. Note that because of a small value of T_K each Kondo state is extending very much in the real space on the conduction electron system. When temperature decreases beyond about T_K a kind of coherence among each Kondo state develops and the system changes the character to the so called coherent Kondo state or Kondo lattice. In particular for pure systems such as CeB_6 the resistivity should become zero at T = 0 K with a peak at around T_K. The merit to study dense Kondo system at higher temperature compared with the dilute system is as follows. Because of dense system, we can study accurately the Kondo properties up to higher temperature region including various dynamical properties which makes it possible to compare in detail with the improved theoretical results. We have now rigorous solution of impurity Kondo problem but it is still limited to the static properties [7] and for dynamical properties we should use some kind of approximation. Note that there are some serious problems on how to determine T_K experimentally. In principle, it should be determined, if possible, at higher temperature where each Kondo state behaves independently. A quasielastic-linewidth for the inelastic neutron scattering may be the best way to evaluate T_K with the help of theoretical analysis [8]. Note that even for a single Kondo system T_K depends on temperature and frequency for dynamical processes because of the crystal field splitting and the spin-orbit splitting even for the simplest system of $4f^1$ in Ce^{+++} [9].

The behavior of coherent Kondo state at low temperature may be classified roughly into three categories. The first category is typically represented by $CeAl_3$ in which nothing particularly happens and the system shows pure Kondo lattice character [10]. The low temperature behavior can be described as a heavy fermion in the sense that the resistivity, $\rho(T) = \rho(0) + AT^2$, the susceptibility, $\chi(T) = \chi(0) + BT^2$, and the specific heat, $C(T)/T = \gamma + CT^2$, behave like Fermi liquid. Note that even the single Kondo system behaves similarly [7]. In $CeAl_3$ the γ-value is about 1.5 J/mole-K^2, nearly one thousand times larger than that for the usual metals. The behavior of the diluted system also changes continuously keeping the Fermi liquid character. This means that the single Kondo character is also kept without any essential modification even in the Kondo lattice. However, the actual electronic state of the Kondo lattice is still hard to achieve theoretically now because of difficulty to introduce the coherency properly. In lowest temperature, one order of magnitude lower than T_K, C(T)/T shows a dip indicating a possibility of pseudo-gap at the Fermi energy [11]. Origin of this phenomenon is also not clear.

In the second category, a kind of magnetic order appears below T_K as is represented typically by CeB_6 in Fig. 1, in which two transition temperatures T_1 and T_2 correspond to the two kink points for resistivity. Because the magnetic order is realized from the competition with the Kondo state, the observed orders are unusual compared with the ordinary magnetic orders. In CeB_6, which crystallizes in the CaB_6 type with the simple cubic Ce lattice, the J = 5/2 level splits into Γ_8 quartet ground level and Γ_7 doublet excited level with 550 K of excitation energy. T_K for the Γ_8 Kondo state is evaluated to be about 2 K. The order below T_1 = 3.3 K is still mysterious because we can not yet obtain a consistent ordering model to explain the results of neutron scattering [12] and NMR [13] but is thought to be anyway a kind of quadrupolar ordering enhanced by a compatible antiferromagnetic dipolar ordering induced by external magnetic field. The ordering below $T_2 = 2.3$ K is now well known from the neutron scattering, a complicated two-Q antiferromagnetic order. However, still the NMR result is not easily explained. The magnetic ordering in this region is understood as an induced magnetic order on the Kondo lattice. Actually, the properties above 0.6 K to T_2 is governed mostly by the magnons but below 0.5 K the Fermi liquid character of the Kondo lattice dominates as shown in Fig. 2 [6a,14]. The same results were also observed in $CeAl_2$, a typical dense Kondo system with Γ_7 ground level [15].

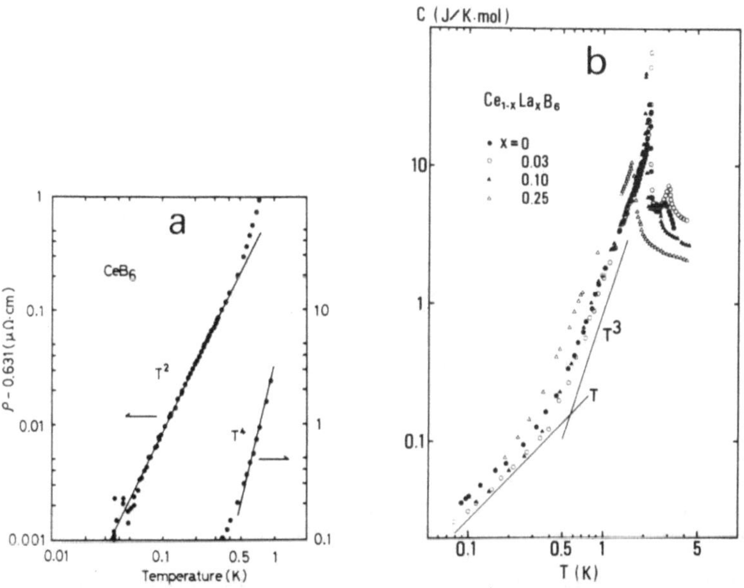

Fig.2 Magnetic part of resistivity $\rho - \rho(0)$ for CeB_6 (a), and the specific heat C for $Ce_{1-x}La_xB_6$, (b), are plotted as functions of temperature in log-log scale. (ref.6a for (a) and ref.14 for (b))

The third category has only one example so far, $CeCu_2Si_2$, in which superconductivity appears at 0.6 K [11,16]. $CeCu_2Si_2$ is a typical dense Kondo system with T_K of about 5 K and thus with a large γ value, the order of 1 J/mole-K^2. This is the first example of heavy fermion superconductivity and attracting much interest. The detail is given in the following paper [11,16]. As the origin of superconductivity, paramagnon effect is thought to be important, because it is very near the magnetic order and then rather a triplet superconductivity like He^3 is thought to be realized. But the Kondo enhanced electron-phonon coupling is also thought to be important [11,16], which will cause usual singlet superconductor, but heavily distorted by the Kondo effect. However, because we still do not know the real character of Kondo lattice, main part is left for future studies. The question is rather why we have so few examples even though we have many dense Kondo systems. Note that every dense Kondo material in category 2 becomes category 1 by dilution. For example in $La_xCe_{1-x}B_6$ no magnetic order appears in zero field for $x > 0.5$, in which γ-value increases with decreasing magnetic moment and becomes largest at the critical region for the vanishing magnetism due to paramagnon enhancement. This is clearly seen in $CeSi_x$ series, which changes from the

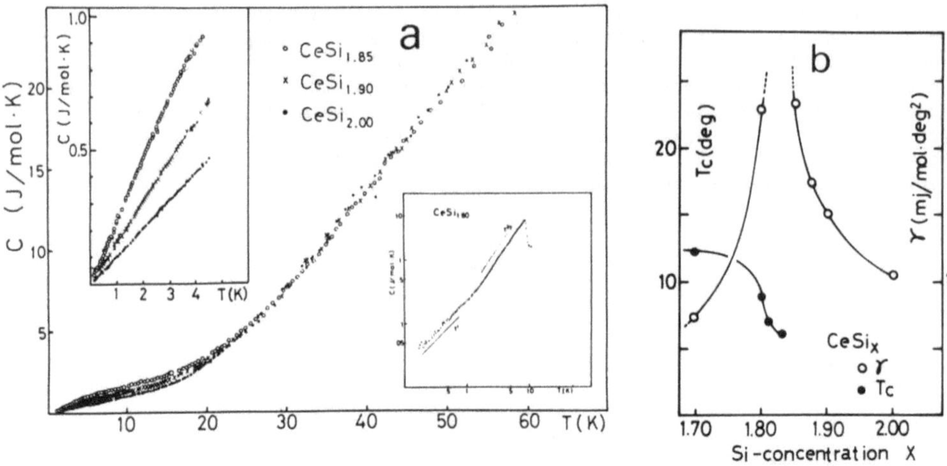

Fig.3 Graphs of specific heat for $CeSi_x$ are shown for x = 2, 1.9 and 1.85 in (a). The samples for x < 1.85 show ferromagnetism at low temperatures. The left hand insertion shows low temperature region for T-linear behavior. The right hand insertion shows the data for x = 1.80 indicating T-linear behavior at lowest temperature. (b) shows the values for γ and T_c, the ferromagnetic Curie temperature. The larger reduction of γ for x = 1.70 is due to larger induced moment. The value of T_K decreases continuously with decreasing value for x. Peak of γ at x = 1.85 is due to paramagnon effect. (ref.17)

category 1 for x = 1 to category 2 in decreasing value of x as seen in Fig. 3 [17]. In this sense we have a lot of heavy fermion systems without magnetic ordering, good candidates for heavy fermion superconductor.

When the $4f^1$ level approaches E_F further, the valence fluctuation state is realized. For simple one impurity problem, there are no essential differences between the Kondo regime and the valence fluctuation regime because they are scaled by T_K. In real materials, however, significant differences are observed. Typical example is the γ-α transition in Ce metal, in which both phases are of the same fcc structure but the volume is about 12% smaller in α-phase and thus the γ-phase belongs to the dense Kondo regime but the α-phase belongs to the valence fluctuation regime [18]. More drastic difference is observed in CeN compared with other Ce-monopnictides, as seen later. In α-Ce, various properties are similar to the usual metals, showing no visible anomalies, so that the promotion model was proposed before in which the $4f^1$ level is above the Fermi energy and thus the 4f electrons are transferred to the 5d conduction bands leaving tetravalent Ce ($4f^0$) configuration [18]. This picture was rejected by the L_{III} absorption measurement mentioned before as well as by the photoemission spectra, PES, and bremsstrahlung isochromat spectra, BIS, or inverse PES [19]. In PES the 4f spectra for α and γ Ce are essentially the same, manifesting the double peak structure as shown in Fig. 4. This means that the occupied 4f spectra seen by PES, which corresponds to the one particle 4f Green's function, show no essential change in the γ-α transition. Therefore, the possible model for the γ-α transition should be the following. The volume decrease causes no essential change in the position of 4f level relative to E_F but changes the c-f mixing matrix substantially, which causes large difference in T_K [20]. Actually T_K in α-Ce is thought to be about 2000 K, much larger than the crystal field splitting and comparable with the spin-orbit splitting. The observed normal properties are simply due to this

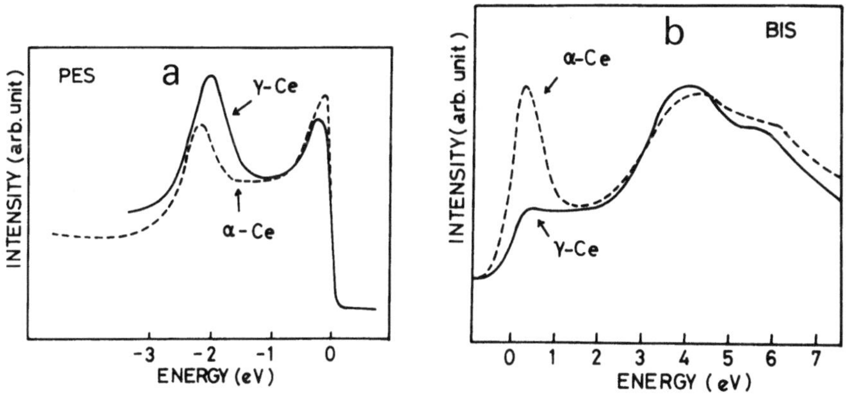

Fig.4 PES (a) and BIS (b) spectra for α, dashed lines, and γ, solid lines, Ce. The similar behavior of PES with Ce-monopnictides shown in Fig.11 indicates the same mechanism for the double peak structure. On the other hand, the large peak near E_F in BIS for α-Ce indicates the VF, or Kondo effect. The peak at 4 eV is due to $4f^2$ level. (ref.19)

large T_K value. Recent detailed study of the polarized inelastic neutron scattering on α-Ce also supports this picture [21]. There are many materials in this category to show superconductivity including α-Ce. The character of superconductivity in these materials is essentially the same as that of usual singlet superconductor [11,16]. Of course the Kondo regime and the VF regime are continuous and there are many materials belonging to the boundary between them in which the γ-value is around 100 mJ/mole-K^2. More systematic studies of these boundary materials are needed.

An interesting and important question for the Kondo lattice, or the heavy fermion systems in more general sense, is the Fermi surface character. So far no experimental results are reported on the real heavy fermion systems because of their heavy masses. In any case a strong magnetic field is necessary but it may cause serious destruction of the Kondo character. Therefore theoretical cooperation is essentially needed. There is one experimental result for $CeSn_3$ which belongs to the boundary of the Kondo and VF regime, or may be better to classify into the VF regime because its γ-value is about 20 mJ/mole-K^2. The result is similar to that of $LaSn_3$ in important parts [22]. There is no accepted theory to calculate the Fermi surface in Kondo lattice system but some preliminary crude models are proposed [23] including the simple band calculation for the trivalent Ce ion [24], which gives, however, the value of about one for the average occupation number of the 4f states. Ce-monopnictides are another interesting example and will be treated later more in detail.

Yb-compounds show similar properties as Ce-compounds in the sense of electron-hole symmetry for the 4f states, in which Yb^{++} ($4f^{14}$) corresponds to Ce^{++++} ($4f^0$). However, the $4f^1$-hole level in Yb-compound situates in general much closer to the Fermi energy than that for $4f^1$ electron level in Ce-compound and thus in Yb-compounds it is possible to have really divalent materials such as YbB_6, which is very similar to SrB_6. Differently from α-Ce, L_{III} spectra, PES and BIS are interpreted as the simple divalent material. This difference is also partly due to the smaller c-f mixing matrix, about a factor two smaller than that in Ce-compounds [25]. Then we have an interesting question of what happens when the 4f level is just on the Fermi energy. A typical material for this category is YbB_{12} [26]. In this material a gap is

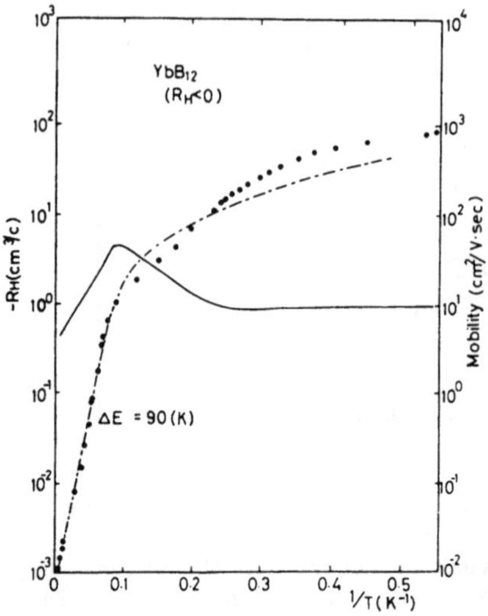

Fig.5 Hall constant R_H and mobility of YbB_{12} are shown as functions of inverse temperature 1/T. The mobility is obtained simply by the product of resistivity ρ and the Hall constant, R_H, from the experimental data. The dot-dashed line is the simple model calculation with the gap ΔE = 90 K. The better fitting is obtained by the two gap model, ΔE_1 = 90 K for the conduction band and ΔE_2 = 20 K for the f band. The lowest temperature region is governed by the impurity band conduction, which always shows metallic conduction with a finite mobility. (ref.26)

really open at the Fermi energy with a width of about 100 K as seen in various transport properties as well as in the specific heat and T_1 for NMR. Some examples are shown in Fig. 5 [26]. For the origin of the gap the Wigner lattice formation due to f-f interatomic interaction and c-f mixing mechanism are proposed [27]. It is clear that the both should be important and thus the issue is which is more important and what properties are really calculated from the models. So far, however, the situation of calculation is still poor and not enough to predict quantitatively [26,27]. In the diluted materials, $Yb_{1-x}Lu_xB_{12}$, the above properties change gradually to that of dense Kondo and dilute Kondo systems [28].

Sm-compounds also show valence fluctuating properties. The most typical example is SmB_6, which shows properties very similar to YbB_{12} mentioned before [27]. However, the dense Kondo-like behavior seen in Ce and Yb-compounds is hard to be seen in Sm-compounds. The reason is thought to be as follows. The value of T_K increases exponentially as the degeneracy ν of the magnetic states to form the Kondo singlet increases, that is, $T_K \propto W\exp[-W/\nu J]$, where W is the band width and J is the effective c-f interaction parameter. In Ce ν is fourteen making T_K large compared with the inter-atomic magnetic interaction, which does not include ν and smallest in Ce. On the other hand in the 3d compounds such as Fe and Mn, the orbital freedom is quenched by the strong crystal field and only the spin freedom remains, which is rather large. This makes the inter-atomic magnetic interaction energy much larger than the intra-atomic Kondo binding energy. This is the reason to reduce the concentration of 3d magnetic impurity far below percent range to observe the Kondo effect. The situations in Sm and Eu are rather similar to the 3d compounds considered above,which makes dense Kondo behavior difficult. The dense Kondo behavior is thus observed only when the 4f level is very near the Fermi energy which makes T_K large. Typical examples may be Sm_4X_3, in which X = Bi, Sb and As. As shown in Fig. 6 [29], the behavior of resistivity is of typical dense Kondo. Both Sm_4As_3 and Sm_4Sb_3 show ferromagnetic transition at about 150 K but its effect on the resistivity is very small showing that the Kondo character changes very little, or the induced moment is small, even though the peak of the specific heat is large. The value of T_K is also thought to be about 150 K. The situation in Eu seems to be similar to Sm because both Eu^{++} ($4f^7$) and Sm^{+++} ($4f^5$) are magnetic, S = 7/2 and J = 5/2, respectively, in the ground states and Eu^{+++} ($4f^6$) and Sm^{++} ($4f^6$) are singlet ground state, J = 0, with the excited J = 1 level at about 300 K. The valence fluctuation behavior in Eu-compounds are, however, very much different from that in Sm-compounds. This difference stems primarily from the non-magnetic character in the trivalent Eu. The valence fluctuating Eu-compounds are usually simple singlet at lowest temperature because of Eu^{+++}. With increasing temperature the valency changes rapidly into divalency because of lattice expansion and entropy term, both of which favor divalency. At high temperature, simple divalent Eu-compounds are realized. It is strange, however, that even in the transient temperature region no anomalous behavior such as SmB_6 type or the dense Kondo type is observed. More studies are needed to clarify this point.

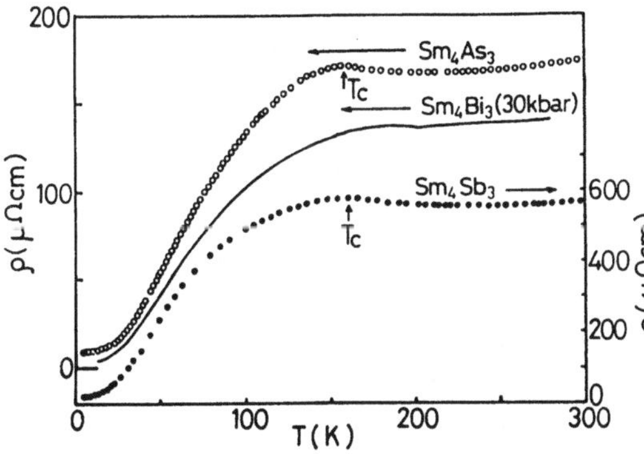

Fig.6 Electrical resistivities for Sm_4As_3, Sm_4Sb_3 and Sm_4Bi_3 under pressure of 30 kbar are shown. In the normal pressure Sm_4Bi_3 is in a typical mixed valence state. Note the very similar behavior among three materials. The ferromagnetic Curie temperature T_c, which is not yet measured in Sm_4Bi_3, is shown by arrows. In low temperature, the T^2 law is applicable for $\rho(T)$. (ref.29)

Actinide compounds are now attracting much attention because they bridge between the 4f and 3d systems. Actually some compounds, in particular heavy actinide compounds, show normal localized 5f character like the normal 4f system, some compounds, in particular light actinides, are like 3d compounds showing band 5f character, and some compounds show VF characters. A typical example to show the dense Kondo behavior is UBe_{13}, which is described in detail in the following article [11]. T_K is evaluated to be several degree Kelvin. However, the feature of PES and BIS for 5f is substantially different from that of Ce-compounds mentioned before. This has only a single peak with a peak position about 1 eV above E_F with a half width of about 4 eV [30]. This structure looks to suggest the 5f band structure but the 5f band width by the usual band calculation gives a width of less than a half of the experimental value as shown in Fig. 7 [31]. A PES experiment performed at low temperature with a better resolution shows that a sharp peak grows at the Fermi energy [32]. This may be related with the Kondo state but the width is still two orders of magnitude larger than that expected from T_K = 5 K. In this sense, U_3X_4, in which X means P, As, Sb and Bi, may be easier to understand.

Various properties of U_3X_4 are similar to those of Sm_4X_3 mentioned before as shown in Fig. 8 for resistivity [33]. The Curie temperature for ferromagnetic ordering with weak moment is also similar. Various properties are understood, therefore, as the weakly induced ferromagnetic state on the Kondo lattice with T_K of order of 100 K for the tetra valence of U. Heavy fermion character is actually observed with a strange magnetic behavior [34]. The feature of PES is also interpreted as the two peaks structure similar to that in Ce-compounds studied before, even though the lower peak is weak [35]. The magnetic ordering is again complicated, not a simple ferromagnetism.

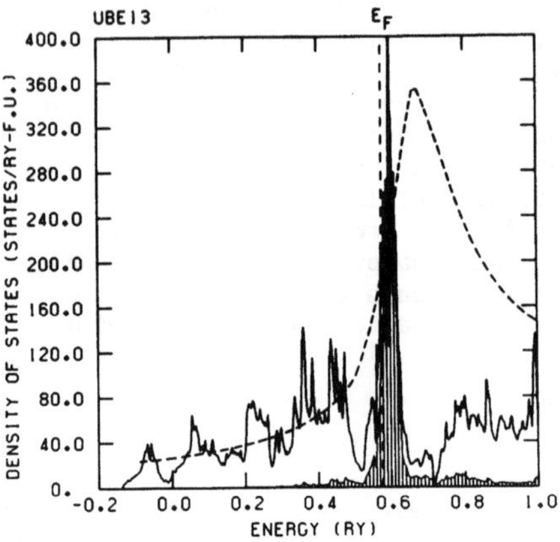

Fig.7 Density of states for UBe_{13} obtained by the usual band calculation is shown together with the PES-BIS experiment, a dashed line (ref.30). The hatched part shows the component of f character. The Fermi energy is shown by the vertical dashed line. The calculation does not include the spin-orbit interaction. The 5f spin-orbit interaction is expected to make the 5f band width wider by about 1 eV. (ref.31)

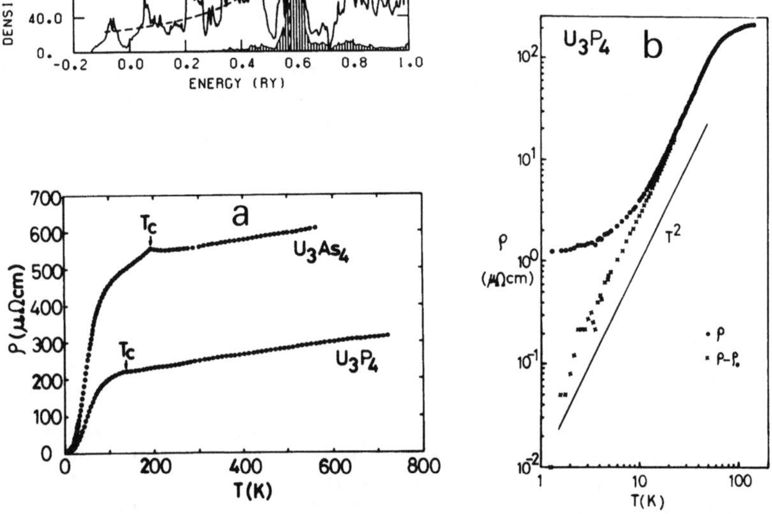

Fig.8 Electrical resistivities for U_3As_4 and U_3P_4 are shown (a). The ferromagnetic Curie temperatures T_c are shown by arrows. The resistivity of U_3P_4 in the ordered region is shown by log-log plot in (b). The T^2 law applies up to 50 K. (ref.33)

To summarize this chapter, the valence fluctuating states have some common features in one aspect but have also rich variety depending on the atomic configuration as well as on various competitions with other interactions such as magnetic order, spin fluctuation, superconductivity, polaron effect etc. In principle, it has the most fundamental interest in the sense that it covers just the transient region from the localization and magnetism to the delocalization and non-magnetism.

3. Intra-atomic Interaction

It is clear that intra-atomic d-f exchange interaction is ferromagnetic, or of positive sign, and thus the negative sign responsible to the Kondo effect should come from the c-f mixing effect. The c-f mixing effect is clearly of inter-atomic character due to the symmetry respect and thus much more complicated than the intra-atomic interactions. At first, therefore, we survey the important fundamental interactions in rare earth compounds.

The fundamental interactions may be classified into intra- and inter-atomic interactions. Among intra-atomic interactions the largest one is the f-f Coulomb interaction. All the parameters for the intra-atomic interactions are known from the analysis of the atomic optical data. For the f-f Coulomb interaction, the spherical part U_{ff}, or the so called correlation energy, changes from about 15 eV in Ce to 17 eV in Yb [36]. These values are nearly equal to the correlation energy in the 3d electrons for Ni because the extension of wave function is similar for 3d and 4f, while the 5f wave function spreads a little bit more than 4f and thus the correlation energy in the 5f electrons around U is several eV smaller than the corresponding one of 4f. As well known, the multipole and exchange f-f Coulomb interactions together with the spin-orbit interaction make the various J-multiplet levels. For single electron or hole, Ce^{+++} or Yb^{+++}, the spin-orbit split between $J = 5/2$ and $7/2$ is 0.27 eV or 1.27 eV, respectively. In the 5f electron the spin-orbit split increases in general and is about 1 eV, for example, in U. In the 4f systems, the L-S coupling scheme is well applicable but in the 5f systems, for example in U, some excited levels are rather well described by the j-j coupling scheme because of the larger spin-orbit and the smaller multipole-exchange interactions. These J level schemes are usually well conserved in the 4f systems even in the crystals as seen by spectroscopies. However, the correlation energy is modified strongly by various screening mechanisms, in particular by the intra-atomic d-f Coulomb interaction.

The intra d-f Coulomb interaction may be written as

$$H_{df} = U_{df} - 2J_0\vec{S}_d\vec{S}_f + \sum_{k=1}^{4}\{A_k - B_k(\frac{1}{2} + \frac{1}{S}\vec{S}_d\vec{S}_f)\}t_k(u_d^{(k)}u_f^{(k)}) \qquad (1)$$

in which the first term is the spherical Coulomb term, the second the spherical exchange, the third the multipole-multipole and the fourth the anisotropic exchange interactions. U_{df} changes from 10 eV in Ce to 12 eV in Yb. When the bare f-f Coulomb energy U_{ff} is screened perfectly by the d-f Coulomb interaction, the corresponding process is $2(f^nd^1) \to f^{n-1}d^2 + f^{n+1}d^0$ and then the required energy, which may be called the effective f-f correlation energy in the atomic model, is about 5 eV both in Ce and Yb. In the crystals, the screening is not so perfect and thus the effective f-f correlation energy U_{eff} obtained naively from PES and BIS is about 5 ~ 8 eV. In any case, these values are sufficiently larger than the c-f mixing energy, or the simple f band width, and thus guarantee the atomic configuration even in the crystals. In the 3d systems, for example in Ni, the energy difference $2(d^9s^1) \to d^{10}s^0 + d^8s^2$ is rather negative, which guarantees the 3d band picture in the usual sense. The 5f system is just in between the 4f and 3d systems in the present problem, too. The all parameters shown in Eq. (1) are known from the fitting of the atomic optical data. Note that the screening process is dynamical. Equation (1) contributes both the scattering processes and the self-energies in various forms including the single site and many sites mutual interaction types.

4. The c-f Mixing Interaction and f-Level

The most important inter-atomic interaction is the so called mixing interaction. This interaction was first introduced in the present problem by ANDERSON [37]. However, the definition of mixing interaction is vague and controversial because it depends on the definition of the basal states, which have further the non-orthogonality problem. We believe that, if it is given by one particle model, the usual band calculation should give the best answer because it is thought to be best for one particle model. To have definite values, the result of the so called first principle band calculation is translated to the tight binding model with the orthogonal plane waves (OPW). Then various transfer integrals, that is the mixing matrices, as well as the level positions are obtained. For the f-level position, however, the band calculation does not give a reliable value because the strong correlation energy is not so well treated in the present stage of the band calculation. Therefore, it is usually better to determine the f level position from PES and BIS measurements. However, because of strong correlation and the related strong many-body effect, these spectra are also not so simple and thus careful analysis is needed.

In many-body systems, in general, the concept such as f level position has no meaning in the rigorous sense because it is the one particle concept. The usual definition for f-level in many-body systems may be obtained from the one particle f-Green function, which is nothing but PES and BIS for f-character. If there is a well defined single peak for each of them, the occupied and unoccupied f levels are defined naturally. Unfortunately, in the f-electron systems in general, the situation is not so simple and very often more than two peaks appear.

The mechanisms for multiple peaks are thought to be as follows [38,48]. When one f electron is removed, various processes to screen the created one positive charge occur. The most important one is the intra-atomic d-f Coulomb interaction as was mentioned before. The probability to have such a well screened state is reduced by the overlapping integral between the initial and the final states. The situation is similar to the zero-phonon transition probability in the well known electron-phonon screening problem. The biggest difference in these two cases is the difference in the electron mass in the former and the ionic mass in the latter, which causes the much smaller reduction in the present problem. The many-phonon process in the latter corresponds to the many-electron-hole pair excitation process in the former making a long tail in the deeper energy side. This peak, well screened by the d-f Coulomb interaction, may be called the effective f level in the crystal. However, usually, the process does not stop at this stage because of the c-f mixing mechanism. Note that as shown in Appendix the d-f screening effect is not restricted to the metals with occupied d-bands but also applicable equally well to a not-so-large band gap insulator with vacant d-band because of the d-characters in the filled valence bands through the strong p-d mixing matrix element at least comparable with the band gap energy. A typical example is seen in Ce-monopnictides as shown later.

The c-f mixing effects may be classified into the real process and the virtual process, or the screening process. The former process is as follows. Because of the c-f mixing interaction, the f-hole is filled in the final state again where a hole is created in the valence band with many-electron-hole pairs. This is a kind of resonant process in which the f-hole state is the intermediate states. When the c-f mixing effect is small, this gives simply a width for the above f-hole peak. However, because of the large degeneracy of the f-hole, in particular 14-fold degeneracy for the case of Ce^{+++}, even if each c-f matrix is small, the resultant f hole life time becomes so short that the life time broadening can be nearly equal to the typical valence band width in many cases. Then the PES peak for the f-hole can not be seen at the f-level position but appears at the both edges of the valence band. This is the origin of the two peak structure observed in various rare earth compounds, in particular for the PES of light rare earth compounds such as Ce, Pr and Nd in which the f-hole degeneracy is large. When the c-f mixing becomes still larger, these two peaks leave the band edges and become real bound states when there are gaps beyond the valence band. Or, when the Fermi energy exists on the valence-conduction bands, the system becomes the mixed valence type valence fluctuating state. Note that for each c-f mixing process, the dynamical d-f Coulomb screening considered before works causing the same reduction on the c-f mixing matrix element. Therefore, in total, the fourth power of the overlapping integral, $|\langle i|f\rangle|^4$, is multiplied in the present process compared with the static d-f screening model. As shown in Appendix, however, this reduction is not so large; in typical cases only about 30%. As shown in the next section, this reduction factor agrees very well with the experimental fact.

The screening effect, or the self-energy term, of the c-f mixing interaction is more complicated than the d-f Coulomb screening, or the usual screening processes, treated before. At first, we consider the initial state in which the f-electrons are occupied. Because of the c-f mixing interaction, one f-electron is transferred to one of the unoccupied conduction electron states, which are situated above the Fermi energy at zero temperature. Then the process in which each occupied conduction electron is transferred into the f hole and comes back as shown in Fig. 9a occurs. This electron-hole screening process is very unique in the sense that, if in the initial process the f electron is transferred just on the Fermi energy, this screening yields logarithmic divergence. This is the origin of the Kondo state caused from the instability at the Fermi energy. Note that this is really a typical many-body mechanism because when the f-electron is not transferred to the conduction band this mechanism does not occur. In the present approximation, the Kondo state is given by the diagram shown in Fig. 9b, in which the self energy of the η-th magnetic state is given by Fig. 9a. In the so called non-crossing approximation (NCA), Σ_η is given by Fig. 9c. In this case the diagrams of Figs. 9b and 9c should be solved simultaneously. NCA may be a better approximation but has a trouble in giving singular Kondo state at zero-temperature. These problems will be discussed in more detail in the following articles [39]. When the temperature is much higher than the Kondo binding energy typically given by the Kondo temperature T_K, this log-divergent problem becomes less important. Then the mixing with the states of the higher energies but higher density of states and stronger mixing matrices becomes important.

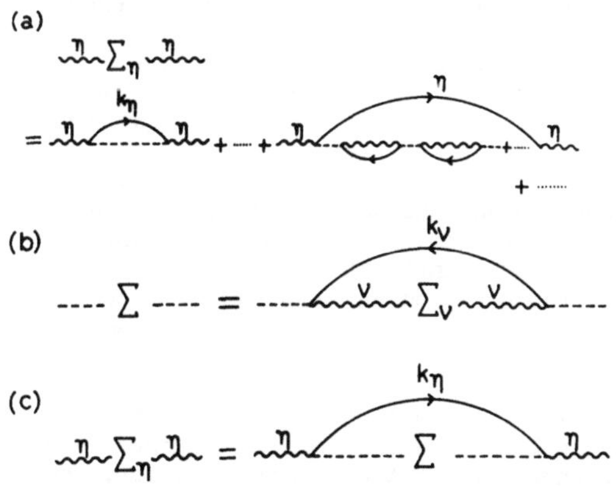

Fig.9 Diagrams for the magnetic state with the η-th symmetry (a) and the Kondo singlet state (b). In (a), the first term of the right side is the usual second order self-energy by the c-f mixing while the higher order terms include the screening by the electron-hole pair shown by a bubble. In NCA scheme (c), this screening is not due to a simple bubble but the Kondo singlet.

This is nothing but the crystal field energy by c-f mixing process in which the above c-f mixing screening process gives a substantial amount of shift up of the effective f-level energy in the usual second order process as shown later. Because the initial state has a substantial amount of states in which one f electron is transferred to conduction electron states about T_K above the Fermi energy, the BIS to put one electron on the f state has a peak at about T_K above the Fermi energy [38,39]. In the same way, PES has also a peak at E_F. The peak at E_F has not yet been observed clearly in PES except in a few cases [40] but is observed clearly by BIS for typical VF materials [41].

To summarize the problem of f level, the natural definition may be the well Coulomb screened f-hole and electron levels. When the c-f mixing is small, these levels correspond to the peak of PES for f-hole and BIS for f-electron. When the c-f mixing effect is strong, PES and BIS structures become complicated. The double peak structure is very common in PES for light rare earth compounds, which is, however, not related with VF directly, while the double peak structure in BIS relates with VF. Then a careful analysis is needed to evaluate the above defined f-levels from PES and BIS. Sometimes, the effective f level is defined as the self-consistent level for various mechanisms including the c-f mixing, such as the band calculation of the Fermi surface for conduction electrons and the lowest order indirect f-f interactions as well as the lowest order crystal field calculation. As became clear from the precedent consideration, thus defined level includes the c-f mixing screening effect which causes more than 1 eV shift up of the effective f-level in the usual Ce-compounds compared with the Coulomb-screened level mentioned before. A typical example will be shown in the next section.

5. Typical Examples

As a typical example, we consider rare earth-monopnictides, REX, in which RE means a rare earth atom and X a pnictogen, N, P, As, Sb, or Bi. In particular we treat CeX. The band calculation shows that in the compounds with the lighter pnictogens a narrow gap opens between the valence bands made mostly by the p-states on the pnictogen with a maximum at Γ-point and the conduction bands, the lower part of which is made mostly by the $5d\text{-}t_{2g}$ state on the rare earth atom with minima at three X-points [42]. The Fermi surfaces obtained from the band calculation agree very well with the de Haas van Alphen measurements on LaSb and LaBi [43]. All the mixing parameters are well estimated from the band calculation as shown in Fig. 10 [44]. The various mixing parameters as well as the level positions obtained by the tight binding model fitting of the APW ab-initio calculation are listed in Table in [38b]. It should be noted that the p-f mixing matrix in CeX is about three-quarters of that in LaX because the f-wave function shrinks as the number of occupied f-electron increases, in particular rapidly from La to Ce [45]. It should be noted, however, that except La-compounds the simple 4f band calculation is not applicable but the difference between the occupied and unoccupied states should be included correctly. In this sense we should be careful to use the values except La.

Two peak structure is seen in PES clearly in all of the Ce-monopnictides [46] except CeN [47], which shows typical valence fluctuation of mixed valence type. Because all the information is known except the f-level position, the PES is fit by changing the f-level position. Then, if we use the static Coulomb screening model and thus use the full mixing values, the upper peak is pushed up above the Fermi energy even in CeAs because of too large mixing matrix. However, when the mixing effect is reduced by 30% as was studied before due to the dynamical d-f screening effect, we obtain very good

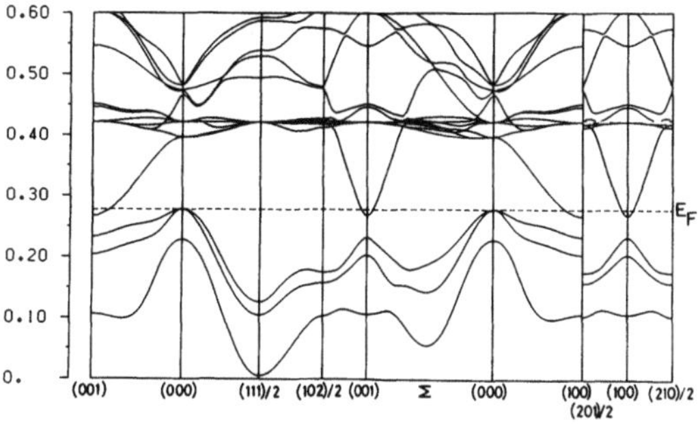

Fig.10 Tight binding band calculation for LaSb. The c-f mixing matrices, for example, are obtained by comparing the result without the c-f mixing. The spin-orbit interactions are included. The top of the valence band with the Γ_8 symmetry is pushed down by the c-f mixing with the f level with the same Γ_8 symmetry, which is shifted up by 0.4 eV at the Γ-point as seen in this figure. In ferromagnetic CeSb, the main part of the Γ_8 f level is occupied, situated below the Fermi energy. Energy is given in Rydberg and the abscissa is given in units of $2\pi/a$. (ref.44)

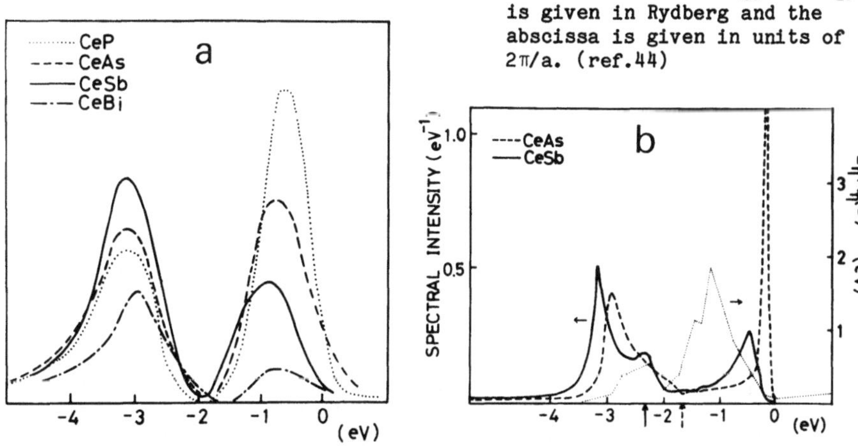

Fig.11 Experimental results of PES for Ce-monopnictides are shown in (a) (ref.46). The spectral intensity in (a) is given in arbitrary unit. The calculated results for CeSb and CeAs are shown in (b) (ref.38b,48). The mixing intensity, the mixing matrix times density of states, for CeSb is also shown by dotted line. The f level positions are shown by the arrows. The peaks of BIS for CeSb and CeAs are at 4.5 eV and 5.5 eV above the Fermi energy, respectively. (ref.49)

agreement with experiment as shown in Fig. 11 [38b,48]. The peak positions do not change from CeBi to CeP because they are fixed at two edges of the valence band while the intensity ratio changes systematically because the f-level approaches to E_F from Bi to P, which is consistent with the peak of BIS [49]. In this case, only in CeN, the upper peak is pushed up above the Fermi energy in agreement with the experimental fact that CeN is the mixed valence type valence fluctuating material. This is due to both the shallowest f-level position and the strongest c-f mixing in CeN. The strong c-f mixing in CeN is mainly due to the strong d-f mixing matrix because of the smallest lattice constant in CeN. In CeBi and CeSb, the d-f mixing is much smaller than the p-f mixing.

Because the f-level position is fixed, we can now calculate everything. At first, the crystal field energies in Nd, Pr and Ce-monopnictides are calculated. As was discussed before, the effective f-level is shifted due to the c-f mixing screening effect for the second order perturbation scheme. This shift is now calculated to be about 1 eV in Ce-monopnictides but much smaller in Pr and Nd because in Pr and Nd the f-level positions are much deeper [50]. The crystal field splittings in rare earth monopnictides and monochalcogenides were studied by neutron inelastic scattering and it was reported that the result for the fourth-order term can be interpreted very well by the point-change model with the effective charge Z of about unity except Ce-compounds, in which the crystal field splitting is anomalously small [51]. For example in CeP and CeAs, which are normal magnetic materials, the crystal field splittings are less than half of the values extrapolated from the values for Pr and Nd compounds. This anomaly is now naturally explained by the c-f mixing model without any adjustable parameter. At first, it is shown quite generally that the contributions form the point charge model and the d-f Coulomb-exchange term, the fourth-order terms in Eq. (1), are nearly cancelled out leaving much smaller values than the experimentally observed values [50]. Therefore, the c-f mixing term should be the main term. The actual calculation shows that in Pr and Nd compounds the main contribution is due to the mixing process from the occupied valence electrons to the vacant f-level making f^2 configuration. The scaling law found in the point charge model is then well reproduced. Furthermore, the ratio between the sixth-order term and the fourth-order term agree very well with experimental results. Note that this ratio is zero for the d-f Coulomb-exchange mechanism because there are no sixth-order term and also very small in the point charge model due to the factor $\langle r_f^2 \rangle / R^2 \sim 0.01$, in which $\langle r_f^2 \rangle$ is the square average extension of the 4f wave function and R the lattice constant. On the other hand in the c-f mixing mechanism there are no such small parameter to cause reduction in the higher order terms [50,52]. Note also that for the conduction electron scattering, for example the analysis of the temperature dependence of electrical resistivity, the d-f Coulomb-exchange terms become main contribution because now the lower-order terms are responsible in which the d-f interaction dominates the c-f mixing interaction [53]. In the Ce-compounds, however, because the effective f-level is shifted up to the Fermi energy, the $f \to c$ transfer contribution becomes important, cancelling out about a half of the $c \to f$ contribution mentioned before causing again excellent agreement with the experiment. It should be noted that the shift of the effective f-level due to the c-f mixing screening is essential, which is, however, not seen by PES directly. In the same way, anomalous magnetic properties in CeBi and CeSb are explained naturally again without any adjustable parameter [50,54]. In these materials the p-f mixing interaction with the valence holes of a few percent per Ce atom created at around the Γ-point due to a weak overlapping of the conduction and the valence bands is essentially important. Because the Fermi energy is comparable with the c-f mixing interaction this p-f mixing effect is strongly non-linear, which is the origin of the anomalous properties in these materials. Again the shift of the effective f-level which is now about 1.5 eV, larger than that for the crystal field calculation because the valence holes are near the Fermi energy, is essentially important. The result of the band calculation for the ferromagnetic state of $J_z = 5/2$ is shown in Fig. 12 [55]. The J = 5/2 level position is the effective level shifted by the c-f mixing screening and thus does not correspond to PES. Detailed correspondence with PES is given in the following article [38b]. The result of Fig. 12 should correspond to the Fermi surface experiment. The top of the valence bands is distorted strongly by the non-linear p-f mixing effect. This distorted Fermi surface was really observed experimentally by de Haas van Alphen effect [43a,56]. The cylindrical part extending along the Γ-X axis is most characteristic.

Both CeBi and CeSb show typical dense Kondo behavior above the ordering temperature with the similar T_K values, evaluated to be about 20 ~ 50 K. These

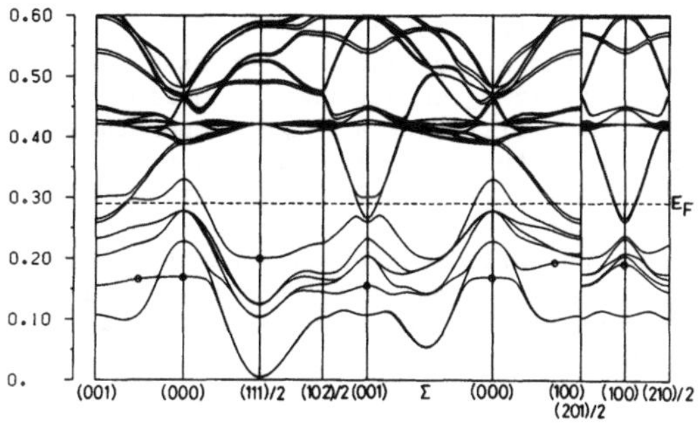

Fig.12 The tight binding band calculation of CeSb in the ferromagnetic state for J_z = 5/2. The effective f level position for J_z = 5/2 is at 1 eV below the Fermi energy. The bands of predominantly J_z = 5/2 f character are marked by the open circles. Note that the fairly large dispersion is due to the p-f mixing interaction. All other f levels are put at 1.5 eV above the Fermi energy. One particular p-band which mixes fully with the J_z = 5/2 f state is pushed up above the Fermi energy at the Γ-point. This particular band extends along the Γ-(001) axis making a characteristic Fermi surface. Energy is given in Rydberg. (ref.55)

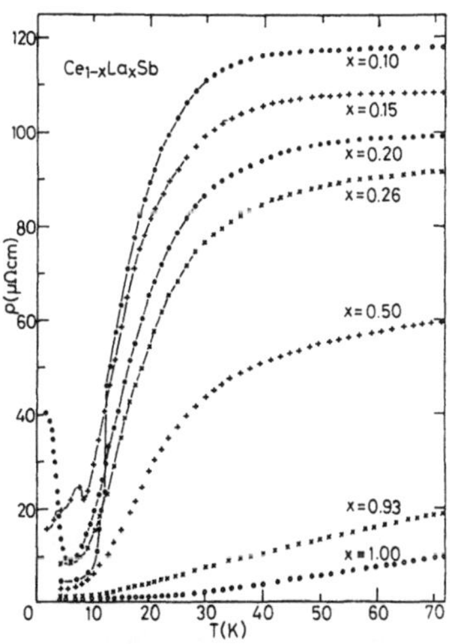

Fig.13 Electrical resistivities of $Ce_{1-x}La_xSb$ for various values of x are shown as functions of temperature. The magnetic resistivity, ρ-ρ(LaSb), shows dense Kondo behavior except x = 0.93 with a peak well above the magnetic ordering temperature, which is 15 K for CeSb and decreases quickly with increase of x. (ref.57)

are unique in the sense that only a few percent of conduction electrons kill all the Ce moments causing a Kondo lattice at low temperature. Actually the resistivity has a peak well above the ordering temperature indicating clearly existence of the coherent Kondo state. In particular in the ordered states the Kondo layers are observed making various types of stacking with the magnetic layers. It is also strange that in the dilute limit, x → 0.03 in $La_{1-x}Ce_xSb$, no dilute Kondo behavior is observed as shown in Fig. 13 [57]. These are explained as follows. In CeSb the ground crystal field level is Γ_7 and the Γ_8 level has an excitation energy of about 38 K. The valence holes at Γ-point mix only with Γ_8 strongly through the p-f mixing while the conduction electrons around X-points mix mostly with Γ_7 through the weak d-f mixing. Due to the strong non-linear p-f mixing effect the Γ_8 ordering appears as the first-order transition. Then, due to the strong anisotropy of the p-f mixing matrix, the moments align ferromagnetically in the plane perpendicular to the z-axis with the moment along the z-axis. Due to the cylindrical dispersion along the z-direction as shown in Fig. 12, the inter-layer interaction is very weak causing very complicated phase diagram due to various different types of stacking. In CeSb, the Kondo layer, which is the Γ_7 ground level Kondo state, also participates the variation of the stacking. The phase diagrams of CeSb and CeBi for low temperature region are explained very well by the p-f mixing model again without any adjustable parameter [50]. The Kondo state seen at higher temperature is clearly due mainly to the Γ_8 levels through the strong p-f mixing. The vanishing Kondo effect in the temperature dependence of resistivity in the dilute alloys is due to the increase of the crystal field splitting which is caused by the weaker p-f mixing effect due to smaller valence hole number with dilution [50]. The resistivity at high temperature is governed by conduction band electrons because they practically do not participate the Γ_8 Kondo state, while the resistivity at low temperature is governed by the valence hole conduction because they do not participate in the Γ_7 Kondo state existing at low temperature. The Hall constant measurement clearly shows the above change of main carrier [58]. Note that the Kondo temperature T_K is given as the binding energy of the singlet state as shown in Fig. 9. Then T_K may be given approximately as [59]

$$T_K = E_f + \sum_\eta \lambda_\eta J_\eta \ln[D/(T_K + \delta_\eta)] \tag{2}$$

$$J_\eta = \rho(E_F)|V_{\eta F}|^2/(1 - \Sigma_\eta') \tag{3}$$

where E_f is the lowest level energy, λ_η the degeneracy for the η-th level, $\rho(E_F)$ the density of states at E_F, $|V_{\eta F}|^2$ the square average of the c-f mixing matrix for η-th level at E_F, D the band width, Σ_η' the derivative of the η-th level self-energy at E_F and $\delta_\eta = \delta_\eta{}^0/(1 - \Sigma_\eta')$ is the corrected level splitting of η-th level measured from the ground level. When δ_η is smaller than T_K, δ_η may be neglected and η-th level may be included in the ground level. Then T_K is further approximated as

$$T_K = D\exp[E_{eff}/\sum_\eta \lambda_\eta J_\eta] \quad \text{where} \tag{4}$$

$$E_{eff} = E_f + \sum_\nu \lambda_\nu J_\nu \ln(D/\delta_\nu) \tag{5}$$

is the effective ground level energy for $\delta_\eta < T_K$, while the ν-summation runs on the levels for $\delta_\nu > T_K$. The above formula means that, even if the d-f (Γ_7) mixing is small, the Γ_7 Kondo temperature can be large because the effective Γ_7 level for the Kondo state is shifted strongly by the p-f(Γ_8) mixing as shown in Eq. (5) making E_{eff} very small. This screening is similar to those considered before, but much larger here, because for the Kondo singlet state the mixing with the states with energies of T_K and δ_η above the Fermi energy is concerned and thus the factor $\ln(D/\delta_\eta)$ appears. Actual calculation agrees well with experimental values. Note that to calculate the detailed Kondo properties a simplified model is necessary and in many cases only the ground doublet level is treated. In such a case the level energy is taken to be only a couple of hundred degrees [60]. This is due to the result in Eq. (5) but we should be very careful whether the same picture is applicable or not for the dynamical problems.

To summarize the detailed treatment of Ce-monopnictides, we should emphasize the fact that CeX is really a typical example in which the c-f mixing is very simple because it follows the symmetry conservation very well due to the holes being at Γ-point and the electrons at X-points. This makes the treatment simple and transparent. It is also important to note that the c-f mixing has really various interesting aspects and all these properties should be treated self-consistently. The Kondo problem, or the valence fluctuation problem, is related with these problems undividedly.

Finally a new material $CeRh_3B_2$ is mentioned briefly. This material shows the dense Kondo behavior above room temperature in the temperature dependences of resistivity and susceptibility, a large negative θ_p value [61]. Below room

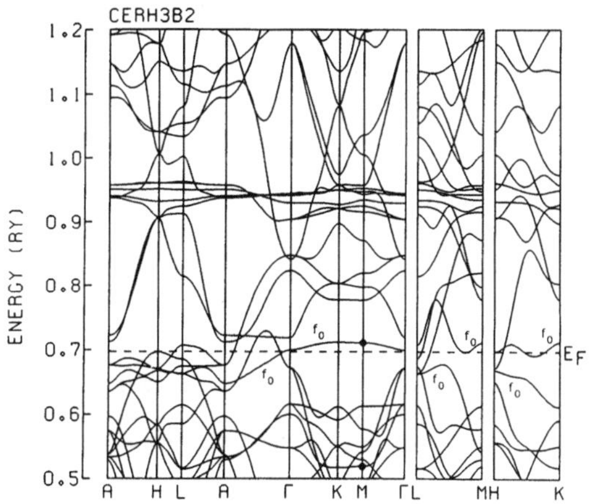

Fig.14 The APW band calculation for $CeRh_3B_2$ in the ferromagnetic state for f_0 (l_z = 0) state occupied. The bands with predominant f_0 character are marked by f_0. Note the large dispersion along the z-axis, which is caused by the direct (ffσ) mixing along the chain. While the weak dispersions on the planes perpendicular to the z-axis, namely the Γ-K-M and A-H-L planes, are caused mostly by the d-f mixing typically shown by the solid circles on the M-point. The bands below E_F are mostly d-bands on Rh sites. (ref.62)

temperature, these behaviors change rapidly and the material becomes ferromagnetic at 120 K, which is unusually high, the highest among the Ce-compounds and two orders of magnitude larger than that expected from other $RERh_3B_2$. In this material Ce atoms form chains in which the nearest neighbour distance along each chain is unusually short. Our band calculation shows that the direct f-f overlap along the chain is large and a real f band with a large dispersion along the chain is formed as shown in Fig. 14 for the $l_z = 0$ level, which is extending along the chain direction [62]. The ferromagnetism, which is stabilized by creating holes at the top of that $l_z = 0$ band, can explain various anomalous properties in this material. This state is also a kind of valence fluctuation but clearly different from those considered so far. This material is interesting, therefore, in the sense to bridge typical VF in the 4f system to VF in the 5f and 3d systems.

Acknowledgements

The author deeply appreciate various supports by O. Sakai, K. Takegahara, T. Saso and M. Takeshige.

Appendix

We consider the case that the valence p-bands are filled and the conduction d-bands separated from the valence bands with a gap are vacant. The model is also applicable to Ce-monopnictides treated in detail in the text. To make the calculation simpler, we pick up only one rare earth site from which an f-electron is removed and the surrounding nearest neighbour anions. We also consider only d-symmetry molecular orbits and assume that the ten-fold degenerate bonding orbits belong to the valence bands and are occupied. Then the initial and the final states, in which an f-electron is removed, may be written as

$$|i\nu\rangle = \alpha\psi_{d\nu} + \sqrt{(1-\alpha^2)}\psi_{p\nu}, \tag{A1}$$

$$|f\nu\rangle = \beta\psi_{d\nu} + \sqrt{(1-\beta^2)}\psi_{p\nu}, \tag{A2}$$

where $\psi_{d\nu}$ is the ν-th d orbital state and $\psi_{p\nu}$ is the molecular orbital of d_ν symmetry made by p-states on anions. The number of d states on the Ce site should increase by one in the final state. Therefore the following equation holds,

$$10\beta^2 = 10\alpha^2 + 1 \tag{A3}$$

and then the total overlapping integral is evaluated to be

$$\langle i|f\rangle = \langle i\nu|f\nu\rangle^{10} \simeq [1 - 0.01/8\alpha^2(1 - \alpha^2)]^{10} \tag{A4}$$

in which $\beta^2 - \alpha^2 = 0.1$ is treated as the expansion parameter in the last equation. When the p-d mixing is of the same order of the energy separation between the d and p levels, α^2 is larger than 0.1. In CeSb, α^2 is evaluated to be about 0.2. Then $\langle i|f\rangle$ becomes 1 - 0.08 and $|\langle i|f\rangle|^4$ about 0.7. A small reduction of $\langle i|f\rangle$ from unity is due to the large degeneracy of d-orbitals and the strong p-d mixing interaction.

References

1. T. Kasuya: in Magnetism, ed. by G. T. Rado and H. Suhl (Academic Press, New York 1966) vol. IIB, p. 215
2. For a review see: G. T. Rado and H. Suhl (ed) Magnetism (Academic Press, New York 1973) vol. V

3a. Valence Instabilities and Related Narrow-Band Phenomena, ed. by R. D. Parks (Plenum Press, New York 1977)

3b. Valence Fluctuations in Solids, ed. by L. M. Falicov, W. Hanke and M. B. Maple (North-Holland, Amsterdam 1981)

3c. Valence Instabilities, ed. by P. Wachter and H. Boppart (North-Holland, Amsterdam 1982)

3d. Proc. of the 4th Int. Conf. on Valence Fluctuations, ed. by E. Müller-Hartmann, B. Roden and D. Wohlleben, published in J. Magn. Magn. Mat. 47&48 (1985)

3e. Proc. of the 5th Int. Conf. on Crystalline Field and Anomalous Mixing Effects in f-Electron Systems, Sendai, 1985, to be published in J. Magn. Magn. Mat.

4. J. Röhler: J. Magn. Magn. Mat. 47&48, 175 (1985), and references therein
5. M. Kasaya, J. M. Tarascon and J. Etourneau: Solid State Commun. 33, 1005 (1980)
   H. Boppart and P. Wachter: Phys. Rev. Lett. 53, 1759 (1984)

6a. N. Sato, A. Sumiyama, S. Kunii, H. Nagano and T. Kasuya: J. Phys. Soc. Jpn. 54, 1923 (1985)

6b. N. Sato, S. Kunii, I. Oguro, T. Komatsubara and T. Kasuya: J. Phys. Soc. Jpn. 53, 3967 (1984)

7. A. Okiji and N. Kawakami: in this volume
   N. Kawakami and A. Okiji: in this volume
   P. Schlottmann: in this volume
8. S. Maekawa, S. Kashiba, S. Takahashi and M. Tachiki: in this volume
9. B. Cornut and B. Coqblin: Phys. Rev. B5, 4541 (1972)
   K. Yamada and K. Yosida: in this volume
10. For a recent review see: N. B. Brandt and V. V. Moshchalkov: Adv. Phys. 33, 373 (1984)
11. F. Steglich: in this volume
12. J. M. Effantin, J. Rossat-Mignod, P. Burlet, H. Bartholin, S. Kunii and T. Kasuya: J. Magn. Magn. Mat. 47&48, 145 (1985)

13. M. Takigawa, H. Yasuoka, T. Tanaka and Y. Ishizawa: J. Phys. Soc. Jpn. 52, 728 (1983)
14. T. Furuno, N. Sato, S. Kunii, T. Kasuya and W. Sasaki: J. Phys. Soc. Jpn. 54, 1899 (1985)
15. F. Steglich, C. D. Bredl, M. Loewenhaupt and K. D. Schotte: J. de Physique 40, C5-301 (1979)
16. P. Fulde: in this volume
17. H. Yashima and T. Satoh: Solid State Commun. 41, 723 (1982)
H. Yashima, N. Sato, H. Mori and T. Satoh: Solid State Commun. 43, 595 (1982)
18. K. A. Gschneidner, Jr.: in [3a], p. 89
19. E. Wuilloud, H. R. Moser, W. -D. Schneider and Y. Baer: Phys. Rev. B28, 7354 (1983)
D. M. Wieliczka, C. G. Olson and D. W. Lynch: Phys. Rev. B29, 3028 (1984)
20. J. W. Allen and R. M. Martin: Phys. Rev. Lett. 49, 1106 (1982)
21. R. M. Galera, D. Givord, J. Pierre, A. P. Murani, C. Vettier and K. R. A. Ziebeck: J. Magn. Magn. Mat. 47&48, 139 (1985), and to be published
22. W. R. Johanson, G. W. Crabtree, D. D. Koelling, A. S. Edelstein and O. D. McMasters: J. Appl. Phys. 52, 2134 (1981)
23. H. Harima and T. Kasuya: to be published in [3e]
24. A. Yanase: J. Magn. Magn. Mat. 31-34, 453 (1983)
25. H. Harima and T. Kasuya: to be published in [3e]
26. M. Kasaya, F. Iga, M. Takigawa and T. Kasuya: J. Magn. Magn. Mat. 47&48, 429 (1985)
27. T. Kasuya, M. Kasaya, K. Takegahara, T. Fujita, T. Goto, A. Tamaki, M. Takigawa and H. Yasuoka: J. Magn. Magn. Mat. 31-34, 447 (1983)
28. F. Iga, M. Kasaya and T. Kasuya: to be published in [3e]
29. A. Ochiai, T. Suzuki and T. Kasuya: to be published in [3e]
30. E. Wuilloud, Y. Baer, H. R. Ott, Z. Fisk and J. L. Smith: Phys. Rev. B29, 5228 (1984)
31. K. Takegahara, H. Harima and T. Kasuya: J. Magn. Magn. Mat. 47&48, 263 (1985) and to be submitted to J. Phys. F:Metal Phys.
32. A. J. Arko, C. G. Olson, D. M. Wieliczka, Z. Fisk and J. L. Smith: Phys. Rev. Lett. 53, 2050 (1984)
33. N. Niitsuma, S. Takagi, T. Suzuki and T. Kasuya: unpublished data
34. S. Takagi, N. Niitsuma, T. Suzuki and T. Kasuya: to be published in [3e]
35. S. Suga, M. Yamamoto, K. Soda, T. Mori, S. Takagi, N. Niitsuma, T. Suzuki and T. Kasuya: to be published in [3e]
36. W. C. Martin, R. Zalubas and L. Hagan: Atomic Energy Levels-The Rare-Earth Elements (National Bureau of Standards, Washington 1978)
37. P. W. Anderson: Phys. Rev. 124, 41 (1961)
38a. O. Sakai, H. Takahashi, M. Takeshige and T. Kasuya: Solid State Commun. 52, 997 (1984)
38b. M. Takeshige, O. Sakai and T. Kasuya: in this volume
39. Y. Kuramoto: in this volume
40. H. Sugawara, A. Kakizaki, I. Nagakura, T. Ishii, T. Komatsubara and T. Kasuya: J. Phys. Soc. Jpn. 51, 915 (1982)
41. F. U. Hillebrecht, J. C. Fuggle, G. A. Sawatzky, M. Campagna, O. Gunnarsson and K. Schonhammer: Phys. Rev. B30, 1777 (1984)
42. A. Hasegawa: J. Phys. C:Solid St. Phys. 13, 6147 (1980)
43a. H. Kitazawa, T. Suzuki, M. Sera, I. Oguro, A. Yanase, A. Hasegawa and T. Kasuya: J. Magn. Magn. Mat. 31-34, 421 (1983)
43b. A. Hasegawa: J. Phys. Soc. Jpn. 54, 677 (1985)
44. O. Sakai and M. Takeshige: private communication
45. H. Harima and T. Kasuya: to be published in [3e]

46. W. Gudat, M. Iwan, R. Pinchaux and F. Hulliger: in [3c] p. 249
47. E. Wuilloud, B. Delley, W. -D. Schneider and Y. Baer: J. Magn. Magn. Mat. 47&48, 197 (1985)
48. M. Takeshige, O. Sakai and T. Kasuya: to be published in [3e]
49. F. U. Hillebrecht, W. Gudat, N. Martensson, D. D. Sarma and M. Campagna: J. Magn. Magn. Mat. 47&48, 221 (1985)
50. H. Takahashi and T. Kasuya: J. Phys. C:Solid St. Phys. 18, 2697, 2709, 2721, 2731, 2745, 2755 (1985)
51. R. J. Birgeneau, E. Bucher, J. P. Maita, L. Passell and K. C. Turberfield: Phys. Rev. B8, 5345 (1973)
52. K. Takegahara, A. Yanase and T. Kasuya: in Crystalline Electric Field Effects in f-Electron Magnetism, ed. by R. P. Guertin, W. Suski and Z. Zolnierek (Plenum, New York 1982) p. 533
53. Y. Aoki: thesis, Tohoku University 1983 (unpublished)
54. K. Takegahara, H. Takahashi, A. Yanase and T. Kasuya: Solid State Commun. 39, 857 (1981)
55. O. Sakai, M. Takeshige, H. Harima, K. Otaki and T. Kasuya: to be published in [3e]
56. H. Aoki, G. W. Crabtree, W. Joss and F. Hulliger: to be published in [3e]
57. M. Sera: thesis, Tohoku University 1982 (unpublished)
58. M. Sera, T. Suzuki and T. Kasuya: J. Magn. Magn. Mat. 31-34, 385 (1983)
59. T. Kasuya: Physics Monthly 5, 350 (1984) (in Japanese)
60. B. Coqblin, A. K. Bhattacharjee, B. Cornut, F. Gonzalez-Jimenez, J. R. Iglesias-Sicardi and R. Jullien: J. Magn. Magn. Mat. 3, 67 (1976)
61. K. N. Yang, M. S. Torikachvili, M. B. Maple and H. C. Ku: J. Low Temp. Phys. 56, 601 (1984)
62. K. Takegahara, H. Harima and T. Kasuya: submitted to J. Phys. Soc. Jpn.

Heavy-Fermion Superconductivity: Experimental Status Report

F. Steglich

Institut für Festkörperphysik, Technische Hochschule Darmstadt
D-6100 Darmstadt, Fed. Rep. of Germany

New observations are reported for the heavy-fermion superconductors $CeCu_2Si_2$, UBe_{13} and UPt_3. These results are discussed with respect to the following issues: (i) the competition between different cooperative phenomena in $CeCu_2Si_2$; (ii) the question whether the whole Fermi surface contributes to the superconducting transition in the three compounds; (iii) the nature of the superconducting order parameter. Taking into account more published data for these systems, we can delineate the present status of the field: $CeCu_2Si_2$ behaves as a singlet superconductor with possibly anisotropic Cooper pairing. UPt_3 appears to possess an anisotropic order parameter, either of the singlet or triplet type. No definite conclusions can yet be drawn for UBe_{13}, mainly because its superconducting properties are severely modified by field and temperature dependences of the normal-state parameters.

1. Introduction

Within the last few years an increasing number of lanthanide- and actinide-based intermetallic compounds have been found to behave as "heavy-fermion" systems [1,2]. Inherent to these materials is the existence of f-derived magnetic moments above some characteristic temperature $T^* \simeq 10K$, whereas below T^* they approach a seemingly "simple" metallic state, however, with quite unusual parameter values. For example, the linear specific-heat coefficient γ is typically of the order of several hundred mJ/K^2mole and can even exceed $1J/K^2mole$, a number that has to be compared with $\gamma \lesssim 1mJ/K^2mole$ for simple metals like Cu. From these extraordinary γ- values the existence of low-lying heavy-mass excitations ("heavy fermions") is commonly concluded. Only a few of the heavy-fermion systems, however, keep these FERMI-liquid properties down to very low temperatures. Famous examples are UAl_2 [3] and $CeAl_3$ [4], recently joined by $CeCu_6$ [5] and $CeCu_4Al_8$ [6]. In most cases, like in $NpSn_3$ [7], $CeAl_2$ [8], CeB_6 [9], U_2Zn_{17} [10] and UCd_{11} [11], the groundstate is a magnetically ordered one, usually of complex nature.

The main impact on the field derived from the discovery of "heavy-fermion superconductivity" (HFS) in the tetragonal high-γ compound $CeCu_2Si_2$ [12]. From gigantic values of the specific-heat jump and the absolute slope of the upper critical magnetic field, B'_{c2}, at T_c [13], which both scale with $\gamma \simeq 1J/K^2mole$, the existence of COOPER-pairs formed by heavy fermions has been concluded. It was very re-assuring that this exotic groundstate was subsequently discovered also in two binary actinide-based compounds, cubic UBe_{13} (OTT, RUDIGIER, FISK and SMITH, Ref.14) and hexagonal UPt_3 (STEWART, FISK, WILLIS and SMITH, Ref.15). Though presently known for only three systems, HFS must be considered a general phenomenon: this is suggested by differences not only in the crystal structures, but also in many normal (n)-state properties of these intermetallics.

Since an extensive review on HFS has recently been published by STEWART [2], we will focus in the present paper on a few selected issues of timely interest: (i) Other cooperative phenomena, e.g. magnetic order [16,17] and the formation of a coherent Fermi-liquid phase [18,19], have been observed in certain (n)-state $CeCu_2Si_2$ samples. Their relationship to HFS is the subject of Sect.2. (ii) FULDE et al. [20] have calculated the quasiparticle bandstructure for $CeCu_2Si_2$ exhibiting both heavy and light masses. To investigate in which way these different types of quasiparticles contribute to the superconducting low-T phase, we shall compare in Sect.3 those terms in the specific heat, $C_S(T)$, and thermal conductivity, $\kappa_S(T)$, which are linear in temperature. Such a comparison is also made for the U-based systems. (iii) $CeCu_2Si_2$, UBe_{13} and UPt_3 have been considered the first triplet (odd-parity) superconductors [21-24], though there are as many arguments in favor of singlet (even-parity) superconductivity [20,25-30]. To gain insight into the nature of the Cooper-pair states in HFS, Sections 3 and 4 are devoted to a discussion of the full T-dependences of $C_S(T)$ and $\kappa_S(T)$ as well as of $B_{c2}(T)$. In Sect.5 we compare the three systems by taking into account other superconducting properties, and we conclude the paper by addressing some major open problems in the field.

2. Systematics in the Low-Temperature Behavior of $CeCu_2Si_2$

Early investigations of HFS in $CeCu_2Si_2$ have been rendered difficult by the following puzzling observations: (1) T_c of stoichiometric polycrystalline samples can vary between 0.7K and < 60mK [12,31,32]. (2) Single crystals grown from stoichiometric melts (hereafter: "stoichiometric" single crystals) do not superconduct down to the mK-regime [18,33-35]. (3) Polycrystals with T_c < 0.5K, though bulk superconductors according to the Meissner effect, do not show a jump in the specific heat at T_c, but exhibit a giant linear term in the zero-field specific heat well below T_c [18]. Obviously, these samples behave as "gapless superconductors" with a density of states (DOS) near the Fermi level, E_F, being almost identical with the DOS in the normal state. By taking into account more recent experimental results we shall suggest below that these "sample dependences" in the low-temperature behavior of $CeCu_2Si_2$ originate in an interesting competition between different cooperative phenomena, i.e. HFS, magnetic ordering and "coherence" effects.

MOSHCHALKOV et al. [36] have combined finite-pressure and alloying experiments to demonstrate the role of the atomic volume in determining the relationship between magnetic order and HFS in $CeCu_2Si_2$. Their results on polycrystalline $Ce_{1-x}La_xCu_2Si_2$ alloys and on "stoichiometric" single crystals measured at varying external pressures are reproduced in Fig. 1. La-concentrations and pressure values are converted [37] into relative volume changes $(V-V_0)/V_0$. Doping with La leads to an expansion of the lattice and, for $x \geq 0.1$, to a complete disappearance of superconductivity [39]. Rather, "spin-glass" type cusps are observed in the low-field ac-susceptibility [36]. The "freezing temperature" T_m passes over a maximum and drops suddenly when V approaches V_0 (Fig. 1). Such a volume dependence of T_m is expected [40] to occur in "KONDO-lattice" systems precursive to the "magnetic instability" at which the Kondo-singlet formation becomes more efficient than the interionic RUDERMAN-KITTEL-KASUYA-YOSIDA (RKKY) interaction. Most interestingly, HFS in $CeCu_2Si_2$ seems to develop right at this magnetic instability; this is inferred from the observation in [36] that superconductivity in "stoichiometric" single crystals sets in, if an external pressure of nothing more than 1 kbar ($|\Delta V|/V_0 \simeq 0.1\%$) is applied. In addition, T_c is found to "saturate" at about 0.5K for $p \gtrsim 4$ kbar [41].

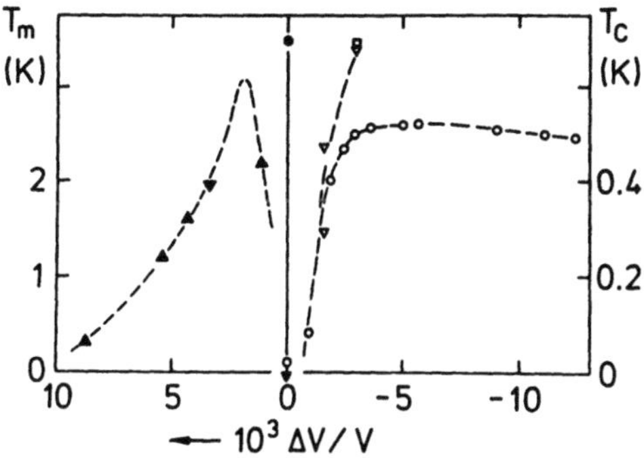

Fig.1 Magnetic-ordering temperatures T_m (closed symbols, left scale) and superconductive transition temperatures T_c (open symbols, right scale) of $CeCu_2Si_2$ and homologs as a function of relative volume change $\Delta V/V_0 = (V-V_0)/V_0$ with $V = a^2c$. $V_0 = 167.43$ Å^3 is the unit-cell volume of "stoichiometric" single crystals at ambient pressure [38]. ●, ○ : "stoichiometric" single crystals at $p \geq 0$ [33,38]. ◻ : single crystal grown from 26at% Cu excess. ▽ : polycrystals prepared either with 10at% Cu excess (T_c = 0.68K) or stoichiometrically (T_c < 0.5K). ▼ : polycrystal of $CeCu_{1.9}Si_2$. ▲ : polycrystals of $Ce_{1-x}La_xCu_2Si_2$, V interpolated between V = 169.25Å^3 for $LaCu_2Si_2$ and V_0 [36]. Dashed lines are intended as guides to the eye

The importance of volume suggested in Ref. 36 is strongly supported by results on $CeCu_2Si_2$ samples with varying stoichiometry. Independent investigations by ISHIKAWA et al. [42] and SPILLE et al. [39] revealed that V can be systematically varied by changing the Cu-concentration x in polycrystalline $CeCu_xSi_2$: Cu-deficit results in an expansion and Cu-excess in a compression of the lattice. Consequently, a $CeCu_{2.2}Si_2$ sample shows bulk superconductivity with high T_c (0.68K), whereas for $CeCu_{1.9}Si_2$ no superconductivity is found; instead, as susceptibility and specific-heat results consistently show (Figs. 2a and 3a), cooperative "freezing" of some (presumably ferromagnetic-cluster) moments occurs at T_m = 1.9K. The correlations discovered for polycrystalline samples could subsequently be utilized to prepare bulk superconducting single crystals [38,39]. Such a crystal, grown by a BRIDGMAN-technique from a melt containing 26at% Cu-excess, exhibits almost the same values for V and T_c (Fig. 1) and very similar susceptibility [43] and specific-heat [44] behavior as the $CeCu_{2.2}Si_2$ polycrystal, cf. Figs. 2c and 3c with Figs. 2b and 3b. On the other hand, a "stoichiometric" single crystal that does not superconduct down to 20mK, shows $\chi(T)$ and C(T) anomalies at T_m = 3.5K signaling an inhomogeneously broadened transition into a long-range antiferromagnetically ordered state [45]. Note that, apart from a higher T_m-value, the anomalies for the single crystal shown in Figs. 2c and 3c differ qualitatively from the corresponding "spin-glass" anomalies for the $CeCu_{1.9}Si_2$ polycrystal.

The above comparison between different samples leads to the following picture: In the Kondo-lattice system $CeCu_2Si_2$, some attractive interaction exists between the heavy fermions, which is masked by magnetic ordering, but becomes dominant via a small compression of the volume, already. The reason for the large T_c-variation in this system is not yet clear. There are indications [18,39] that for "gapless-superconducting" samples (i.e. those with T_c <0.5K) the transition temperature might track the volume dependence of the Kondo temperature T_K. Furthermore, volume compression in $CeCu_2Si_2$ is found to be associated with a remarkable reduction of the

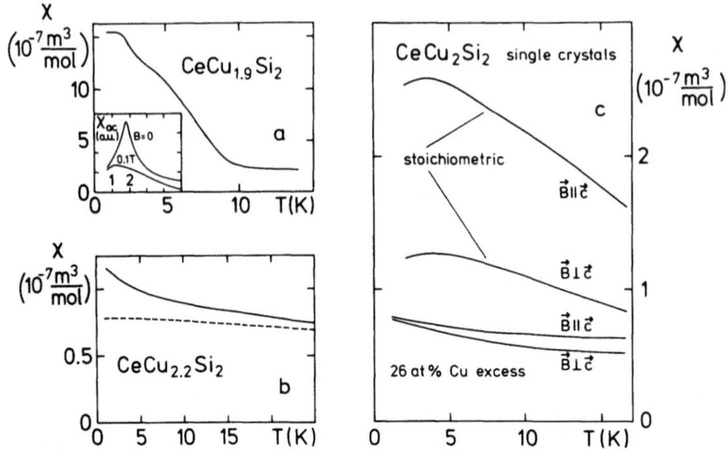

Fig.2 dc-susceptibility χ_{dc} as a function of temperature for $CeCu_2Si_2$. a: $CeCu_{1.9}Si_2$ polycrystal [17]; inset shows χ_{ac} vs. T at B_{dc} = 0 and 0.1T, respectively. b: $CeCu_{2.2}Si_2$ polycrystal [17]; solid line: raw data; dashed line: intrinsic χ, after correction for "impurity contribution". c: single crystals grown from either stoichiometric melt [38] or with 26at% Cu-excess [43]. Measurements were done with different orientations of the field relative to the tetragonal c-axis

residual resistivity [46]. This raises the question of the relationship between HFS and spatial coherence.

In a periodic Kondo lattice, like $CeAl_3$ [47,19,48-50], n-state $CeCu_2Si_2$ [18,19,48,49], $CeCu_6$ [51], $CeRu_2Si_2$ [52] or $CeCu_4Al_8$[6], distinct low-temperature maxima exist in the specific heat, divided by T, as well as in the thermopower. These effects which can be removed by moderate alloying [19,48,49], were ascribed to a "conerence-derived" pseudogap in the resonant DOS of a Kondo lattice near E_F [19,53]; its half-width being measured by the temperature T_0 at which the peak in C(T)/T occurs. For $CeCu_2Si_2$, T_0 ranges between 0.3 and 0.4K, see the B = 0 result of two gapless-superconducting samples in Fig. 4 and the data taken for $CeCu_{2.2}Si_2$ in an overcritical field B = 2T in Fig. 6a.

In the following we discuss some polycrystal results which suggest a distinct influence of the "coherence-gap" formation on superconducting properties of $CeCu_2Si_2$: (i) In the absence of a maximum in $C_n(T)/T$ (cf. Fig. 4), no superconductivity is found down to 20mK. Thus, the existence of a coherence gap appears as a necessary condition for HFS in $CeCu_2Si_2$. On the other hand, (ii) the pronounced C(T)/T variation as observed for the gapless-superconducting sample No. 10 (Fig. 4) suggests a rather sharp gap structure in the DOS at E_F which, in turn, may be the cause for the relatively low T_c (0.47K) of this sample. One may, therefore, speculate that the maximum T_c ($\lesssim$ 0.7K) found for samples like $CeCu_{2.2}Si_2$ results from a compromise between the formation of a coherence gap on the one hand and some smearing of this DOS structure due to the Cu-excess on the other. (iii) Since "gapless superconductivity" is found for samples with $T_c \lesssim T_0$ only, formation of a superconducting gap is inhibited in the presence of a well defined coherence gap.

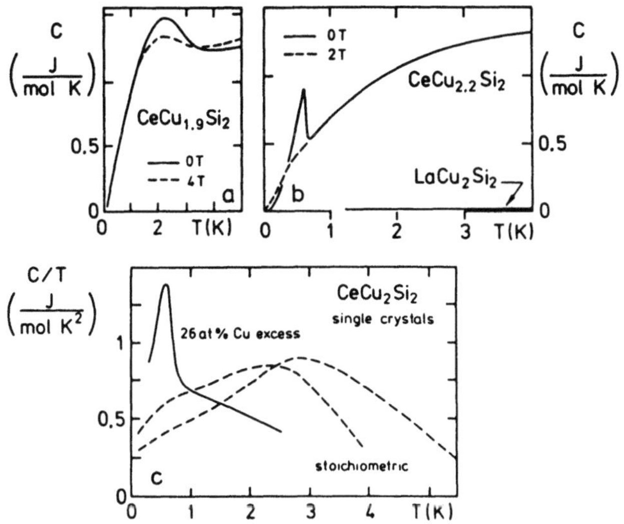

Fig.3 Specific heat C vs. T for $CeCu_2Si_2$ [44]. a: $CeCu_{1.9}Si_2$ polycrystal at B = 0 and 4T. b: $CeCu_{2.2}Si_2$ polycrystal at B = 0 and 2T as well as $LaCu_2Si_2$ at B = 0. c: C/T vs. T plot for single crystals grown from stoichiometric melts and with 26at% Cu excess

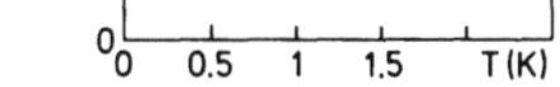

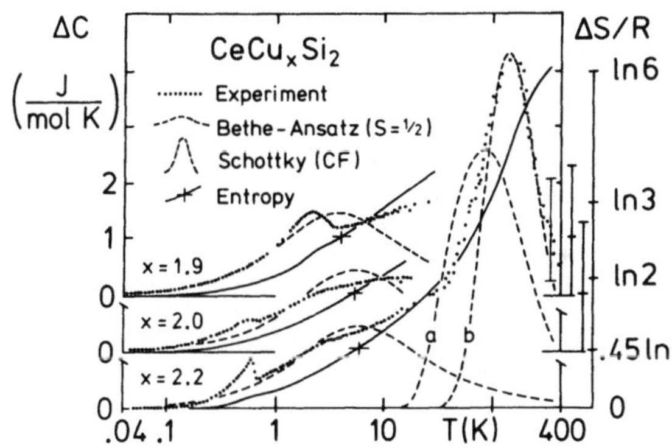

Fig.4 C/T vs. T for polycrystals of $CeCu_{1.9}Si_2$ [44] and $CeCu_2Si_2$ [18]

Fig.5 Specific heat ΔC and entropy ΔS for $CeCu_xSi_2$ polycrystals with differing x against a logarithmic temperature scale [44]. ΔC = C[$CeCu_xSi_2$] - C[$LaCu_xSi_2$] displays the 4f-contribution of Ce. Error bar indicates the uncertainty of high-temperature data. Dashed lines give specific heats calculated for the single-ion S = 1/2 Kondo problem [55] and for the crystal-field level scheme with three doublets at 0K, 140K and 364K (a, Ref. 44) as well as 0K, 360K and 360K (b)

The relationship between the three phenomena HFS, magnetic order and coherence-gap formation is illustrated in Fig. 5, in which the Ce-derived specific heats are compared for the three $CeCu_xSi_2$ samples discussed above [44]. Below, we wish to focus on the low-T results only [54]. In case of

$CeCu_{1.9}Si_2$, a shift of entropy from the "single-ion anomaly" as obtained for an S = 1/2 Kondo impurity by BETHE-Ansatz calculations [55,56] to the "spin-glass anomaly" at T = 2.5K is clearly visible. Surprisingly enough, no significant deviations of the measured data points from the single-ion theory can be resolved below 1K. Turning now to the superconducting $CeCu_{2.2}Si_2$ sample with a pronounced specific-heat jump, one recognizes again some entropy deficit above 2K, which becomes compensated at lower temperatures. Since the entropy transfer is much larger for the sample No. 10 with a rather sharp coherence gap, we consider this feature another indication for the DOS-change when going from a dilute Kondo ion to a periodic Kondo lattice.

Although the comparative study presented in this section could unravel some of the puzzles posed by $CeCu_2Si_2$, more questions remain open and require further investigations: Why does Cu-excess lead to volume compression? Why does it, even on a 10at% level, not remove the coherence gap-in contrast to the effect of 5at% Cu-deficit? Why do low-volume samples show low resistivities? What is the reason for the saturation of T_c near 0.5K in pressurized single crystals [33], whereas transition temperatures up to 0.7K can be obtained for polycrystals? In this context, more detailed knowledge of the ternary phase diagram [42],as well as progress in the single-crystal characterization [34,38],are urgent tasks.

3. Thermal Properties in the Superconducting State

In this section we discuss low-temperature results of the specific heat and thermal conductivity for $CeCu_2Si_2$ [49,57], UPt_3 [58] and UBe_{13} [59]. We concentrate on the specific-heat data of Figs. 6a-c first. For all three systems the transition to superconductivity is indicated by a "jump" in the zero-field data. While this feature is rather sharp for UBe_{13} and $CeCu_2Si_2$, only a broadened hump is found in case of UPt_3. Replacing the measured transitions in the C/T vs. T representation in the usual entropy-conserving way by discontinuous ones, we obtain good agreement with literature results [32,15,22] concerning both T_c and the jump height, ΔC. When ΔC is normalized to the n-state specific heat at T_c, $C_n(T_c)$, this ratio is found to be 2.4 for UBe_{13}, typical for strong-coupling superconductors, but much greater than 1.43, the value predicted by the theory of BARDEEN, COOPER and SCHRIEFFER (BCS). While for $CeCu_2Si_2$, $\Delta C/C_n(T_c) \simeq 1.1$ comes out somewhat lower than the BCS-value, it is particularly small (0.6) for UPt_3. Note that a field as high as 8T is not sufficient to suppress the specific-heat jump in UBe_{13} (Fig. 6c).

For a superconductor in zero magnetic field the entropy at T = T_c equals that in the normal state. From the B = 0 results of Figs. 6a-c the entropy balance $\Delta S = {}_0\int^{T_c} T^{-1}(C_s(T) - C_n(T))dT = 0$ is not met if both, $C_s(T)$ and the n-state results for T > T_c, are extrapolated to T = 0 in a straightforward way: An entropy deficit for $CeCu_2Si_2$, but an excess in case of the U-based compounds, would be obtained. A possible reason for this mismatch is fine structure in the DOS at E_F. For $CeCu_2Si_2$ the results in an overcritical field of 2T are, in fact, apt to resolve the entropy problem completely: They reveal a minimum in the DOS at E_F. For the U-based systems, on the other hand, a very narrow maximum of the DOS right at E_F would have to be assumed in order to achieve $\Delta S = 0$. The corresponding $\gamma_n(T)$-upturns at very low temperatures are schematically indicated by the thin solid lines in Figs. 6b and c.

For UPt_3 such an increase in γ_n seems to start already above T_c = 0.35K when measured at B = 0.5T. It is further supported by a quadratic, rather

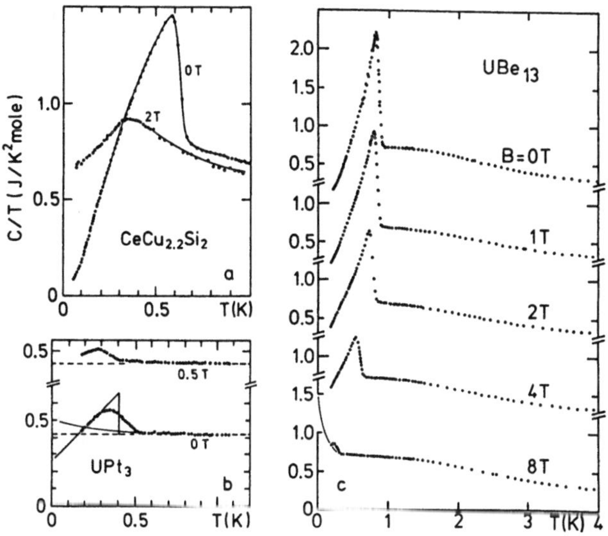

Fig.6 C/T vs.T for polycrystals of $CeCu_{2.2}Si_2$ (a), UPt_3 (b) and UBe_{13} (c) at $B = 0$ and $B > 0$. Solid lines in (a) are intended as guides to the eye. Dashed line in (b) corresponds to $B = 0$ results obtained for the same sample between 1.3 and 16K (J.J.M. Franse, P.H. Frings, A. de Visser and A. Menovsky: Ref. 49, p. 116). Thick solid lines in (b) indicate idealized sharp jump, see text, and extrapolation of results in the superconducting state to $T = 0$. Thin solid lines in (b) and (c) are schematic illustrations of how to fulfill the entropy requirement, as explained in the text

than a linear, T-dependence of the n-state thermopower [58,60]. For UBe_{13}, measured at $B = 8T$, there is no upturn down to the corresponding transition temperature of 0.26K. If one assumes that the B-dependence of $C_n(T)/T$ below T_c is as weak as observed for $T \gtrsim T_c$, $\Delta S = 0$ would require a steep upturn in the zero-field $\gamma_n(T)$, see Fig. 6c. It should be noted, however, that other than electronic degrees of freedom might, in principle, be involved. Thus, for instance, if a magnetic or structural phase transition at $T \lesssim T_c$ occurs near the onset of superconductivity, the entropy requirement could be fulfilled without relying on a γ_n-upturn.

We now turn to the temperature dependence of C_s, the specific heat in the superconducting state. Taking the increase of the n-state specific-heat coefficient in Fig. 6b as granted, the C_s/T data for UPt_3, which lie on a straight line down to the experimental temperature limit of 0.15K, can be extrapolated linearly to $T = 0$. Consequently, $C_s(T)$ contains a large linear term $\gamma_s T$ with $\gamma_s \simeq 0.26 J/molK^2$, besides a contribution $\alpha_s T^2$ [61]. Extrapolation of the C_s/T results for $CeCu_2Si_2$ to $T = 0$ reveals, if any, an only very small intercept, i.e. $\gamma_s \lesssim \gamma_n/30$ (Fig. 6a). No power law can be fitted to these data in any substantial T-window. It was already noticed earlier [49] that $C_s(T)$ follows approximately a T^m-dependence with $m \simeq 2$ somewhat below T_c, but gains an increasing exponent towards lower T. As suggested by the plot of Fig. 7a, there might exist a power-law dependence with $m \simeq 2.8$ below $T = 0.1K$. A $C_s \sim T^3$ dependence had been reported for UBe_{13} [22]. The C_s/T^m vs. T fits in Fig. 7b indicate that a very good power-law behavior exists in a large T-range (0.2 to 0.8K), although with $m = 2.86$ rather than 3. No hint at any linear term can be found from these data.

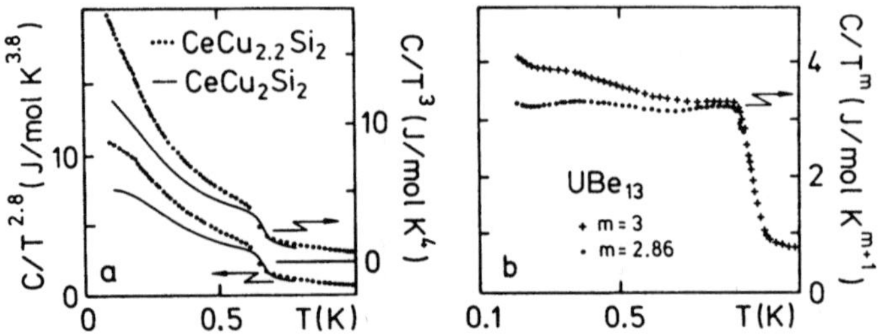

Fig.7 C/T^m vs. T plots for polycrystals of $CeCu_{2.2}Si_2$ and $CeCu_2Si_2$ (No. 4, Ref. 18) with m = 2.8 and 3 (a) as well as for a UBe_{13} polycrystal with m = 2.86 and 3 (b)

Before we analyze these different temperature dependences of $C_s(T)$, we wish to address the results of the thermal conductivity, $\kappa_s(T)$: At sufficiently low temperatures, where the electrons are dominantly scattered from static lattice imperfections, the electronic parts C_{es} and κ_{es} should show the same temperature dependence, provided that their Fermi velocity does not depend on temperature. Whereas the specific heat is dominated by the electronic contribution in the whole range of accessible temperatures, this is not obvious for the thermal conductivity: Above T_c UPt_3 [58] and, in particular, $CeCu_2Si_2$ [57] show considerable phonon heat transport. In a conventional superconductor, the electronic part falls below the n-state values, while the phonon contribution κ_{ps} shows a relative increase below T_c. κ_{ps} becomes more important for lower temperatures where less quasiparticles, the main phonon scatterers, exist. At sufficiently low temperatures κ_{ps} must be limited by boundary scattering ("size effect"). This upper limit of κ_{ps} can be estimated via $\tilde{\kappa}_{ps} = 1/3\bar{\Lambda}v_0C_p$, where $\bar{\Lambda}$ is the mean grain size in the polycrystalline sample, v_0 the average sound velocity and C_p the phonon specific heat. This thermal-conductivity behavior is clearly observed in Fig. 8 for the conventional superconductor HfV_2 [62]. In parti-

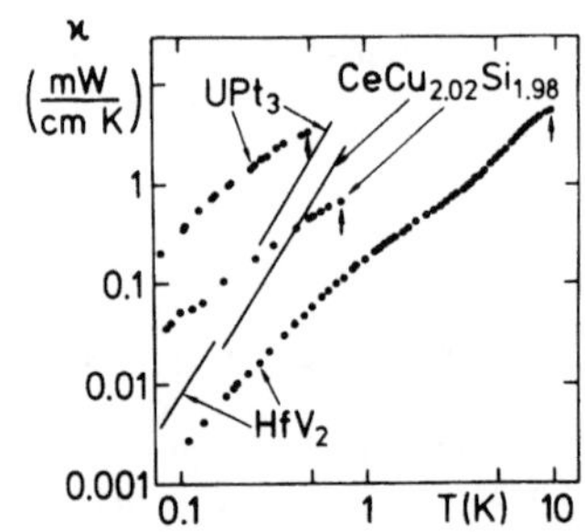

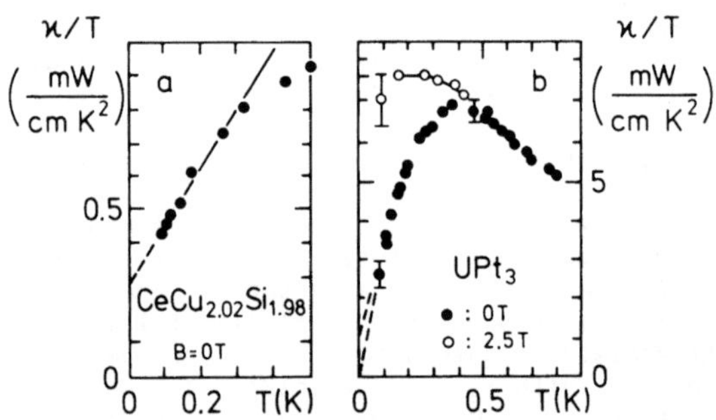

Fig.8 Temperature dependence of the thermal conductivity κ vs. T (in a log-log plot) for UPt_3 and $CeCu_{2.02}Si_{1.98}$ as well as for the conventional superconductor HfV_2 [62]. Straight lines represent $\tilde{\kappa}_{ps}$ vs. T, the upper bound of phonon conductivity limited by the "size effect", see text. Thick arrows mark transitions to superconductivity

Fig.9 κ/T vs.T for polycrystals of $CeCu_{2.02}Si_{1.98}$ at B = 0 (a) and UPt_3 at B = 0 and 2.5T (b). Dashed lines denote possible extrapolations of B = 0 results to T = 0

cular, the measured conductivity κ_s is smaller than $\tilde{\kappa}_{ps}$ in the whole T-range, with a tendency to approach it asymptotically at very low temperatures. Here, $\kappa_s(T)$ appears dominated by the phonon contribution. Turning back to UPt_3 and $CeCu_2Si_2$, we find the total thermal conductivity to exceed the upper limit of κ_{ps} at low temperatures considerably. Therefore, we conclude that in these systems the dominant contribution to the heat transport well below T_c stems from the electrons, rather than phonons.

Figure 9 shows the same results in a representation κ/T vs. T. Below T = 0.25K the data are well fitted by a straight line displaying a substantial quadratic term $\beta_s T^2$ to $\kappa_{es}(T)$. However, whereas a relatively large intercept λ_s (about 30% of the n-state value $\lambda_n = \kappa_{en}/T$ (as $T \to 0$) [57]) is noticed for $CeCu_2Si_2$, λ_s amounts to at most 15% of λ_n in the case of UPt_3 [61]. The value $\lambda_s = 0.55$ mW/cmK² recently reported by other authors [60] is in good agreement with this result. We finally note that a T^2-law for $\kappa_s(T)$ below T = 0.3K with no indication of an additional linear term was found for UBe_{13} by JACCARD et al. [50].

Contributions linear in T to the thermal properties of heavy-fermion superconductors may either indicate a gapless excitation spectrum in the superconducting phase due to pair-breaking or portions of the Fermi surface which stay normal in the accessible range of temperatures $T < T_c$. Adopting this last point of view in the following and taking into account the large $\gamma_s T$-contribution to the specific heat, we conclude that in UPt_3 normal electrons with heavy mass exist down to $T = 0.3T_c$, at least. In fact, if we subtract $\gamma_s T_c$ from the total n-state specific heat at T_c, a more realistic height of the specific-heat jump is obtained: Instead of $\Delta C/C_n(T_c) \simeq 0.6$ we find $\Delta C/[C_n(T_c) - \gamma_s T_c] \simeq 1.5$, close the BCS-value.

By comparing the ratios λ_s/γ_s and λ_n/γ_n, which refer to the superconducting and normal states, we estimate in the limit of a T-independent electronic mean free path (as $T \to 0$) that the Fermi velocity v_{Fs} of those electrons remaining normal below T_c is given by $v_{Fs}/\bar{v}_{Fn} \simeq (\lambda_s/\gamma_s)/(\lambda_n/\gamma_n)$ where $\bar{v}_{Fn}$ is the average Fermi velocity in the normal state. Inserting our estimates $\lambda_s \simeq 1$mW/cmK² and $\gamma_s = 0.26$ J/K²mole, we find $v_{Fs} \simeq 0.3\ \bar{v}_{Fn}$ in the case of UPt_3. This comparison suggests a variation in the Fermi velocity by a factor of about three, in accord with a recent bandstructure calculation for UPt_3 [64]. The less heavy quasiparticles seem to form Cooper pairs with large gap, whereas the heaviest ones do not pair down to rather low temperatures. FREEMAN et al. [30] who have recently calculated the electron-phonon coupling for UPt_3, based upon an LDA treatment, come to a similar conclusion: Those states on the Fermi surface with dominating d-character form Cooper pairs at $T = T_c$, while the more f-like states remain normal down to well below T_c. A simple explanation for this result makes use of the low average Fermi velocity in UPt_3 [65,66] which exceeds the average sound velocity by only a factor of two. These phase-space arguments [61] strongly disfavor Cooper pairing when the Fermi velocity becomes further reduced. A similar reasoning may explain the strong T_c-variation among polycrystalline $CeCu_2Si_2$ samples: here, a 40% reduction of T_K (and, thus, presumably of v_F) was found to lead to a complete suppression of superconductivity [18].

For $CeCu_2Si_2$, comparison of the specific-heat and thermal-conductivity results leads to a completely different picture: The very small $\gamma_s T$-term (Fig. 6a) suggests that those normal electrons existing well below T_c must have rather light masses. The compensating effect of their large Fermi velocities would then explain the large linear contribution $\lambda_s T$ to the electronic thermal conductivity (Fig. 9a) [57]. In fact, a similar estimate as above yields $v_{Fs} \geq 10\ \bar{v}_{Fn}$. Hence, one can assume non-renormalized light electrons with very low T_c to coexist with the heavy quasiparticles, presumably conduction electrons dressed by their strong interaction with the localized 4f-electrons. Only the latter ones are responsible for the observed HFS. Again, these conclusions are in general agreement with recent theoretical results [20]: A calculation of the renormalized bandstructure of $CeCu_2Si_2$, which modifies LDA [67] by taking into account proper phase shifts for the Kondo scattering, shows that there remain portions of the Fermi surface which do not develop heavy masses and probably do not participate in the Cooper-pair formation. The large variation in effective mass may be interpreted by the small spatial extent of the 4f-wavefunctions which does not allow for hybridization with all orbitals of neighboring atoms contributing to the Fermi surface [61]. Finally, since no linear contributions to $C_s(T)$ and $\kappa_s(T)$ have been resolved for UBe_{13}, the whole Fermi surface seems to be involved in the superconductivity of this material.

The temperature dependence of the thermal properties beyond the linear contributions yields information about the excitation spectrum of the superconducting condensate. Simple T-power laws are obtained in group-theoretic classification schemes for the superconducting order parameter, as developed by VOLOVIK and GOR'KOV [68] and UEDA and RICE [69], see also VARMA [21]. For example, it was suggested that, in the case of gap zeros on points on the Fermi surface, both $C_{es}(T)$ and $\kappa_{es}(T)$ ought to be proportional to T^3. This situation would be realized for an axial state, like the ANDERSON-BRINKMAN-MOREL (ABM) state in liquid ^3He. On the other hand, $C_{es} \sim \kappa_{es} \sim T^2$ would characterize gap zeros on lines as expected for, e.g., the polar state [21]. The results of Figs. 6 and 7 reveal a strikingly non-universal specific-heat behavior for the three heavy-fermion superconductors: Only for UPt_3 a simple power law is found, whereas a power law with non-integral exponent exists for UBe_{13} and no power law can be resolved for $CeCu_2Si_2$ down to 0.1K.

An approximate T^3-dependence of the electronic specific heat over quite a large temperature regime does not necessarily involve an anisotropic pairing: This is illustrated in Fig. 10 where recent results on the conventional superconductor HfV_2 are compared with UPt_3, $CeCu_2Si_2$ and UBe_{13}. Reduced units are used in this C_e/T vs. T^2 plot in order to account for

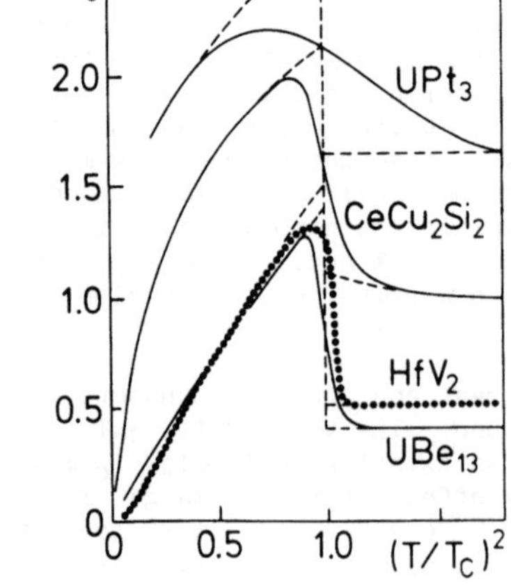

Fig.10 Electronic specific heats divided by T and normalized by $\Delta C/T_c$, as function of $(T/T_c)^2$ for three heavy-fermion superconductors and the conventional superconductor HfV_2 [62]. Thick solid lines are used instead of data points. Dashed lines indicate idealized sharp jumps, see text

different gap sizes. For HfV_2, $C_{es}(T)$ follows roughly a T^3-dependence in the same range of reduced temperatures as for UBe_{13}. This behavior, already observed earlier for ZrV_2 [70], was ascribed to the strong-coupling nature of HfV_2 [62], substantiated by the enhanced value of $\Delta C/\gamma_n T_c \simeq 1.9$.

One of the most striking results of this section is the T^2-term in the electronic contribution to the thermal conductivity for $CeCu_2Si_2$ and UPt_3 (Figs. 9a and b), confirming a previous observation on UBe_{13} [50]. It remains to be explained by future work, however, why for $CeCu_2Si_2$ and UBe_{13} the thermodynamic quantity $C_s(T)$ shows a T-dependence more complex than the transport property $\kappa_s(T)$.

4. Upper Critical Magnetic Fields

In order to gain insight into the type of Cooper pairing in heavy-fermion superconductors, we compare in this section recent experimental results on the temperature dependences of their upper critical magnetic fields with theoretical models [71].

The experimental $B_{c2}(T)$ curves for single crystal and polycrystalline samples of $CeCu_2Si_2$, a UBe_{13} polycrystal and a single crystal of UPt_3 are displayed in Figs. 11a-f. In general, our new results confirm previous observations: (i) In the case of $CeCu_2Si_2$, $B_{c2}(T)$ starts with a large slope at T_c, $B'_{c2} = -(dB_{c2}/dT)_{T_c} \simeq 20T/K$ [13,38]. At lower T, the critical-field curve flattens off distinctly and even shows a shallow maximum near $T = 0.2K$ [38]. (ii) The $B_{c2}(T)$ curve of UBe_{13} begins with an extremely large $B'_{c2}(T_c)$ [72]. In Fig. 11d continuously recorded $B_{c2}(T)$ values in the vicinity of T_c [71] are compared with results obtained from discrete measurements. According to this plot, $|dB_{c2}/dT|_{T_c} > 100$ T/K and $(dB_{c2}/dT)_{T_c}$ may even assume positive values. At somewhat lower temperatures, $B_{c2}(T)$ shows an approximately linear T-dependence. Here, B'_{c2} is $\simeq$ 25T/K, compared to $\simeq$ 42T/K as found for a UBe_{13} single crystal by MAPLE et al. [72]. (iii) For UPt_3, a much smaller $B'_{c2}(T_c)$ is observed [65]. In addition, measurements with the field both parallel ($B_{c2\parallel}$) and perpendicular ($B_{c2\perp}$) to the hexagonal c-axis confirm the anisotropy in $B_{c2}(T)$ earlier found by CHEN et al. [65] and WILLIS et al. [73], $B_{c2\parallel}/B_{c2\perp}$ being as large as $\simeq 1.5$ near T_c, but decreasing to $\simeq 1$ as $T \rightarrow 0$.

Figure 11 contains some new findings for the U-based heavy-fermion superconductors and, thus, reveals interesting "sample dependences" of the critical fields: For UBe_{13} we observe two inflection points in $B_{c2}(T)$ near 4T and 7T. This contrasts both the conspicuous linear behavior discovered in [72] and the $B_{c2}(T)$ upturn to $B \simeq 13T$ communicated in [74]. Furthermore a discontinuous change in the slope near 0.4T is clearly visible for B_{c2} of UPt_3, when measured with B applied perpendical to the c-axis. Instead, a continuous change in $B'_{c2}(T)$ was reported for this configuration in [65].

In view of these various deviations from the standard shape of a critical-field curve, comparison of the experimental results with existing theories for $B_{c2}(T)$ appears, at first glance, not very promising. However, we will use weak-coupling theory which covers, in general, all possible types of conventional and non-conventional pairing as discussed in [68,69]. In addition, weak-coupling theory can take into account complex properties of the conduction electrons, e.g. anisotropies, multi-band structure or substantial spin-orbit coupling. The major drawback of weak-coupling theory in describing $B_{c2}(T)$ of heavy-fermion superconductors may be found in the fact that k_BT_c is not small compared with the typical energies of the

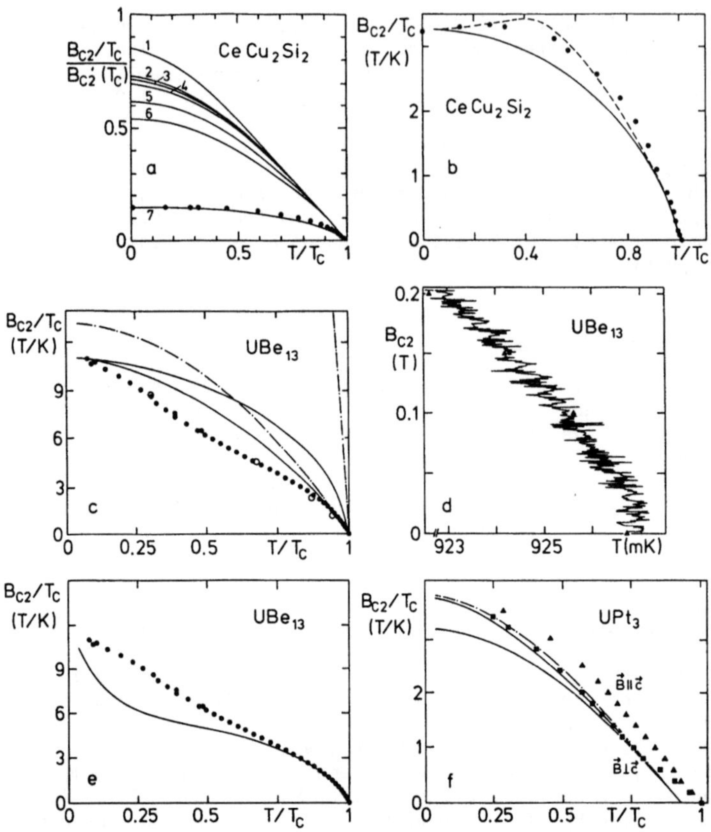

Fig.11 Temperature dependence of the upper critical magnetic field for a single crystal of nominal composition $CeCu_{2.52}Si_2$ with T_c = 0.65K, $B'_{c2}(T_c)$ = 21 T/K (relative orientation $B \perp c$) (a), a $CeCu_2Si_2$ polycrystal (No. 4, Ref. 18) with T_c = 0.66K (b), a UBe_{13} polycrystal with T_c = 0.93K (c-e, Ref. 71), and two UPt_3 samples, cut perpendicular and parallel to the hexagonal c-axis from the same single crystal (T_c = 0.49K) (f, Ref. 71). A plot B_{c2}/T_c vs. T/T_c is used, except for (a) where the ordinate is normalized by $B'_{c2}(T_c)$ and for (d) where a continously recorded critical-field curve in the vicinity of T_c is plotted as B_{c2} vs. T and compared with results of discrete measurements (triangles), see text. All data points are obtained from midpoints of resistive transitions, except for the open circles in (c) which indicate the midpoints of calorimetric transitions (T_c = 0.87K). Curves in the figures refer to various fits of the "dirty-limit model" and the "p-wave model" as explained in the text

heavy-fermion normal state. This can lead to a violation of the basic assumption of weak-coupling theory that all n-state properties should be temperature- and field-independent in the parameter ranges set by T_c and B_{c2}. We will return to this problem later in this section.

Below we compare our experimental results with two models, the "dirty-limit model" and the "p-wave model" which represent two very simple limits of the general weak-coupling theory. Both models assume an isotropic Fermi surface, a requirement which appears to be not perfectly, but reasonably well fulfilled [38,65]: anisotropy effects observed in the upper critical fields of heavy-fermion superconductors do not exceed about 20% (with the notable exception of UPt_3 in the vicinity of T_c, see Fig. 11f).

The dirty-limit model [75] is adequate for any conventional superconductor with $\ell/\xi_0 \ll 1$ (ℓ: mean free path, $\xi_0 = \hbar v_F/2\pi k_B T_c$: coherence length). It exhibits PAULI-limiting, i.e. a reduction of $B_{c2}(T)$ by the action of the B-field on the spins of the quasiparticles forming the Cooper pairs. Pauli limiting is absent in the p-wave model [76], representative for triplet superconductors with equal-spin pairing along the direction of the field. Well-known examples are the ABM state and the polar state. The p-wave model was originally applied to the pure limit ($\ell/\xi_0 \gg 1$) only, but has recently been generalized by RAINER et al. [71], who have included impurity scattering.

In the dirty-limit model $B_{c2}(T)$ is determined by four n-state parameters: (i) The electronic diffusion constant $D = \pi^2 k_B^2/(3e^2\gamma_n\rho_n)$, where γ_n and ρ_n are the n-state linear specific-heat coefficient and resistivity, respectively; (ii) The pair-breaking parameter $\alpha = \hbar/(\pi k_B T_c \tau_s)$; (iii) The spin-orbit parameter $b = 1.78\ \lambda_{so} = 0.19\hbar(k_B T_c \tau_{so})$, and the susceptibility-enhancement factor $R = N(\chi_n)/N(\gamma_n)$. D and α already fix the slope $B'_{c2}(T_c)$, whereas b and R determine the magnitude of the Pauli-limiting effect. Usually, an appropriate choice of b also allows to fix $B_{c2}(T)$ as $T \to 0$. (In non-cubic materials the diffusion constant can be anisotropic which results in an anisotropic $B_{c2}(T)$ dependence). Deviations from the dirty limit can be accounted for, to a very good approximation, by $D^* = D/(1+1.2\ell/\xi_0)$ [77].

The p-wave model contains only two parameters to fit $B_{c2}(T)$, the diffusion constant D and the pair-breaking parameter, which reads $\alpha = \xi_0/\ell$ for a triplet superconductor with anisotropic orbital part of the order parameter. D is determined by ρ_n and γ_n, and the measured slope $B'_{c2}(T_c)$ fixes α.

In Fig. 11 results from fits of these two models to the data are also shown. For the dirty-limit model, D and b are used to adjust the theory to the experimental values of $B'_{c2}(T_c)$ and $B_{c2}(0)$. The $B_{c2}(T)$ curves obtained in this way are not very sensitive to α and R, unless one uses unreasonably large α's. Note that $\alpha > 1$ would imply unreasonably high intrinsic transition temperatures which exceed the measured ones by more than a factor of four. We have set $\alpha = 0$ in the dirty-limit analysis. Since the susceptibility-enhancement factor R is not accurately known for the U-based systems, we have used "reasonable values" [2], namely $R = 0.4$ for UBe_{13} and $R = 0.1$ for UPt_3. (Changing R-values at fixed $B'_{c2}(T_c)$ and $B_{c2}(0)$ does not cause significant changes in the overall shape of $B_{c2}(T)$). In Fig. 11a, $R = 0.8$ was used as deduced for $CeCu_2Si_2$ at $T = T_c$ [49]. The p-wave model was adjusted to the measured $B'_{c2}(T_c)$. The latter, along with the diffusion constant, determines the pair-breaking parameter ξ_0/ℓ.

The critical field according to the p-wave model vastly exceeds the measured low-T values for $CeCu_2Si_2$. This demonstrates a rather strong Pauli-limiting effect and rules out triplet superconductivity with only equal-spin pairing along the magnetic-field direction. In Fig. 11a we compare the experimental results with results of several different models all of which include no Pauli limiting. These are (i) the p-wave model for the polar state (1), the SCHARNBERG-KLEMM (SK) state [78] (3), and the ABM state (5), (ii) the s-wave model without Pauli limiting (R = 0) for the "pure limit" (2) and for the "dirty limit" (4), as well as (iii) the same models as for (1) and (4), but with pair-breaking parameter $\alpha = 2$ (6).

Obviously, all these different models lead to $B_{c2}(T)$ curves of very similar shapes. On the other hand, a reasonable fit of the dirty-limit model to the data can be achieved by choosing $\alpha = 0$, $R = 0.8$ and $b = 0.65$. The latter two numbers are representative of a strong Pauli-limiting effect.

The remaining deviations at intermediate temperatures can be reduced considerably by taking into account corrections to the weak-coupling results originating from the temperature and/or field dependences of n-state properties, e.g. the resistivity, susceptibility and specific-heat coefficient γ_n. To demonstrate this improvement we have chosen a polycrystalline sample with sufficiently high residual resistivity. For such a sample the major corrections to weak-coupling theory stem from the T-dependence of γ_n, which was measured for this same sample in an overcritical field of 2T [18]. In Fig. 11b we compare measured $B_{c2}(T)$ data for this sample with results of the dirty-limit model with (i) $b = 0.75$ and a fixed value of $R = R(T_c) = 0.75$ (solid line) as well as (ii) $b = 1$ and a T-dependent $R(T) = (\chi_n/\mu_0\mu_{eff}^2)/(\gamma_n(T)/\pi^2 k_B^2)$ as determined by the measured T-dependence of γ_n (dashed line). Here, the low-T susceptibility $\chi_n = 8 \cdot 10^{-8}$ m³/mole [79] and the low-T effective moment $\mu_{eff} = 1.65\mu_B$ [32] have been used [80]. It is clearly seen in the figure that the corrections caused by $\gamma_n(T)$ are of the proper size to understand the observed "enhancement" of the measured $B_{c2}(T)$ at intermediate temperatures.

There is one other interesting result concerning $B_{c2}(T)$ of $CeCu_2Si_2$. Comparing the single-crystal data of Fig. 11a with those for the polycrystal of Fig. 11b, we notice an almost identical extrapolated value $B_{c2}(0) \simeq 2.1$ T. This indicates that the Pauli-limiting effect is almost independent of the orientation of the field relative to the crystal axes. More precisely, a direct measurement on another single crystal with B both perpendicular ($B_{c2\perp}$) and parallel ($B_{c2\parallel}$) to the tetragonal c-axis revealed that $B_{c2\parallel}(0)$ exceeds $B_{c2\perp}(0)$ by only 20% [38]. This somewhat reduced Pauli- limiting effect may be due to a moderately enhanced spin-orbit scattering in the "parallel orientation" [38].

For UBe_{13} the dirty-limit model and the p-wave model have been adjusted to both $B'_{c2}(T_c) \simeq 200$ T/K, representing the extremely steep $B_{c2}(T)$ slope right at T_c, and to $B'_{c2} \simeq 25$T/K as found somewhat below T_c (Fig. 11c). Our "dirty-limit" curves (solid lines) were obtained by choosing $b = 1.1$ and $b = 7$, respectively. The results of the p-wave model (dash-dotted lines) invoke a respective pair-breaking parameter $\xi_0/\ell = 0.44$ and $\xi_0/\ell = 5.4$. No good fit is obtained with any of these models. The inapplicability of the p-wave model is particularly striking: This holds even for the lower dash-dotted curve, since the large ξ_0/ℓ implies unrealistically high intrinsic values of both T_c ($\simeq$ 18K) and ΔC ($\simeq 16\gamma_n T_c$). In Fig. 11e, we have employed into the dirty-limit model all T- and B-dependences of n-state parameters known. The most pronounced of these effects is the gigantic negative magnetoresistance, which also depends on temperature. The resulting solid curve was adjusted to $B'_{c2} = 25$T/K. We can achieve agreement with the experimental results only down to 0.8 T_c. The decrease of ρ_n with increasing field (and decreasing temperature) leads to a flattening of $B_{c2}(T)$ that disagrees with the data points. Furthermore, taking into account the low-T upturn in $\gamma_n(T)$ as inferred from entropy arguments (see Fig. 6c) we find a marked increase of $B_{c2}(T)$ which does not agree with the data either. A very good fit of the $B_{c2}(T)$ data of [72] could be obtained within the dirty-limit model by TACHIKI et al. [81], who have taken into account the B-dependence of the (usually field-independent) weak-coupling parameters, including the BCS coupling constant. We have refrained to include a magnetic-field dependence of the latter, since we cannot measure it independently. Of course, using a

B-dependent coupling constant as an adjustable parameter, we would be able to account for the remaining discrepancies of order 30%. Though the origin of the anomalous features in the $B_{c2}(T)$ curve remains unknown, our analysis yields one important result: Because of the failure of the p-wave model to describe the experimental results, the upper critical field of UBe_{13} appears to be substantially Pauli limited, too.

If we finally consider UPt_3 (Fig. 11f), the steeper part of the $B_{c2}(T)$ curves can be well fitted either by ignoring Pauli limiting in the dirty-limit model almost completely (with $\alpha = 0$, $R = 0.1$ and $b = 0.5$, upper solid line) or by the p-wave model ($\xi_0/\ell = 0.37$, dash-dotted line). This might indicate either equal-spin pairing along B in a triplet state or strong spin-orbit scattering in a singlet state. It should be noted that the spin-orbit parameter cannot exceed $b = 1$ in order to keep the spin-orbit scattering time τ_{so} larger than the total transport scattering time τ_{tr}. An equally good fit as the one in Fig. 11f can be achieved with $b < 1$ and $R = 0.3$, the most likely value for the susceptibility-enhancement factor [2]. It is worth to note that a vanishing pair-breaking parameter is necessary to fit the dirty-limit model to the data. If we use $\alpha \simeq 0.9$ as estimated from the reduced specific-heat jump as measured, $\Delta C/C_n(T_c) \simeq 0.6$ (Fig. 6b), no satisfactory fit is possible, even if $R = 0$ and $b = \infty$, i.e. in the absence of Pauli limiting (lower solid line in Fig. 11f). However, $\alpha \simeq 0$ is in accord with $\Delta C/[C_n(T_c)-\gamma_s T_c] = 1.5$ as obtained if the effect of normal electrons well below T_c is corrected for (Sect. 3).

The most peculiar observation for UPt_3 is the discontinuous change of the $B_{c2}(T)$ slope near 0.4T, if B is oriented perpendicular to the c-axis. According to [82] such a "kink" in the critical-field curve may occur in non-conventional superconductors. More specifically, it originates in a competition between a weakly anisotropic pairing interaction in a superconductor with non-cubic crystal structure and the orientational effect of a magnetic field: The interaction may favor an orientation of the order parameter which differs from the orientation as induced by the field, for which zeros of the anisotropic gap favorably lie in the direction of the shielding currents around vortices. In this case the "interaction effect" wins at low B-field, and a relatively flat slope $B'_{c2}(T_c)$ is observed. B'_{c2} then shows a sudden increase at a certain field value when the orientation in accord with the flux lines becomes energetically more favorable. When the magnetic field and the interaction favor the same orientation of the order parameter, however, B_{c2} starts with the larger slope right at T_c ($B \parallel c$, see Fig. 11f).

5. Summary and Outlook

A comparison between the three heavy-fermion superconductors, based upon results of the preceeding sections and from the literature, is given in Table 1. The literature results concern (i) the dc-JOSEPHSON effect between $CeCu_2Si_2$ and Al as well as between UPt_3 and Al, Nb or UPt_3 as counterelectrodes [52,83], (ii) the normalized ultrasonic attenuation α_s/α_n for UPt_3 [84] and UBe_{13} [85] and (iii) the spin-lattice relaxation rates of ^{63}Cu in $CeCu_2Si_2$ [86] and of ^9Be in UBe_{13} [87]. Obviously, there exist as many differences as similarities between the three systems:

(1) Entropy considerations suggest the existence of a low-energy scale with $T_0 < 0.1T^*$, $T^* \approx 10K$ being the characteristic (Kondo or spin-fluctuation) temperature. In $CeCu_2Si_2$ (like in $CeAl_3$), the quasiparticle - DOS exhibits a narrow relative minimum in the vicinity of E_F, whereas the possible existence of a very narrow DOS-peak centered at E_F was inferred for the two U-based compounds.

Table 1 Comparison between some low-temperature properties of $CeCu_2Si_2$, UPt_3 and UBe_{13}. Hyphen: no experiment reported. a: [52,83]. b: [84]. c: [85]. d: [86]. e: [87]. f: [50]. Dashed frame as explained in the text

	$CeCu_2Si_2$	UPt_3	UBe_{13}
low-energy scale ($T_0<0.1T^*$)	$\gamma_n(T)$ peak at 0.3-0.4K	$\gamma_n(T)$ upturn as $T \to 0$?	$\gamma_n(T)$ upturn as $T \to 0$?
variations of m^*			
Cooper pairs:	heavy quasipart.	less heavy qu.	whole FS
normal states:	light electrons	heaviest qu.	
anomalies in $B_{c2}(T)$	flattening at low T shallow peak	anisotropy in $B'_{c2}(T_c)$ "kink" for $B\perp c$	$\lvert B'_{c2}(T_c)\rvert \gtrsim$ 100T/K 2 inflection points
ℓ/ξ_0	0.5	3	0.2
Pauli limiting	strong	none	substantial
Josephson effect	large	none(a)	-
T-power laws:			
$C_s(T) - \gamma_s T$	complex, eventually $T^{2.8}$ for T<0.1K	T^2	$T^{2.9}$
α_s/α_n	-	T^2(b)	T^2(c)
T_1^{-1}	T^3(d)	-	T^3(e)
$\kappa_s(T) - \lambda_s T$	T^2	T^2	T^2(f)

(2) For UPt_3 and $CeCu_2Si_2$, comparison between the linear contributions to both the specific heat and electronic thermal conductivity supports the idea of variations in the effective mass of the quasiparticles. In the former system, m^* varies by a factor of order three only, and the formation of Cooper pairs in the accessible T-range appears restricted to the less heavier quasiparticles. On the other hand, in $CeCu_2Si_2$ normal electrons with light masses seem to coexist with Cooper pairs formed by heavy fermions. Though a variation of effective masses cannot be ruled out for UBe_{13}, from the absence of such linear contributions to $C_s(T)$ and $\kappa_s(T)$ it is inferred that the whole Fermi surface participates in the superconductivity of this material.

(3) The outward shape of $B_{c2}(T)$ shows several anomalous features, which are distinctly different for the three systems. From the fit of the p-wave model to the measured $B_{c2}(T)$ slope at T_c, the ratio ℓ/ξ_0 is obtained and found to exceed one only in the case of UPt_3. A standard s-wave analysis yields quite similar results [71]. It is worth to note that the extremely short mean free path in UBe_{13} is, to a large part, due to intrinsic (T-dependent) scattering:The $B = 0$, n-state resistivity which is $\simeq 130\mu\Omega$cm at $T = T_c$ is extrapolated to decrease to $\simeq 50\mu\Omega$cm as $T \to 0$ [88]. Finite-field

measurements up to B = 10T strongly suggest a further reduction of ρ_n (T = 0) by a factor of about 7 as B → ∞ [88], a fascinating property of UBe_{13} that has yet to be explained. In $CeCu_2Si_2$, variations of the residual resistivity between 3.5 and 60μΩcm were observed in superconducting samples with T_c > 0.5K, see Sect. 2. Here, ℓ/ξ_0 may come close to one for particularly well annealed samples.

(4) A Pauli-limiting effect in $B_{c2}(T)$ is well established for $CeCu_2Si_2$ and UBe_{13}, being especially strong and only weakly orientation dependent for the former system. The existence of Pauli limiting rules out the ABM state as a possible superconducting state [71]. The critical-field curve of UPt_3 is nearly free of Pauli limiting, allowing for any S = 0 or S = 1 state.

(5) POPPE [52,83] could recently demonstrate the existence of a dc-Josephson current through a weak link between $CeCu_2Si_2$ and a clear-cut singlet superconductor, Al. Since this Josephson current is found to be of the same large size as in conventional junctions between two singlet superconductors, $CeCu_2Si_2$ was concluded to be a superconductor with dominant S = 0 pairing. No Josephson effect could, on the other hand, be detected for UPt_3 with Al, Nb and UPt_3 as counterelectrodes [52,83]. The absence of a Josephson effect was considered to be highly consistent with an anisotropic pairing interaction, either S = 0 or S = 1.

(6) Information about the possible type of pairing anisotropy may be derived from the simple T-power laws as found in certain superconducting properties. While VARMA [21] takes those powers in the dashed frame of Table 1 as evidence for gap zeros along lines on the Fermi surface, RICE et al. [22] and RODRIGUEZ [89] argue that a T^3-law in $C_s(T)$, as approximately observed for UBe_{13}, and the T^2-dependence of the ultrasonic attenuation ratio α_s/α_n prove the existence of gap zeros on points. Group-theoretic arguments [68,69] would necessarily yield S = 0 pairing in the former case, while both S = 0 and S = 1 pairing would be compatible with the latter.

This brings us to the last topic of this section, namely a listing of some major questions in the field, of which only a few can, so far, be answered:

- Which mechanism causes the formation of heavy fermions?
- Is the attractive interaction between them phonon-mediated as in conventional superconductors or is it of electronic origin?
- What are the actual superconducting pair states in the different systems?

The compound $CeCu_2Si_2$ is presently best understood. Here, the formation of heavy fermions can be explained in the framework of the Kondo-lattice concept [20,52]. For a discussion of the microscopic background, the reader is referred to GREWE [90]. Elastic anomalies in $CeCu_2Si_2$ point to the importance of phonons [91]. In particular, the rather strong volume dependence of the Kondo energy k_BT_K [33,41] provides a specific electron-phonon coupling mechanism which can serve to explain a $T_c \lesssim$ 1K, see FULDE et al. [20] and GREWE [26]. This kind of superconductivity represents an S = 0 state, in agreement with the observation of a large Josephson effect. In some theoretical treatments [27,28,92] an anisotropic S = 0 pairing interaction is obtained for Kondo lattices. Thus, for instance, OHKAWA [92] finds gap zeros along lines, in accord with the power laws in the spin-lattice relaxation rate [86] and thermal conductivity. Arguments against an anisotropic pairing are based on (i) the complex T-dependence of $C_s(T)$ and (ii) the fact that there is virtually no change in T_c if the mean free path ℓ (always $\leq \xi_0$) is varied by more than one order of magnitude.

As far as the U-based compounds are concerned, one can at present only speculate about the origin of their heavy-fermion low-T phases. In the case of UPt_3, one may assume an intermediate-valent ground state close to a magnetic instability. The latter would explain the various paramagnon-derived anomalies reported for this material [93,15,66,94]. According to Table 1, UPt_3 is most likely to exhibit anisotropic Cooper pairing, to which new support is lent by the finding of a "kink" in $B_{c2}(T)$ for the $B \perp c$ configuration. As to the spin state of the order parameter, no definite conclusions can be drawn from the available data. For example, FRANSE et al. [95] have recently observed for dilute $U(Pt_{1-x}Pd_x)_3$ alloys ($x \leq 0.05$) an enhancement of "paramagnon effects", but a suppression of superconductivity. From this result, they have ruled out paramagnon-mediated triplet superconductivity [96]. Consistent with such a mechanism, however, a depression of T_c - along with an increase of T^* - was observed under external pressure [97].

In order to explain HFS in UBe_{13}, OTT et al. [22] favor a paramagnon-mediated S=1 pairing, so far safely known to exist in liquid ^3He only [98]. In addition, RICE et al. [99] have generalized the BRINKMAN-RICE theory [100] to finite temperatures to describe the n-state thermodynamics of the heavy fermions in UBe_{13}. A more conventional view was adopted by OVERHAUSER and APPEL [29], who have explained most of the anomalous properties of UBe_{13} within a one-electron picture, leading to S = 0 pairing superconductivity. In view of the experimental results listed in Table 1 and of the controversial conclusions drawn from them [21,22,89], the nature of both the spin and orbital state of the order parameter remains unclear. The major difficulty with this system lies in the extreme field and temperature dependences of its n-state parameters within B- and T-ranges comparable with B_{c2} and T_c. Hence, before one can hope to understand the origin of HFS in UBe_{13}, these n-state anomalies have to be unravelled.

Acknowledgments

It is a great pleasure for me to thank the members of the experimental solid-state physics group of the Fachgebiet Technische Physik at TH Darmstadt, especially U. Ahlheim, C.D. Bredl, H.M. Mayer, U. Rauchschwalbe, L. Pawlak, G. Sparn and H. Spille for an enjoyable collaboration. I have greatly benefitted from many enlightening discussions and a stimulating cooperation with N. Grewe, D. Rainer and H. Rietschel, and I should like to acknowledge several stimulating conversations with P. Fulde and K.D. Schotte. Special thanks are due to R. Lamatsch for her expert typing of the manuscript and to K. Sachse-Schneider for her skilful drawings. This work was supported by the Deutsche Forschungsgemeinschaft under the auspices of the Sonderforschungsbereich 65 Frankfurt/Darmstadt.

References

1. For an early report, see: F. Steglich, J. Aarts, C.D. Bredl, W. Lieke, D. Meschede, W. Franz and H. Schäfer: J. Magn. Magn. Mat. 15-18, 889 (1980)
2. For a recent review, see: G.R. Stewart: Rev. Mod. Phys. 56, 755 (1984)
3. R.J. Trainor, M.B. Brodsky and H.V. Culbert: Phys. Rev. Lett. 34, 1019 (1975)
4. K. Andres, J.E. Graebner and H.R. Ott: Phys. Rev. Lett. 35, 1779 (1975)
5. Y. Onuki, Y. Shimizu and T. Komatsubara: J. Phys. Soc. Jpn. 53, 1210 (1984)
   G.R. Stewart, Z. Fisk and M. Wire: Phys. Rev. B30, 482 (1984)
   H.R. Ott, H. Rudigier, Z. Fisk, J.O. Willis and G.R. Stewart: Solid State Commun. 53, 235 (1985)

6. U. Rauchschwalbe, U. Gottwick, U. Ahlheim, H.M. Mayer and F. Steglich: J. Less Common Metals, in press
7. R.J. Trainor, M.B. Brodsky, B.D. Dunlap and G.K. Shenoy, Phys. Rev. Lett. 37, 1511 (1976)
8. B. Barbara, J.X. Boucherle, J.L. Buevoz, M.F. Rossignol and J. Schweizer: Solid State Commun. 24, 481 (1977)
C.D. Bredl, F. Steglich and K.D. Schotte: Z. Phys. B29, 327 (1978)
9. K. Winzer and W. Felsch: J. Phys. (Paris) 39, C6-832 (1978)
T. Komatsubara, T. Suzuki, M. Kawakami, S. Kunii, T. Fujita, Y. Ishikawa, A. Takase, K. Kojima, M. Suzuki, Y. Aoki, K. Takegahara and T. Kasuya: Ref. 1, p. 963
10. H.R. Ott, H. Rudigier, P. Delsing and Z. Fisk: Phys. Rev. Lett. 52, 1551 (1984)
11. Z. Fisk, G.R. Stewart, J.O. Willis, H.R. Ott and F. Hulliger: Phys. Rev. B30, 6360 (1984)
12. F. Steglich, J. Aarts, C.D. Bredl, W. Lieke,D. Meschede, W. Franz and H. Schäfer: Phys. Rev. Lett. 43, 1892 (1979)
13. U. Rauchschwalbe, W. Lieke, C.D. Bredl, F. Steglich, J. Aarts, K.M. Martini and A.C. Mota: Phys. Rev. Lett. 49, 1448 (1982)
14. H.R. Ott, H. Rudigier, Z. Fisk and J.L. Smith: Phys. Rev. Lett. 50,1595 (1983)
15. G.R. Stewart, Z. Fisk, J.O. Willis and J.L. Smith: Phys. Rev. Lett. 52, 679 (1984)
16. B. Batlogg, J.P. Remeika, A.S. Cooper and Z. Fisk: J. Appl. Phys. 55, 2001 (1984)
B. Batlogg, J.P. Remeika, A.S. Cooper, G.R. Stewart, Z. Fisk and J.O. Willis: J. Magn. Magn. Mat. 47&48, 42 (1985)
17. U. Rauchschwalbe, W. Baus, S. Horn, H. Spille, F. Steglich, F.R. de Boer, J. Aarts, W. Assmus and M. Herrmann: J. Magn. Magn. Mat. 47&48, 33 (1985)
18. C.D. Bredl, H. Spille, U. Rauchschwalbe, W. Lieke, F. Steglich, G. Cordier, W. Assmus, M. Herrmann and J. Aarts: J. Magn. Magn. Mat. 31-34, 373 (1983)
19. C.D. Bredl, S. Horn, F. Steglich, B. Lüthi and R.M. Martin: Phys. Rev. Lett. 52, 1982 (1984)
20. H. Razafimandimby, P. Fulde and J. Keller: Z. Phys. B54, 111 (1984)
N. d'Ambrumenil and P. Fulde: Ref. 17, p. 1
21. C.M. Varma: Bull. Am. Phys. Soc. 29, 404 (1984); Comments in Solid State Physics, in press
22. H.R. Ott, H. Rudigier, T.M. Rice, K. Ueda, Z. Fisk and J.L. Smith: Phys. Rev. Lett. 52, 1915 (1984)
23. P.W. Anderson: Phys. Rev. B30, 4000 (1984)
24. O. Valls and Z. Tesanovic: Phys. Rev. Lett. 53, 477 (1984)
25. M. Tachiki and S. Maekawa: Phys. Rev. B29, 2497 (1984)
26. N. Grewe, Z. Phys. B56, 111 (1984)
27. K. Miyake, T. Matsuura, H. Jichu and J. Nagaoka: Progr. Theor. Phys. 72, 1063 (1984)
28. F.J. Ohkawa and H. Fukuyama: J. Phys. Soc. Jpn. 53, 4344 (1984)
29. A.W. Overhauser and J. Appel: Phys. Rev. B31, 193 (1985)
30. T. Oguchi, A.J. Freeman and G.W. Crabtree: preprint (1985)
31. G.W. Hull, J.H. Wernick, T.H. Geballe, J.V. Waszcak and J.E. Bernardini: Phys. Rev. B24, 6715 (1981)
32. W. Lieke, U. Rauchschwalbe, C.D. Bredl, F. Steglich, J. Aarts and F.R. de Boer: J. Appl. Phys. 53, 2111 (1982)
33. F.G. Aliev, N.B. Brandt, V.V. Moshchalkov and S.M. Chudinov: Solid State Commun. 45, 215 (1983)
34. G.R. Stewart, Z. Fisk and J.O. Willis: Phys. Rev. B28, 172 (1983)
35. Z. Kletowski: J. Less Common Metals 95, 127 (1983)

36. F.G. Aliev, N.B. Brandt, V.V. Moshchalkov and S.M. Chudinov: J. Low Temp. Phys. 57, 61 (1984)
37. Here, a bulk modulus B = 1.1 · 10^3 kbar for $CeCu_2Si_2$ was used, see: I.L. Spain, F. Steglich, U. Rauchschwalbe and H.D. Hochheimer: Proc. X^{th} AIRAPT Conference, Amsterdam (1985); Physica B, in press
38. W. Assmus, M. Herrmann, U. Rauchschwalbe, S.Riegel, W. Lieke, H. Spille, S. Horn, G. Weber, F. Steglich and G. Cordier: Phys. Rev. Lett. 52, 469 (1984)
39. H. Spille, U. Rauchschwalbe and F. Steglich, Helv. Phys. Acta 56, 165 (1983)
40. S. Doniach: Physica 91B, 231 (1977)
M. Lavagna, C. Lacroix and M. Cyrot: Phys. Lett. 90 A, 210 (1982)
41. A more complex $T_c(p)$ dependence for p $\gtrsim$ 20 kbar was recently reported by B. Bellarbi, A. Benoit, D. Jaccard, J.M. Mignot and H.F. Braun: Phys. Rev. B30, 1182 (1984)
42. M. Ishikawa, H.F. Braun and J.L. Jorda: Phys. Rev. B27, 3092 (1983)
43. W. Baus: Diploma Thesis, TH Darmstadt (1985), unpublished
44. C.D. Bredl, W. Lieke, R. Schefzyk, M. Lang, U. Rauchschwalbe, F. Steglich, S. Riegel, R. Felten, G. Weber, J. Klaasse, J. Aarts and F.R. de Boer: Ref. 17, p. 30
45. The $\chi(T)$-results on "stoichiometric" single crystals grown by other techniques are similar to those in Fig. 2b, see Ref. 16 and Y. Onuki, Y. Furukawa and T. Komatsubara: J. Phys. Soc. Jpn. 53, 2197 (1984). One should also note that such crystals show a large magnetoresistivity, which is almost absent in Cu-excess crystals [17]
46. ρ_0-values for superconducting $CeCu_2Si_2$ single-crystal and polycrystalline samples range between 3.5 μΩcm [13] and 60 μΩcm [38], while for non-superconducting ones $\rho_0 \gtrsim$ 100 μΩcm [38]
47. J. Flouquet, J.C. Lasjaunias, J. Peyrard and M. Ribault: Ref. 32, p. 2127
48. C.D. Bredl, N. Grewe, F. Steglich and E. Umlauf: Proc. Int. Conf. on Low Temperature Physics, LT 17, U. Eckern, A. Schmid, W. Weber, H. Wühl, eds., North-Holland, Amsterdam (1984),p. 327
49. F. Steglich, C.D. Bredl, W. Lieke, U. Rauchschwalbe and G. Sparn: Physica 126B, 82 (1984)
50. D. Jaccard and J. Flouquet: Ref. 17, p. 45
51. T. Fujita, T. Satoh, Y. Onuki and T. Komatsubara: Ref. 17, p. 66
52. F. Steglich, U. Rauchschwalbe, U. Gottwick, H.M. Mayer, G. Sparn, N. Grewe, U. Poppe and J.J.M. Franse: J. Appl. Phys. 57, 3054 (1985)
53. R.M. Martin: Phys. Rev. Lett. 48, 362 (1982)
54. The Schottky-type anomaly at elevated temperatures is not compatible with the crystal-field level scheme of three Kramers doublets at 0K, 140K and 360K (curve a) as derived from neutron-scattering results [S. Horn, E. Holland-Moritz, M. Loewenhaupt, F. Steglich, H. Scheuer, A. Benoit and J. Flouquet: Phys. Rev. B23,3171 (1981)]. Rather, it suggests [44] a scheme of three doublets at 0K, 360K and 360K (curve b), similar to what was recently deduced from the anisotropy of the static bulk susceptibility [Y. Onuki, T. Hirai, T. Komatsubara, S. Takayanagi, A. Sumiyama, A. Furukawa, Y. Oda and H. Nagano: see Ref. 61; S. Maekawa, S. Kashiba, S. Takahashi and M. Tachiki: these proceedings]
55. H.U. Degranges and K.D. Schotte: Phys. Lett. 91A, 240 (1982)
56. For a treatment of crystal-field split Ce^{3+} ions, see: N. Kawakami and A. Okiji: these proceedings
57. G. Sparn, W. Lieke, U. Gottwick, F. Steglich and N. Grewe: Ref. 17, p. 521
58. J.J.M. Franse, A. Menovsky, A. de Visser, C.D. Bredl, U. Gottwick, W. Lieke, H.M. Mayer, U. Rauchschwalbe, G. Sparn and F. Steglich: Z. Phys. B59, 15 (1985)

59. H.M. Mayer, U. Rauchschwalbe, C.D. Bredl, F. Steglich, H. Rietschel, H. Schmidt and H. Wühl: Phys. Rev. B, submitted
The specific heat of UBe_{13} at finite B-field has also been determined by H.R. Ott, H. Rudigier, E. Felder, Z. Fisk and J.L. Smith: preprint (1985)
60. D. Jaccard, J. Flouquet, P. Lejay and J.L. Tholence: Ref. 52, p. 3082
61. F. Steglich, U. Ahlheim, J.J.M. Franse, N. Grewe, D. Rainer and U. Rauchschwalbe: Proc. 5th Int. Conf. on Crystal Field Effects, ICCF5, Sendai (1985), T. Kasuya, ed.; J. Magn. Magn. Mat., in press
62. B. Lüthi, M. Herrmann, W. Assmus, H. Schmidt, H. Rietschel, H. Wühl, U. Gottwick, G. Sparn and F. Steglich: Z. Phys. B, in press
63. The T^2-dependence in $C_S(T)$ of UPt_3 was found to hold even to somewhat lower temperatures with a γ_S-value being about two times smaller compared to that in Fig. 6b; see: A. Sulpice, J. Gandit, J. Chaussy, J. Flouquet, D. Jaccard, P. Lejay and J.L. Tholence, to be published
64. J. Sticht and J. Kübler: Solid State Commun. 54, 389 (1985)
65. J.W. Chen, S.E. Lambert, M.B. Maple, Z. Fisk, J.L. Smith, G.R. Stewart and J.O. Willis: Phys. Rev. B30, 1583 (1984)
66. A. de Visser, J.J.M. Franse, A. Menovsky, T.T.M. Palstra: J. Phys. F. 14, L191 (1984)
67. T. Jarlborg, H.F. Braun and M. Peter: Z. Phys. B52, 295 (1983)
68. G.E. Volovik and L.P. Gor'kov: ZhETF, in press
69. K. Ueda and T.M. Rice: Phys. Rev. B31, 7114 (1985)
70. H. Keiber, C. Geibel, B. Renker, H. Rietschel, H. Schmidt, H.Wühl, G.R. Stewart: Phys. Rev. B30, 2542 (1984)
71. U. Rauchschwalbe, U. Ahlheim, F. Steglich, D. Rainer and J.J.M. Franse: Z. Phys. B, in press
72. M.B. Maple, J.W. Chen, S.E. Lambert, Z. Fisk, J.L. Smith, H.R. Ott, J.S. Brooks and M.J. Naughton: Phys. Rev. Lett. 54, 477 (1985)
73. J.O. Willis, Z. Fisk, J.L. Smith, J.W. Chen, S.E. Lambert and M.B. Maple: see Ref. 48, p. 245
74. J.W. Chen, S.E. Lambert, M.B. Maple, M.J. Naughton, J.S. Brooks, Z. Fisk, J.L. Smith and H.R. Ott: see Ref. 52, p. 3076
75. E. Helfand and N.R. Werthamer: Phys. Rev. 147, 112 (1966)
K. Maki: Phys. Rev. 148, 362 (1966)
76. K. Scharnberg and R.A. Klemm: Phys. Rev. B22, 5233 (1980)
77. G. Eilenberger in: Ferienkurs Supraleitung, Kernforschungsanlage Jülich (1977), unpublished
78. K. Scharnberg and R.A. Klemm: Phys. Rev. Lett. 54, 2445 (1985)
79. J. Aarts: PhD Thesis, University of Amsterdam (1984), unpublished
80. A change of the crystal-field level scheme as discussed in [54] would result in μ_{eff} close to $1.24\mu_B$, the effective moment of the cubic Γ_7-doublet. This would lead to $R(T_c)$ = 1.4 and 1.3 for the single crystal and polycrystal, respectively. The corresponding b values of 2.8 (Fig. 11a), 3.0 (solid line in Fig. 11b) and 3.9 (dashed line in Fig. 11b) would still imply $\tau_{so} > \tau_{tr}$, where τ_{so} is spin-orbit scattering time and τ_{tr} the total transport scattering time
81. M. Tachiki, T. Koyama and S. Takahashi: these proceedings
82. R. Klemm and K. Scharnberg: Proc. Int. Conf. on Materials and Mechanisms in Superconductivity, Ames (1985), D.K. Finnemore, ed., in press
83. U. Poppe: Ref. 61
84. D.J. Bishop, C.M. Varma, B. Batlogg, E. Bucher, Z. Fisk and J.L. Smith: Phys. Rev. Lett. 53, 1009 (1984)
85. C.M. Varma: private communication (1985)
86. Y. Kitaoka, K. Ueda, T. Kohara, K. Asayama, Y. Onuki and T. Komatsubara: Ref. 61
87. D.E. MacLaughlin, Cheng Tien, W.G. Clark, M.D. Lan, Z. Fisk, J.L. Smith and H.R. Ott: Phys. Rev. Lett. 53, 1833 (1984)

88. U. Rauchschwalbe: unpublished results
89. J.P. Rodriguez: preprint (1985)
90. N. Grewe: Solid State Commun. 50, 19 (1984)
see also, Y. Kuramoto: these proceedings
91. R. Mock and G. Güntherodt: J. Phys. C17, 5635 (1984)
92. F.J. Ohkawa: these proceedings
93. P.H. Frings, J.J.M. Franse, F.R. de Boer and A. Menovsky: Ref. 18, p. 240
94. G. Aeppli, E. Bucher and G. Shirane: preprint (1985)
W.J.L. Buyers, J.K. Kjems and J.D. Garret: preprint (1985)
95. J.J.M. Franse, A. de Visser, A. Menovsky and P.H. Frings: Ref. 61
96. It should be noted that the lack of superconductivity even for x = 0.01 (A. de Visser, J.C.P. Klaasse, M. van Sprang, J.J.M. Franse and A. Menovsky: Proc. Int. Conf. on Magnetism, San Francisco (1985); to be published in J. Magn. Magn. Mat.) might exclusively be caused [15] by the Pd-derived reduction of the electronic mean free path. I wish to thank J.O. Willis for bringing this point to my attention
97. J.O. Willis, J.D. Thompson, Z. Fisk, A. de Visser, J.J.M. Franse and A. Menovsky: Phys. Rev. B31, 1654 (1985)
98. For a recent review, see: D. Vollhardt: Rev. Mod. Phys. 56, 99 (1984)
99. T.M. Rice, K. Ueda, H.R. Ott and H. Rudigier: Phys. Rev. B31, 594 (1985)
100. W.F. Brinkman and T.M. Rice: Phys. Rev. B2, 4302 (1970)

Part II

Exact Solution for Single Impurity

Exact Solution of the Anderson Model and Its Thermodynamics I*

A. Okiji and N. Kawakami

Department of Applied Physics, Osaka University, Suita, Osaka 565, Japan

Bethe ansatz method enables us to treat the Anderson model exactly. The exact solution of the Anderson model with a single orbital and its thermodynamic properties obtained by the Bethe ansatz method are reviewed in this article. Some of the results are included for the degenerate Anderson model with a strong correlation.

1. Introduction

The Anderson Hamiltonian has been considered to be one of the most useful Hamiltonian in the field of the solid state physics [1]. In fact this Hamiltonian has been adopted to explain the magnetic and electronic phenomena in dilute alloys including the Kondo effect, the valence-fluctuation phenomena in the rare-earth compounds [2] and the dynamical processes of the adsorbed atom on metal surfaces [3]. Therefore, lots of theoretical papers appeared with the aim of investigating this model more precisely, in which the renormalization-group approach was done very successfully [4]. Furthermore, a recent study of the Anderson model by means of the Bethe ansatz method enables us to treat this model exactly [5,6]. In this article, the results obtained by this method are reviewed [5-13].

The Anderson Hamiltonian with a single localized orbital is [1]

$$H_A = \sum_{k,\sigma} \varepsilon_k C^+_{k\sigma} C_{k\sigma} + \sum_{k,\sigma} V_k (C^+_{k\sigma} d_\sigma + d^+_\sigma C_{k\sigma}) + \varepsilon_d \sum_\sigma d^+_\sigma d_\sigma + U d^+_\uparrow d_\uparrow d^+_\downarrow d_\downarrow \quad . \tag{1}$$

Here ε_k is the energy of a conduction-electron state; ε_d is the unperturbed energy of the d state on an impurity atom; U is the Coulomb repulsive energy between d electrons; and V_k is the matrix element of the hybridization. Supposing that V_k does not depend on k and the values of U, $|\varepsilon_d|$ and $|V|$ are small compared with the Fermi energy, the following deductions can be made: the system reduces to a one-dimensional one; the impurity effect of electron states far from the Fermi surface on the physical quantities can be neglected; then the energy spectrum ε_k may be expressed as a linear function of k. It was shown by Wiegmann that with the above conditions this Hamiltonian can be diagonalized by means of the Bethe ansatz method [5]. The following results can be deduced:

$$\exp[i(k_j L - \phi_j)] = \prod_{\beta=1}^{M} \frac{B(k_j) - \Lambda_\beta + iUV^2/2}{B(k_j) - \Lambda_\beta - iUV^2/2} \quad , \qquad j=1,2,\cdots,N, \tag{2a}$$

---

* The review article for the Anderson model is published in Adv. in Phys. 32, 453 (1983) by A.M. Tsvelick and P.B. Wiegmann and also the review article for its thermodynamics in J. App. Phys. 55, 1931 (1984) by A. Okiji and N. Kawakami.

$$-\prod_{j=1}^{N}\frac{B(k_j)-\Lambda_\alpha-iUV^2/2}{B(k_j)-\Lambda_\alpha+iUV^2/2}=\prod_{\beta=1}^{M}\frac{\Lambda_\alpha-\Lambda_\beta+iUV^2}{\Lambda_\alpha-\Lambda_\beta-iUV^2}\quad,\quad \alpha=1,2,\cdots,M, \tag{2b}$$

where $\phi_j=2\tan^{-1}[V^2/2(k_j-\varepsilon_d)]$, $B(k)=k(k-U-2\varepsilon_d)$. Here L is the length of the system, k_j denotes a quasimomentum, Λ_α is a variable related to the spin, N(M) is the total (up-spin) electron number and the energy is measured from the Fermi energy.

2. Ground-state properties

In this section we describe the properties of this model at zero temperature briefly. The solutions of Eqs.(2) describing the ground and excited states can be found in the complex plane in general. We found [6] from Eqs.(2) that the solution for the singlet ground-state can be expressed by the complex $k_\alpha^\pm$ which satisfy $B(k_\alpha^\pm)=\Lambda_\alpha\pm iUV^2/2$, where Λ_α is real, and the magnetic excitations can be expressed by the real k. Then we describe the physical quantities at zero temperature with the use of the real Λ and k. In the thermodynamic limit, by introducing the distribution functions for the real Λ and k, Eqs.(2) can be rewritten in the following formulae [6-8],

$$\sigma(\Lambda)-\int_{-\infty}^{a}R(\Lambda-\Lambda')\sigma(\Lambda')d\Lambda'+\int_{-\infty}^{b}Q[\Lambda-B(k)]\rho(k)dk=\sigma_s(\Lambda)\quad, \tag{3a}$$

$$\rho(k)+B'(k)\int_{-\infty}^{b}R[B(k)-B(k')]\rho(k')dk'+B'(k)\int_{-\infty}^{a}Q[\Lambda-B(k)]\sigma(\Lambda)d\Lambda=\rho_s(k)\quad, \tag{3b}$$

where

$$\sigma_s(\Lambda)=\int_{-\infty}^{\infty}Q[\Lambda-B(k)][1/2\pi+\Delta(k)/L]dk\quad, \tag{4a}$$

$$\rho_s(k)=1/2\pi+\Delta(k)/L+B'(k)\int_{-\infty}^{\infty}R[B(k)-B(k')][1/2\pi+\Delta(k')/L]dk'\quad. \tag{4b}$$

Here $B'(k)=dB(k)/dk$ and $\Delta(x)$, $R(x)$ and $Q(x)$ are introduced as the Fourier transforms of $\exp(-V^2|\omega|/2+i\omega\varepsilon_d)$, $[1+\exp(UV^2|\omega|)]^{-1}$, and $(1/2)\mathrm{sech}(UV^2\omega/2)$. Note that the width of the virtual bound state is $V^2/2$ and the width of the conduction band is infinity in this calculation. The distribution functions $\sigma(\Lambda)$ and $\rho(k)$ are divided into two terms, that is, $\sigma=\sigma^c+\sigma^i/L$ and $\rho=\rho^c+\rho^i/L$ where σ^c and ρ^c express the distribution functions for conduction electrons and σ^i and ρ^i those for the impurity parts. Furthermore, it is to be noted that the parameters a and b appearing in the above expressions can be determined by minimizing the energy of this system where the values of U,ε_d and the applied magnetic field H are given. Consequently, it is found that the following conditions [7] must be used for determining the values of a and b,

$$\int_{-\infty}^{a}\sigma^c(\Lambda)d\Lambda=(U+2\varepsilon_d)/2\pi\quad,\quad\int_{-\infty}^{b}\rho^c(k)dk=H/2\pi\quad, \tag{5}$$

where $g\mu_B=1$. It is easily understood that the impurity part of the magnetization S_z and the average number of the localized electron n_d can be written as

$$S_z=\frac{1}{2}\int_{-\infty}^{b}\rho^i(k)dk\quad,\quad n_d=1-\int_{-\infty}^{a}\sigma^i(\Lambda)d\Lambda\quad. \tag{6}$$

2.1 Ground-state energy

The ground-state energy can be expressed as

$$E^i = 2\mathrm{Re}\int_a^\infty k_\alpha^+ \sigma^i(\Lambda_\alpha)d\Lambda_\alpha + \int_{-\infty}^b (k-H/2)\rho^i(k)dk \quad .$$

In particular, the ground-state energy for the symmetric case ($U=-2\varepsilon_d$) in a zero field can be calculated analytically and the following expression is deduced [6]:

$$E_s^i = E_s^i(U=0)+(V^2/\pi)\mathrm{Re}(1+iu)\log(1+iu)-\frac{V^2}{2\pi}\int_0^\infty dx\int_{-\infty}^\infty dy\,\frac{\sqrt{x}\,\mathrm{cosech}\,\pi(x+y^2)}{(y+\sqrt{u}/2)^2+1/4u} \quad , \tag{7}$$

where $u=U/V^2$. Using the above expression, we can obtain the perturbational results [14] for $u \ll 1$ and the energy gain proportional to $T_K[=(UV^2/\pi^2)^{\frac{1}{2}}\exp(-\pi u/4)]$ for $u \gg 1$ (Kondo model) [15]. It is noteworthy to point out that the exact expressions for the ground-state energy and for T_K cannot be obtained for the Kondo model by using the Bethe ansatz method [16]. However, these expressions for the Kondo model can be obtained by calculating them from the symmetric Anderson model for $u \gg 1$.

2.2 Magnetic and charge susceptibilities

With the use of Eqs.(5) and (6), the magnetic susceptibility which is defined as $\chi_m=dS_z/dH$ can be expressed by the following form [11]

$$4\pi\chi_m = [C_aD_b\rho^i(b)/\rho^c(b)-C_bD_a\sigma^i(a)/\sigma^c(a)][C_aD_b-C_bD_a]^{-1} \quad , \tag{8}$$

where

$$C_x = [\partial/\partial x\int_{-\infty}^a \sigma^c(\Lambda)d\Lambda]/\sigma^c(a) \quad , \quad D_x = [\partial/\partial x\int_{-\infty}^b \rho^c(k)dk]/\rho^c(b) \quad , \quad x=a,b.$$

In the symmetric Anderson model at H=0, the magnetic susceptibility is written down explicitly [8]. That is

$$\tilde{\chi}_m = \tilde{\chi}_P + \tilde{\chi}_K \quad ,$$
$$\tilde{\chi}_P = \int_{-\infty}^\infty \frac{(1+u^2-4ux^2)e^{-\pi x^2}dx}{(1+u^2-4ux^2)^2+16u^3x^2} \quad , \tag{9}$$
$$\tilde{\chi}_K = \pi(2\sqrt{u})^{-1}\exp[\pi(u-u^{-1})/4] \quad ,$$

where $\tilde{\chi}_m=\pi V^2\chi_m$. It is thought that $\tilde{\chi}_P$ corresponds to the susceptibility of the Pauli paramagnetism and $\tilde{\chi}_K$ to the susceptibility of the Kondo model [16,17]. For $u \gg 1$, $\tilde{\chi}_P$ tends to zero and $\tilde{\chi}_m$ becomes $\tilde{\chi}_K(=1/2\pi T_K)$ as has been expected (see Ref.8).

The charge susceptibility defined as $\chi_c=-dn_d/d\varepsilon_d$ can be expressed as [11]

$$\pi\chi_c = [C_aD_b\sigma^i(a)/\sigma^c(a)-C_bD_a\rho^i(b)/\rho^c(b)][C_aD_b-C_bD_a]^{-1} \quad . \tag{10}$$

For the symmetric Anderson model [7] at H=0,

$$\tilde{\chi}_c = \int_{-\infty}^\infty \frac{(1+u^2+4ux^2)e^{-\pi x^2}dx}{(1+u^2+4ux^2)^2-16u^3x^2} \quad , \tag{11}$$

where $\tilde{\chi}_c=\pi V^2\chi_c/4$. The numerical results of $\tilde{\chi}_c$ are given in Ref.9. For $u \gg 1$

(Kondo model), $\tilde{\chi}_c$ tends to zero because the charge fluctuation is depressed in this limit. The numerical results for the magnetic and charge susceptibility in the asymmetric Anderson model are given in Ref.7.

2.3 Magnetoresistance

From Eqs.(6), the number of the localized electron with up spin, $n_{d\uparrow}=(n_d-2S_z)/2$ and that with down spin, $n_{d\downarrow}=(n_d+2S_z)/2$, can be obtained. Following the Friedel theory [18], we may write the phase-shift as $\delta_\uparrow=\pi n_{d\uparrow}=\pi(n_d-2S_z)/2$, $d_\downarrow=n_{d\downarrow}=\pi(n_d+2S_z)/2$, and the magnetoresistance as [18]

$$R(U,\varepsilon_d,V^2,H) = \frac{1}{2} R_0 \frac{[\cos\pi n_d-\cos 2\pi S_z]^2}{[1-\cos\pi n_d \cos 2\pi S_z]} ,$$

where

$$R_0 = R(U=0,\ \varepsilon_d=0,\ H=0) \quad .$$

The behaviour patterns of the magnetoresistance are shown in Fig.1 [7].

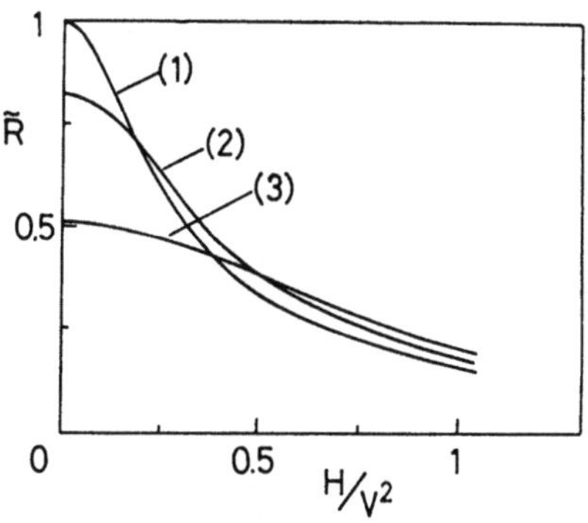

Fig.1. The magnetoresistance $\tilde{R}=R/R_0$ as a function of H. $U/V^2=2.0$: (1) $\varepsilon_d/V^2=-2.12$ and 0.12, (2) −1.65 and −0.35, (3) −1.0 .

3. Thermodynamic properties

In order to examine the thermodynamic properties of the Anderson model, we must seek all the solutions of Eqs.(2) in the complex plane. In the thermodynamic limit, these solutions are grouped into the complex Λ, the complex k and the real k. The solution of the complex Λ has the following form [9,19]:

$$\Lambda_\alpha^{n,\ell} = \Lambda_\alpha^n+i(n+1-2\ell)UV^2/2 \quad , \quad \ell=1,2,\cdots,n, \quad n=1,2,\cdots,\infty,$$

where Λ_α^n is real and the sets of $\Lambda_\alpha^{n,\ell}$ are the nth order strings [19].

The solution of the complex k relates to $\Lambda_\alpha^{n,\ell}$ through the relation $B(k_\alpha^{n,\ell})=\Lambda_\alpha^{n,\ell}\pm iUV^2/2$. The distribution functions for above solutions should be determined by minimizing the thermodynamic potential. From this procedure we can obtain the expression for the impurity contribution to the thermodynamic potential [9-13]:

$$\Omega^i = E_s^i - T\int_{-\infty}^{\infty} \sigma_s^i G(\varepsilon_1^+)d\Lambda - T\int_{-\infty}^{\infty} \rho_s^i G(-\kappa)dk \quad , \tag{12}$$

where $G(x)=\log[1+\exp(x/T)]$ and $\sigma_s^i(\rho_s^i)$ is given by the impurity part of Eq.(4). Note that Ω^i is expressed in terms of the pseudoenergies κ and ε_1^+ instead of the distribution functions. These pseudoenergies are determined by the following set of the integral equations [9-13]:

$$\begin{aligned}
\kappa(k) &= \kappa_s + T\int_{-\infty}^{\infty} Q(B(k)-\Lambda)[G(\varepsilon_1^+)-G(\varepsilon_1^-)]d\Lambda \quad , \\
\varepsilon_1^{\pm}(\Lambda) &= TQ*G(\varepsilon_2^{\pm}) - T\int_{-\infty}^{\infty} B'(k)Q[B(k)-\Lambda]G(\pm\kappa)dk \quad , \\
\varepsilon_n^{\pm}(\Lambda) &= TQ*[G(\varepsilon_{n-1}^{\pm})+G(\varepsilon_{n+1}^{\pm})] \quad , \quad n>>2 \quad , \\
\lim_{n\to\infty} \varepsilon_n^+/n &= U+2\varepsilon_d \quad , \quad \lim_{n\to\infty} \varepsilon_n^-/n = H \quad ,
\end{aligned} \tag{13}$$

where $*$ means convolution and

$$\kappa_s(k) = k-\tfrac{1}{2}U - \varepsilon_d + 2\int_{-\infty}^{\infty}(x-\tfrac{1}{2}U-\varepsilon_d)^2 R[B(k)-B(x)]dx \quad .$$

3.1 Low temperature properties

Although the analytical treatment of Eqs.(13) is quite difficult for an arbitrary temperature, the low temperature quantities can be calculated analytically with the use of the functions at zero temperature [9-11]. For instance the calculation of the thermodynamic potential up to T^2 gives the expression for the coefficient of the T-linear specific heat as follows [11]

$$\tilde{\gamma} = (V^2/8)[\sigma^i(a)/\sigma^c(a)+\rho^i(b)/\rho^c(b)] \quad , \tag{14}$$

where $\tilde{\gamma}$ is normalized at $U=H=\varepsilon_d=0$. From Eqs.(8),(10), and (14) we derive the exact relation between the coefficient of the T-linear specific heat and susceptibilities at T=0 [9-11,14],

$$\tilde{\gamma} = (\tilde{\chi}_m+\tilde{\chi}_c)/2 \quad , \tag{15}$$

for arbitrary values of U, ε_d, and H. This simple relation, which was proposed firstly by YAMADA [14] for the symmetric case by means of the perturbational method, shows that $\tilde{\chi}_m$ and $\tilde{\chi}_c$ are basic quantities which characterize the low temperature thermodynamics of the Anderson model. Now we consider the ratio $W=\tilde{\chi}_m/\tilde{\gamma}$ for the symmetric case in a zero field. It is seen from Fig.2 that as U is increased, the value W increased from unity to 2 rather rapidly. The limiting value of 2, which is called Wilson's ratio, was obtained a decade ago by Wilson numerically [16], and also by Nozieres

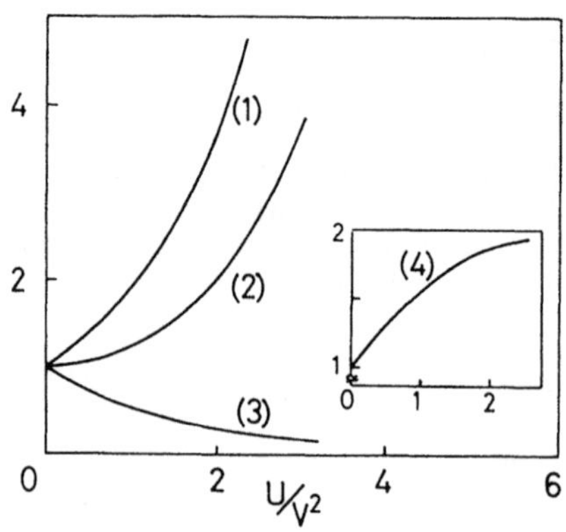

Fig.2. Curves (1),(2),(3) and (4) represent $\tilde{\chi}_m$, $\tilde{\gamma}$, $\tilde{\chi}_c$ and $\tilde{\chi}_m/\tilde{\gamma}$ as a function of U/V^2 in the symmetric case.

with Fermi liquid theory [20], both in the Kondo model. In Fig.2 the numerical results of χ_m, $\tilde{\gamma}$ and $\tilde{\chi}_c$ are also shown for the symmetric Anderson model. The results of $\tilde{\gamma}$ for the asymmetric case in an arbitrary H are given in Ref.11.

3.2 Temperature-dependent susceptibility and specific heat

We show below the numerical results for the magnetic susceptibility and the specific heat over a physically relevant temperature range [12,13]. The results are discussed in two interesting cases: (1) Symmetric case, in connection with the Kondo effect; (2) asymmetric case ($U \gg |\varepsilon_d|, V^2/2$), as a simplified model for the valence-fluctuation phenomena [2]. The numerical calculation is carried out by truncating the integral equations (13) at finite numbers n and by converting the integral equations into a set of matrix equations. These matrix equations are iterated until the results attain stability. The numerical results are thought to be accurate to within a few percent.

(a) Symmetric case

The results obtained for the symmetric Anderson model [12] are given in Figs. 3-5. The curves for the magnetic susceptibility are shown in Fig.3 as a function of the temperature for various values of U. In the high temperature region $T \gg \max(U, V^2/2)$, all the impurity configurations of the valence number $n_d=0,1,2$ are equally populated and effectively degenerate, so that the effective Curie constant $T\chi_m$ takes the value 1/8. At U=0, $T\chi_m$ changes monotonically from 1/8 to zero as the temperature is decreased, which means a transition from a high temperature region to the nonmagnetic region. As U is increased, the value of $T\chi_m$ increases and a localized moment is developed gradually in the temperature region T<U. The localized moment is, however, quenched due to the correlation effect of U in the low temperature region (strong coupling region). This phenomenon means nothing other than the Kondo effect. In this case the magnetic susceptibility takes the value $\chi_m \sim 1/2\pi T_K$ at T=0, where the Kondo temperature T_K is defined as $(UV^2/\pi^2)^{\frac{1}{2}}\exp[-(\pi/4)(U/V^2-V^2/U)]$ in this calculation. The results for temperatures $T \lesssim T_K$ are in accordance with those of the Kondo model [16] (curve 3 in Fig.3). The whole nature of the above results for the magnetic

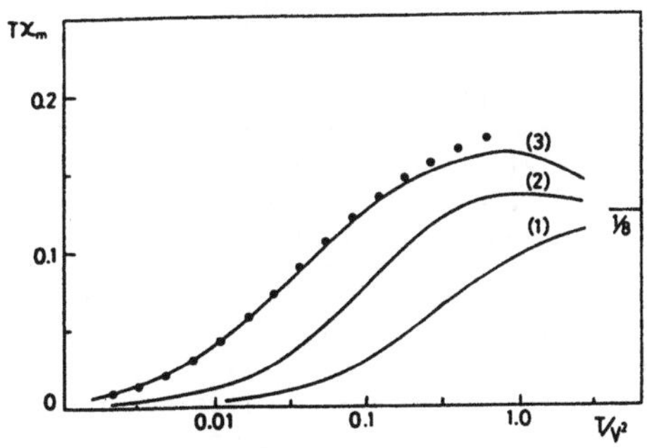

Fig.3. Plots of $T\chi_m$ vs $\log(T/V^2)$ for the symmetric cases where (1) $U/V^2=0$, (2) 2.0 and (3) 4.0. The points are the renormalization-group calculation results for the Kondo model [4,16].

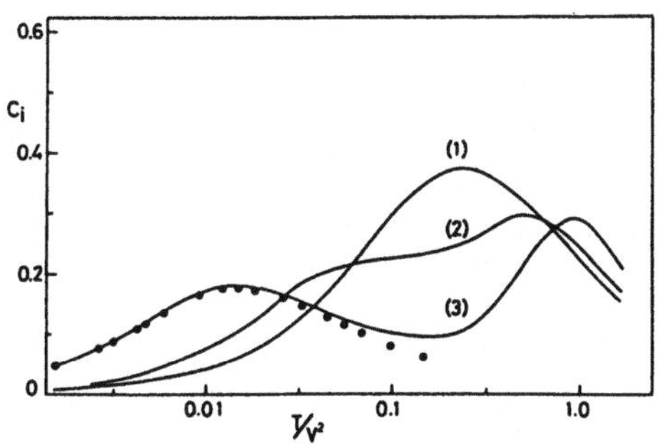

Fig.4. Plots of specific heat C_i vs $\log(T/V^2)$ for the symmetric cases where (1) $U/V^2=0$, (2) 2.0 and (3) 4.0. The points are the values of the exact calculation of the Kondo model [16].

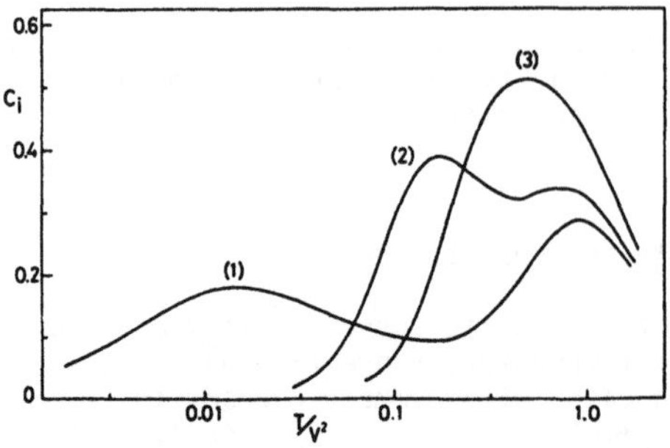

Fig.5. Plots of specific heat C_i vs $\log(T/V^2)$ for the symmetric case U/V^2=4.0 and the magnetic field (1) H/V^2=0 (2) 0.5 and (3) 1.0.

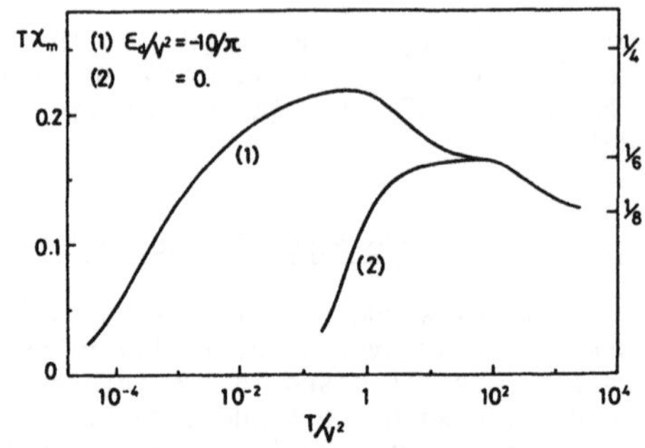

Fig.6. Plots of $T\chi_m$ vs $\log(T/V^2)$ for $U/V^2=10^3/\pi$. The values of ε_d/V^2 are (1) $-10/\pi$ and (2) 0 .

susceptibility shows close agreement with the results of the renormalization-group calculation [4].

The curves for the specific heat as a function of the temperature are given in Fig.4. As U is increased, the peak caused by the Kondo effect appears on the low temperature side. In the case of U/V^2=4.0 the results on the low temperature side agree with those obtained by the Kondo model [16]. The effect of a magnetic field on the specific heat at U/V^2=4.0 is shown in Fig.5. As the magnetic field is increased, the peak of the specific heat caused by the Kondo effect shifts [12,16], just as the Schottky specific heat for a free-spin $C_{FS}=(H/2T)^2\text{sech}^2(H/2T)$ does, and is adsorbed into the peak on the high temperature side. These results show that the many-body bound state (Kondo state) caused by the strong correlation at the impurity site is destroyed by the magnetic field in the same way as it is by the temperature.

(b) Asymmetric case

The results for the asymmetric Anderson model of $U \gg \frac{1}{2}V^2$, $|\varepsilon_d|$ are shown in Figs.6-8. First we discuss the results for H=0. In the high temperature region T>>U, which is characterized by $T\chi_m\sim 1/8$, all the impurity configurations are equally populated and effectively degenerate, so that the value of the specific heat almost vanishes and the impurity occupancy $n_d(T)\sim 1$. As the temperature is decreased, high energy excitations to the configuration space of valence number n_d=2 are gradually suppressed and consequently a transition occurs from the high-temperature region to the valence-fluctuation region where $T\chi_m\sim 1/6$ and $n_d(T)\sim 2/3$. Since in the valence-fluctuation region the n_d=0 and n_d=1 subspaces are still equally populated, the specific heat is not sensitive to the change of temperature and takes the value $C_i\sim 0$ [13]. It is seen from Figs.7 and 8 that the specific heat exhibits the Schottky-type behaviour reflecting the transition from the high temperature region to the valence-fluctuation region. This behaviour can be described by the formula at V=0, $C_{HT}=3(U/T)^2\exp(-U/T)/[3+\exp(-U/T)]^2$. In the low temperature region, the situations are sensitive to the position of ε_d. If $|\varepsilon_d|(\varepsilon_d<0)$ is suitably large as compared with $V^2/2$, the contribution to the thermal excitations from the charge fluctuations gradually vanishes as the temperature is decreased. Consequently the $T\chi_m$ value approaches 1/4, characteristic of the local moment region, and the specific heat shows the Schottky-type behaviour reflecting a transition from the valence-fluctuation region to the local moment region (curve 1 in Fig.7). In this case the ground-state has nearly integral valence $n_d(T)\sim 1$. Therefore we observe the Kondo effect in the temperature region

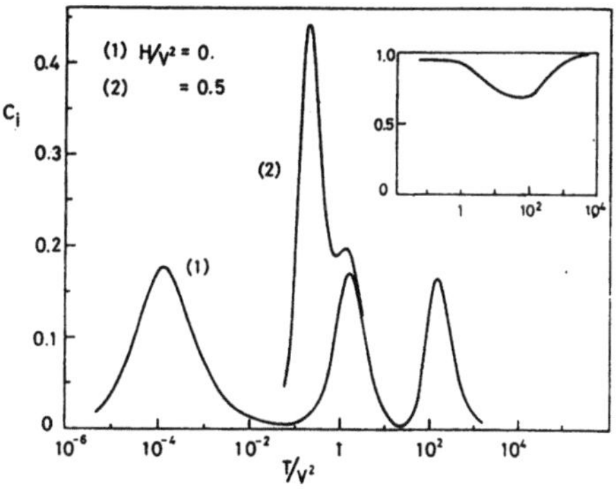

Fig.7. Plots of specific heat C_i vs $\log(T/V^2)$ for $U/V^2=10^3/\pi$ and $\varepsilon_d/V^2=-10/\pi$. The impurity occupancy $n_d(T)$ for H=0 is plotted vs $\log(T/V^2)$ in the inset.

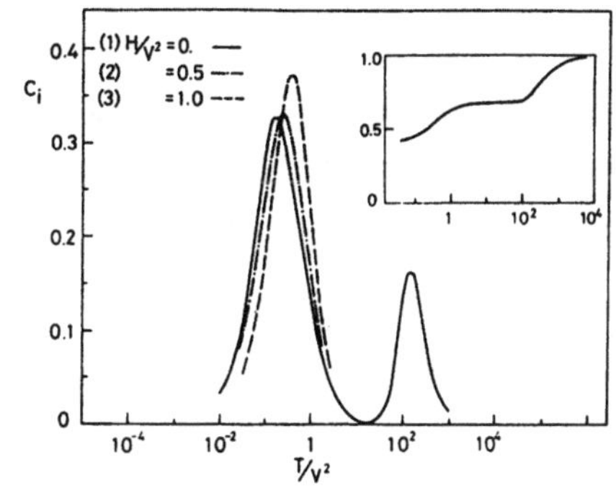

Fig.8. Plots of specific heat C_i vs $\log(T/V^2)$ for $U/V^2=10^3/\pi$ and $\varepsilon_d=0$. The impurity occupancy $n_d(T)$ for H=0 is plotted vs $\log(T/V^2)$ in the inset.

$T \ll V^2/2$. The relevant Kondo temperature [4,10,22] is given by $T_K=(UV^2/\pi^2)\exp[\pi\varepsilon_d(\varepsilon_d+U)/UV^2]$. If the impurity level ε_d approaches the Fermi level, the value of T_K increases and the universal curves for the susceptibility and the specific heat shift to the high temperature side. When $|\varepsilon_d|\lesssim(1/2)V^2$, the Kondo effect disappears and a direct transition occurs from the valence-fluctuation region to the strong coupling region (ground-state with intermediate valence). The variation of the susceptibility and the specific heat corresponding to this transition is shown in the low temperature behaviour of the curve 2 in Fig.6 and the curve 1 in Fig.8. Note that the results obtained for the magnetic susceptibility show close agreement with those of the renormalization-group calculation [4].

The effect of a magnetic field on the specific heat is also shown in Figs.7 and 8. As has been discussed for the symmetric case, the Schottky-type specific heat corresponding to the Kondo effect leads gradually to the specific heat for a free-spin C_{FS} as the magnetic field is increased (Fig.7). It is seen from Fig.8 that the effect of a magnetic field on the specific heat is not so drastic for $|\varepsilon_d|\lesssim(1/2)V^2$(intermediate valence state) as compared with $|\varepsilon_d|>(1/2)V^2$(Kondo state). In this case as the magnetic field is increased the Schottky-type specific heat at the low temperature side leads asymptotically to that for a high magnetic field $H\gg(1/2)V^2,|\varepsilon_d|$,

$$C_{HF} = \frac{(H/2T)^2[\exp(H/2T)+\exp(-H/2T)+4]}{[\exp(H/2T)+\exp(-H/2T)+1]^2}\,, \quad \text{as far as } H\ll U.$$

4. The degenerate Anderson model with a strong correlation

The thermodynamic properties are investigated for the degenerate Anderson model with a strong correlation ($U\to\infty$). Since the detailed formulation for this model in a crystalline field is given in the next article [23], only the obtained results without the effect of a crystalline field are summarized here. The T-linear specific heat C_i can be written as [26],

$$C_i/T = \frac{2\pi^2}{(J+1)(2J+1)}\chi_m + \frac{\pi^2}{3(J+1)}\chi_c\,, \tag{16}$$

where J is the total angular momentum for the ground multiplet of the impurity, χ_m and χ_c are the magnetic and charge susceptibilities at zero temperature [24,25]. Relation (16) is considered as a generalized Fermi-liquid relation for arbitrary J and ε_f. If ε_f lies far above the Fermi level, the value of the ratio $W=T\chi_m/C_i$ tends to $J(J+1)/\pi^2$, which agrees with the results obtained by perturbation theory [27] and the Fermi-liquid theory [28]. On the other hand, if ε_f lies far below the Fermi level, W tends to a generalized Wilson's value $(J+1)(2J+1)/2\pi^2$ [29]. Here, ε_f is the energy of a localized electron.

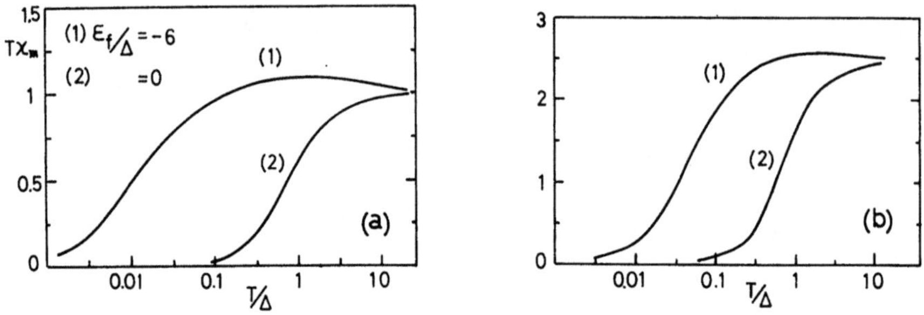

Fig.9. Plots of the effective Curie constant $T\chi_m$ vs $\log(T/\Delta)$ for (a) J=3/2 and (b) J=5/2.

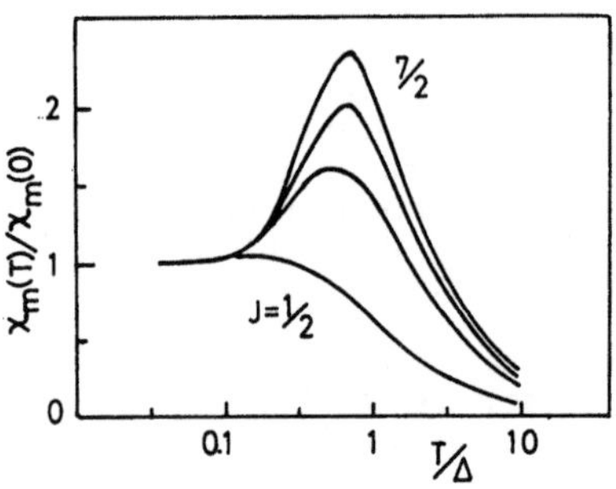

Fig.10. Plots of $\chi_m(T)/\chi_m(0)$ vs $\log(T/\Delta)$ for $\varepsilon_f=0$. The values of $\Delta\chi_m(0)$ for J=1/2, 3/2, 5/2 and 7/2 are 0.17, 0.45, 0.89 and 1.5, respectively.

The numerical results of the magnetic susceptibility are shown in Figs.9 and 10. Firstly we plot the results of the effective Curie constant $T\chi_m$ in Fig.9. In the high-temperature regime $T\gg\Delta(=V^2/2)$, in which both the spin and charge excitations are thermally populated, $T\chi_m$ takes the value of $J(2J+1)/6$. This temperature regime has been referred to as the valence-fluctuation regime in the J=1/2 Anderson model with finite U [4,12,13]. As the temperature is decreased, the value of $T\chi_m$ increases and a local moment is gradually developed (local moment regime) when ε_f lies suitably below the Fermi level (curve (1) in Fig.9a,b). In this case, the Kondo effect is observed in the low-temperature regime and the behaviour of $T\chi_m$ in this regime is essentially the same as that for the Coqblin-Schrieffer model [30]. On the other hand, when ε_f lies near the Fermi level (curve (2) in Fig.9a,b), no local moment is developed and a direct transition occurs from the high-temperature regime to the nonmagnetic regime ($T\chi_m\to 0$). From the results of $T\chi_m$, we can say that the general aspect of this kind of crossover is independent of J and is essentially the same as that for the J=1/2 case [4,12,13

The effect of degeneracy, however, appears rather significantly in the behaviour of the susceptibility curve at low temperatures. In order to see this explicitly we plot the magnetic susceptibility χ_m for $\varepsilon_f=0$ as a function of the temperature in Fig.10. It is seen that the susceptibility curve has a maximum χ_{max} at a finite temperature and the value of χ_{max}/χ_m (T=0) increases as J increases. Such a susceptibility maximum has appeared previously in the theoretical calculation of the Anderson model [4,13,27,28] and the Coqblin-Schrieffer model [30]. In the Coqblin-Schrieffer model ($-\varepsilon_f>>\Delta$), the susceptibility curve is universal and the value of χ_{max}/χ_m (T=0) is also universal for a fixed value of J. The numerical results for this model are given in ref.[30]. As ε_f approaches the Fermi level, the effect of the charge fluctuation becomes conspicuous. In this case the value of χ_{max}/χ_m (T=0) gradually increases from the value for the Coqblin-Schrieffer model [30] (while χ_m(T=0) decreases) and the susceptibility curves lead to those shown in Fig.10. Hence the effect of the charge fluctuation causes the increase of χ_{max}/χ_m(T=0) [31].

5. Summary

We have investigated the ground-state and thermodynamic properties of the Anderson model precisely by means of the Bethe-ansatz method. The Kondo effect has been discussed and its thermodynamic properties have been clarified completely from the standpoint of the correlation effect due to U with the use of the symmetric Anderson model. The nature of the asymmetric Anderson model has been discussed in connection with the valence-fluctuation phenomena.

References

1. P.W. Anderson: Phys. Rev. 124, 41 (1961)
2. See, e.g., Valence Instabilities ed. by P. Wachter and H. Boppart (North-Holland, Amsterdam 1982)
3. See, e.g., Dynamical Processes and Ordering on Solid Surfaces ed. by A. Yoshimori and M. Tsukada (Springer-Verlag, Berlin 1985)
4. H.R. Krishnamurthy, J.W. Wilkins and K.G. Wilson: Phys. Rev. B21, 1003, 1044 (1980)
5. P.B. Wiegmann: Phys. Lett. 80A, 163 (1980)
6. N. Kawakami and A. Okiji: Phys. Lett. 86A, 483 (1981)
7. N. Kawakami and A. Okiji: J. Phys. Soc. Jpn. 51, 1145, 2043 (1982); A. Okiji and N. Kawakami: J. Phys. Soc. Jpn. 51, 3192 (1982); P.B. Wiegmann and A.M. Tsvelick: Sov. Phys. JETP Lett. 35, 120 (1982); J. Phys. C16, 2281 (1983)
8. A. Okiji and N. Kawakami: Solid State Commun. 43, 365 (1982); P.B. Wiegmann, V.M. Filyov and A.M. Tsvelick: Sov. Phys. JETP Lett. 35, 92 (1982)
9. N. Kawakami and A. Okiji: Solid State Commun, 43, 467 (1982); V.M. Filyov, A.M. Tsvelick and P.B. Wiegmann: Phys. Lett. 89A, 157 (1982)
10. A.M. Tsvelick and P.B. Wiegmann: Phys. Lett 89A, 368 (1982); J. Phys. C16, 2321 (1983)
11. N. Kawakami and A. Okiji: J. Phys. Soc. Jpn. 52, 1119 (1983)
12. A. Okiji and N. Kawakami: Phys. Rev. Lett. 50, 1157 (1983)
13. N. Kawakami and A. Okiji: Phys. Letts 98A, 54 (1983); Phys. Rev. Lett. 51, 2011 (1983)
14. K. Yamada: Prog. Theor. Phys. 53, 970 (1975)
15. See, e.g., K. Yosida and A. Yoshimori: Magnetism V ed. by H. Suhl (Academic, New York 1973) p.253
16. K.G. Wilson: Rev. Mod. Phys. 47, 773 (1975); N. Andrei, K. Furuya and J.H. Lowenstein: Rev. Mod. Phys. 55, 331 (1983)

17. It seems that the expression for χ_K is singular at U=0. It is, however, shown by V. Zlatic and B. Horvatic (Phys. Rev. B28, 6904 (1983)) that the expression for χ_m is regular at U=0. This means that the physical quantities calculated by the perturbational expansion by U are correct.
18. J. Friedel: Nuovo Cimento VII, 287 (1958);
A. Sakurai and K. Yosida: Prog. Theor. Phys. 61, 1597 (1979)
19. See, e.g., H.B. Thacker: Rev. Mod. Phys. 53, 253 (1981)
20. P. Nozieres: J. Low Temp. Phys. 17, 31 (1974)
21. F.D.M. Haldane: Phys. Rev. Lett. 40, 416 (1978)
22. F.D.M. Haldane: J. Phys. C11, 5015 (1978)
23. N.Kawakami and A. Okiji: in this volume
24. P. Schlottmann: Phys. Rev. Lett. 50, 1697 (1983); Z. Phys. B51, 49 (1983)
25. E. Ogievetski, A.M. Tsvelick and P.B. Wiegmann: J. Phys. C16, L797 (1983)
26. N. Kawakami, S. Tokuono and A. Okiji: J. Phys. Soc. Jpn. 53, 51 (1984)
27. H. Lustfeld and A. Bringer: Solid State Commun. 28, 119 (1978)
28. D.M. Newns and A.C. Hewson: J. Phys. F10, 2429 (1980)
29. P. Nozieres and A. Blandin: J. Phys. (Paris) 41, 193 (1980)
30. V.T. Rajan: Phys. Rev. Lett. 51, 308 (1983)
31. N. Kawakami and A. Okiji: Phys. Lett. 103A, 205 (1984)

Exact Solution of the Anderson Model and Its Thermodynamics II — Including Crystalline Field and Spin-Orbit Coupling

N. Kawakami and A. Okiji

Department of Applied Physics, Osaka University, Suita, Osaka 565, Japan

The thermodynamic properties of the highly correlated degenerate Anderson model are discussed on the basis of the Bethe-ansatz method, in the presence of the crystalline field and the spin-orbit coupling. The properties of the specific heat are investigated in detail, as a typical example, for the case of Ce impurities in a cubic crystalline field. The analytic expressions are given for the Fermi-liquid relation, the coefficient of the T-linear specific heat and the effective Kondo temperature. The curve of the temperature-dependent specific heat is shown to become asymmetric and have a shoulder structure when the splitting of the crystalline field is increased.

1. Introduction

In the article I [1], the Bethe-ansatz treatment for the thermodynamic properties of the Anderson model has been reviewed in connection with the Kondo problem (symmetric case) and also as a simplified model for the valence-fluctuation phenomena (asymmetric case). This method has been already extended to the more realistic cases, the Coqblin-Schrieffer model [2-4] and the highly-correlated degenerate Anderson model [5-8] (latter model includes former as a special case). These models have been intensively studied in order to understand the properties of the dilute impurity systems such as Ce (Yb) impurities in a metal and recently to clarify the local properties of the valence-fluctuation phenomena in the concentrated systems. In the calculation of the above orbital-degenerate models, the effect of the degeneracy has been found to play an important role at low-temperatures [2-8]. As has been shown in I, for example, the temperature-dependent susceptibility has a maximum at a finite temperature, which did not appear in the original Kondo problem, and the peak structure of the susceptibility curve becomes more prominent with the increase of the degeneracy [4,7-9]. These results have explained some experimental data of Ce (Yb) compounds rather successfully [8,9]. However, the crystalline field, which has been neglected in the above calculation, is expected to come into play and, if it is large, to affect the thermodynamic properties considerably. The investigation of the crystalline field effect on the problem has been done with the various theoretical methods, perturbation [10-12], scaling [13] and more recently Bethe-ansatz [14-16].

In this article, the results obtained by means of the Bethe-ansatz method are reviewed, for the thermodynamic properties of the highly-correlated degenerate Anderson model in the presence of the crystalline field and the spin-orbit coupling [15,16] (ground-state properties have been discussed in ref.14). In the next section, we summarize the procedure of the diagonalization briefly and obtain the basic algebraic equations. Then, we derive the equilibrium thermodynamic equations and discuss the properties of some limiting cases. In the following section the present theory is applied, as a typical example, to the case of Ce impurities in a cubic crystalline field. A brief summary is given in the last section.

2. Diagonalization

We consider the highly correlated degenerate Anderson model described by the Hamiltonian,

$$H_C = H_A + H_W \quad , \tag{2.1a}$$

$$H_A = \sum_{k,m} \epsilon_k C^+_{km} C_{km} + V \sum_{k,m} [C^+_{km} C_{fm} + C^+_{fm} C_{km}]$$

$$+ \epsilon_f \sum_m C^+_{fm} C_{fm} + U \sum_{m \neq m'} C^+_{fm} C_{fm} C^+_{fm'} C_{fm'} \quad , \tag{2.1b}$$

and $U \to \infty$, where $C^+_{fm}(1 \leq m \leq \nu)$ is the creation operator for the ν-fold degenerate electron states of the impurity atom and C^+_{km} is for the conduction electron in the partial wave representation. The second term H_W in (2.1a) represents the sum of the additional contributions from the magnetic field, the crystalline field and the spin-orbit coupling, which may cause the splitting of the ν-fold degenerate states in different ways. First, we consider the electron states of the impurity atom when V=0. We denote the energy splitting of the ν-fold degenerate states due to H_W, as $W_1, W_2, \cdots, W_\nu$, where the value of W_m $(1 \leq m \leq \nu)$ is chosen to satisfy the condition,

$$W_1 \leq W_2 \leq \cdots \leq W_\nu \quad , \quad \sum_{m=1}^{\nu} W_m = 0 \quad . \tag{2.2}$$

The value of W_m and the corresponding eigenfunctions can be obtained in the standard way if the form of H_W is given explicitly (see §4). In the present system including the additional field H_W, we should use the index m which is specified by W_m in the Hamiltonian H_C.

With the condition of the linear dispersion of the conduction electron (constant density of states), the Bethe-ansatz diagonalization of H_C is carried out in two steps. In the first step, we seek for a solution of H_C in a superposition of plane waves characterized by the quasimomentum $k_p^{(1)}$ ($p=1,2,\cdots,N_1$; N_1 is a number of total electrons), the so-called Bethe-ansatz [17]. On applying the antisymmetry and the periodic boundary condition, the problem is reduced to the ancillary problem within the "spin space" in which the permutation symmetry is characterized by Young's tableau with ν-rows (Fig.1). In the present system the particle-state belonging to the m-th row should be considered to correspond to the state specified by W_m. In order to complete the diagonalization of the ancillary problem with the "generalized Bethe-ansatz" [17,18], we necessarily introduce the additional quasimomentum $k_p^{(j)}$ $(2 \leq j \leq \nu)$. Consequently, the basic algebraic equations for $k_p^{(j)}$ are obtained as follows [5,14,15].

$$\exp(ik_p^{(1)}L)\,t(k_p^{(1)}-\epsilon_f) = \prod_{q=1}^{N_2} t(k_p^{(1)}-k_q^{(2)}) \quad , \tag{2.3a}$$

$$-\prod_{q=1}^{N_j} t\{(k_p^{(j)}-k_q^{(j)})/2\} = \prod_{r=1}^{N_{j-1}} t(k_p^{(j)}-k_r^{(j-1)}) \prod_{s=1}^{N_{j+1}} t(k_p^{(j)}-k_s^{(j+1)}) \quad , \quad 2 \leq j \leq \nu, \tag{2.3b}$$

where $t(x)=(x-i\Delta)/(x+i\Delta)$, $\Delta(=V^2/2)$ is the resonance width of the virtual bound-state and L is the length of the system. Here, $N_j - N_{j+1}$ $(N_{\nu+1}=0)$ is the number of electrons belonging to the j-th row specified by W_j in Young's tableau.

3. Thermodynamics

3.1 Thermodynamic equations

We have to consider all the possible solutions of Eq.(2.3) in order to investigate the equilibirum-thermodynamic properties of (2.1). The solution of (2.3) is classified in the complex plane as follows,
(a) n-th order charge complex $k_q^{(j)}$ which makes bound-states with the other solutions within the same column of Young's tableau (Fig.1):

$$k_q^{(j)} = k_q^{(j+1)} \pm i\Delta \quad ,$$

where $1 \le j \le n-1$, $1 \le n \le \nu$ and $k_q^{(n)}$ is real.
(b) m-th order spin complex $k_s^{(j)}$ which makes bound-states within the same row of Young's tableau:

$$k_s^{(j)} = k_{sm}^{(j)} + i(m+1-2\ell)\Delta \quad ,$$

where $2 \le j \le \nu$, $1 \le \ell \le m$ $(m=1,2,\cdots)$ and k_{sm} is real. We introduce the density functions for the real part of the above solutions in the thermodynamic limit: $\rho_n^{(j)}$ $(\tilde{\rho}_0^{(j)})$ for the electron (hole) of the j-th order charge complex and $\rho_m^{(j)}$ $(\tilde{\rho}_m^{(j)})$ for the electron (hole) of the m-th order spin complex. The expression for the various quantities can be written down with these density functions. For instance, the contribution from H_W in (2.1) to the thermodynamic potential is given by [15]

$$W = \sum_{j=1}^{\nu} W_j(N_j - N_{j+1})$$

$$= \sum_{j=1}^{\nu} \{ \sum_{i=1}^{j} W_i \int_{-\infty}^{\infty} \rho_0^{(j)}(k)dk + (W_j - W_{j-1}) \sum_{n=1}^{\infty} n \int_{-\infty}^{\infty} \rho_n^{(j)}(k)dk \} \quad . \qquad (3.1)$$

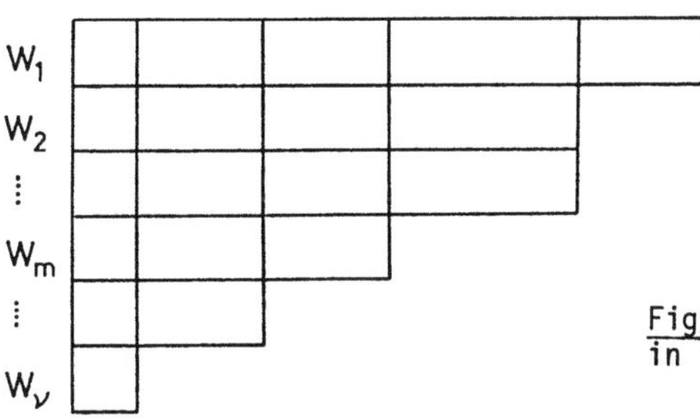

Fig.1. Young's tableau in the "spin space".

The above density functions should be determined by minimizing the thermodynamic potential. The resulting equations are expressed as follows in terms of the pseudoenergies defined by $\varepsilon_0^{(j)} = T\ln(\tilde{\rho}_0^{(j)}/\rho_0^{(j)})$ for $1 \le j \le \nu$ and $\varepsilon_n^{(j)} = -T\ln(\tilde{\rho}_n^{(j+1)}/\rho_n^{(j+1)})$ for $1 \le j \le \nu-1$, $1 \le n$:

$$\varepsilon_0^{(\nu)}/T = k/T + R*G(\varepsilon_0^{(\nu)}) + \sum_{j=1}^{\nu-1} P_j*G(-\varepsilon_0^{(j)}) \quad , \qquad (3.2a)$$

$$G(\varepsilon_n^{(j)}) = P_j*G(\varepsilon_0^{(\nu)})\delta_{n0} + \sum_{\ell=1}^{\nu-1} \{V_{j\ell}*G(-\varepsilon_n^{(\ell)}) - U_{j\ell}*[G(-\varepsilon_{n+1}^{(\ell)}) + G(-\varepsilon_{n-1}^{(\ell)})]\} \quad ,$$

$$\text{for } 1 \le j \le \nu-1 \ , \ n=0,1,2,\cdots, \qquad (3.2b)$$

$$\lim_{n\to\infty} \varepsilon_n^{(j)}/n = W_j - W_{j+1} \quad , \tag{3.2c}$$

where $*$ means convolution, T represents the temperature ($k_B=1$) and $G(x)=\ln(1+\exp(x/T))$. Here P_j and R_j are functions defined by $\sinh(j\Delta\omega)/\sinh(\nu\Delta\omega)$ and $\exp(-\Delta|\omega|)\sinh[(\nu-1)\Delta\omega]/\sinh(\nu\Delta\omega)$ in Fourier space and $U_{j\ell}$ and $V_{j\ell}$ are given by

$$U_{j\ell} = \sum_{i=1}^{z} P_{|j-\ell|+2i-1} \quad , \quad z = \min(j,\ell)$$

$$V_{j\ell} = \sum_{i=1}^{z} (P_{|j-\ell|+2i} + P_{|j-\ell|+2i-2}) \quad .$$

The impurity contribution to the thermodynamic potential is also expressed in terms of the pseudoenergies:

$$\Omega^{i} = -T \sum_{j=1}^{\nu} \int_{-\infty}^{\infty} \Phi_j(k) G(-\varepsilon_0^{(j)}) dk \quad , \tag{3.3}$$

where $\Phi_j(k) = (j\Delta/\pi)/\{(k-\varepsilon_f)^2+(j\Delta)^2\}$.

The formula (3.3) with the ancillary equations (3.2) can describe the equilibrium-thermodynamic properties of (2.1) completely if the form of H_W and the corresponding value of W_j are given explicitly [15]. It is seen from Eqs.(3.2) that the effect of H_W is reflected only in the boundary condition (3.2c) through the value of W_j. This simple result is of much help for the further analytical and numerical calculations. It should be noticed that an ultraviolet cut-off is necessarily introduced for $\varepsilon_0^{(\nu)}$, if one intends to calculate the physical quantities explicitly [7,8].

3.2 Coqblin-Schrieffer limit

If the impurity level ε_f lies far below the Fermi level ($-\varepsilon_f >> \Delta$), the thermal excitation related to the charge fluctuation is depressed. Hence the function $\varepsilon_0^{(\nu)}$ in Eq.(3.2a), which represents the charge fluctuation, is not connected with the other $\varepsilon_0^{(j)}$ and is given by the following equation at T=0:

$$\varepsilon_0^{(\nu)}(k) = k + \int_{Q_\nu}^{\infty} R(k-k')\varepsilon_0^{(\nu)}(k')dk' \quad , \tag{3.4}$$

where the cut-off parameter Q_ν is defined by $\varepsilon_0^{(\nu)}(Q_\nu)=0$ (see also Eq.(3.8)). In this case $\varepsilon_0^{(\nu)}$ enters into Eq.(3.2b) as a driving term, that is,

$$P_j * G(\varepsilon_0^{(\nu)}) = 2\sin(\pi j/\nu) T_0 e^{\pi k/\nu\Delta} \quad , \tag{3.5a}$$

where the Kondo temperature T_0 is given by

$$T_0 = \frac{1}{2\nu\Delta} \exp(\pi\varepsilon_f/\nu\Delta) \int_{Q_\nu}^{\infty} \exp(-\pi k/\nu\Delta)\varepsilon_0^{(\nu)} dk \quad . \tag{3.5b}$$

Note that T_0 defined by (3.5b) is related to the impurity contribution to the T-linear specific heat, C_i, through $C_i=\pi(\nu-1)T/6T_0$. As a result, we can see that Eqs.(3.2b) and (3.2c) describe the equilibrium-thermodynamic properties of the Coqblin-Schrieffer model completely in the presence of the magnetic field, the crystalline field and the spin-orbit coupling [15]. In the absence of H_W, the thermodynamic properties in this model have been already investigated numerically for arbitrary temperatures [4].

3.3 Low-temperature limit

We treat here Eqs.(3.2) analytically at low temperatures and obtain the general expression for the T-linear specific heat. Firstly, we expand the pseudoenergy in Eqs.(3.2) as $\varepsilon_0^{(j)}\sim\eta_j(k)+T^2\lambda_j(k)$. The contribution from $\varepsilon_n^{(j)}$ ($n\geq1$) to the thermodynamic potential can be neglected within the accuracy of T^2. Therefore the expression for Ω^i up to T^2 has the form:

$$\Omega^i = \Omega^i(T=0)+T^2\sum_{j=1}^{\nu}\{\int_{-\infty}^{Q_j}\Phi_j\lambda_j dk-\frac{\pi^2}{6}\Phi_j(Q_j)/(d\eta_j/dQ_j)\} \quad , \tag{3.6}$$

where the cut-off parameter Q_j is defined by $\eta_j(Q_j)=0$, which corresponds to the "Fermi level" of each elementary excitation. The formula (3.6) can be rewritten in a simpler form with the use of the zero-temperature density functions $\rho_I^{(j)}$ (impurity part) and $\rho_C^{(j)}$ (host part). Consequently, the expression for the coefficient of the T-linear specific heat is obtained as [15],

$$\gamma = \frac{\pi}{6}\sum_{j=1}^{\nu}\{\rho_I^{(j)}(Q_j)/\rho_C^{(j)}(Q_j)\} \quad , \tag{3.7a}$$

$$\sum_{\ell=1}^{j}\sum_{i=\ell}^{\nu}[i+j-2\ell]_i^*\rho_I^{(i)} = \Phi_j(k) \quad , \tag{3.7b}$$

$$\sum_{\ell=1}^{j}\sum_{i=\ell}^{\nu}[i+j-2\ell]_i^*\rho_C^{(i)} = \frac{j}{2\pi} \quad , \quad \text{for } 1\leq j\leq\nu \quad , \tag{3.7c}$$

where

$$[n]_i^*f = \int_{-\infty}^{Q_i}dx\,\frac{(n\Delta/\pi)f(x)}{(k-x)^2+(n\Delta)^2} \quad , \quad [0]_i^*f = f \quad .$$

The "Fermi level" Q_j is determined by the following conditional equations,

$$\int_{-\infty}^{Q_j}\rho_C^{(j)}dk = (W_{j+1}-W_j)/2\pi \quad , \quad \text{for } 1\leq j\leq\nu-1 \quad , \tag{3.8a}$$

$$\sum_{j=1}^{\nu}\int_{-\infty}^{Q_j}j\rho_C^{(j)}dk = N_1/L \quad , \tag{3.8b}$$

This general expression for γ suggests that the low-temperature behaviour of the present system shows the Fermi-liquid properties [9-13], because γ is represented only with the density function at the "Fermi level". We can derive the Fermi-liquid relation among the physical quantities at low-temperatures when the form of H_W is given explicitly, which is discussed in the next section.

4. Ce impurities in a cubic crystalline field

For a Ce atom, the large spin-orbit coupling leads to the ground-state with the total angular momentum J=5/2 and the excited state of J=7/2 lying about 300meV higher up. Below, we neglect the effect of the J=7/2 state, which is not expected to play an important role at low temperatures. The calculation including the effect of the J=7/2 state, if it is necessary, can be carried out in a similar way done below. The J=5/2 ground multiplet splits into the Γ_7-doublet and the Γ_8-quartet in a cubic crystalline field and these states split further in the presence of the external magnetic field. In this case the energy splitting $W_j (1 \leq j \leq 6)$ in (3.2) is given by,

$$W_j = \begin{cases} \frac{1}{3}\Delta_c \pm \frac{1}{2}H , \\ -\frac{1}{6}\Delta_c + \frac{1}{2}H \pm \{\frac{1}{4}\Delta_c{}^2 + \frac{4}{3}\Delta_c H + 4H^2\}^{\frac{1}{2}} , \\ -\frac{1}{6}\Delta_c - \frac{1}{2}H \pm \{\frac{1}{4}\Delta_c{}^2 - \frac{4}{3}\Delta_c H + 4H^2\}^{\frac{1}{2}} , \end{cases} \tag{4.1}$$

where Δ_c represents the energy splitting due to a cubic crystalline field ($\Gamma_7(\Gamma_8)$ is ground-state for $\Delta_c>0$ ($\Delta_c<0$)) and H is the magnetic field applied in the z-direction ($g\mu_B=1$). Note that W_j given in (4.1) should be chosen to satisfy the condition (2.2) according to the value of Δ_c and H. With the use of Eqs.(3.2),(3.3) and (4.1), we can evaluate the thermodynamic quantities of the present model for Ce, for arbitrary values of T, Δ_c and H. In the following subsections, we show the results obtained analytically at low-temperatures and numerically at arbitrary temperatures [15,16].

4.1 Fermi-liquid relation

We put forward the calculation of Eqs.(3.7) and (3.8) for a small magnetic field. Henceforth we are concerned with the case for $\Delta_c>0$ (Γ_7 ground state) (case for $\Delta_c<0$ is treated in a similar way). The expansion of the energy W in (3.1) up to H^2 gives the expression for the magnetic susceptibility χ_m at T=0 [14,15],

$$\chi_m = \chi_7 + \chi_8 + \chi_V , \tag{4.2a}$$

$$\chi_7 = (25/36\pi) \lim_{Q_1 \to -\infty} \{\rho_I^{(1)}(Q_1)/\rho_C^{(1)}(Q_1)\} , \tag{4.2b}$$

$$\chi_8 = (65/54\pi) \sum_{j=3}^{5} \lim_{Q_j \to -\infty} \{\rho_I^{(j)}(Q_j)/\rho_C^{(j)}(Q_j)\} , \tag{4.2c}$$

$$\chi_V = (80/9\Delta_c) \int_{-\infty}^{Q_2} \rho_I^{(2)}(k)dk , \tag{4.2d}$$

where χ_7, χ_8 and χ_V are the contributions to χ_m from the Γ_7 state, the Γ_8 state and the Van Vleck paramagnetism. The explicit expression for χ_m is given in ref.14. Next we introduce the "charge susceptibilities" defined by $\chi_c = -\partial(n_7+n_8)/\partial\varepsilon_f$ and $\chi_\Delta = \partial(n_7-n_8)/\partial\Delta_c$ where $n_7(n_8)$ is the occupation number of the Γ_7 state (Γ_8 state). For H=0, χ_c and χ_Δ are calculated as follows,

$$\chi_c = \frac{3}{\pi} \frac{C_2 D_6 \rho_I^{(6)}(Q_6)/\rho_C^{(6)}(Q_6) - C_6 D_2 \rho_I^{(2)}(Q_2)/\rho_C^{(2)}(Q_2)}{C_2 D_6 - C_6 D_2} , \tag{4.3}$$

$$\chi_\Delta = \frac{1}{\pi}\;\frac{C_2D_6\rho_I^{(2)}(Q_2)/\rho_C^{(2)}(Q_2) - C_6D_2\rho_I^{(6)}(Q_6)/\rho_C^{(6)}(Q_6)}{C_2D_6 - C_6D_2} ,$$

where $C_j(D_j)$ is a function of Q_j [15].

We obtain from (3.7),(4.2) and (4.3) the expression for the Fermi-liquid relation which characterizes the elementary excitations at low temperatures as [15],

$$\gamma = \frac{6\pi^2}{25}\chi_7 + \frac{9\pi^2}{65}\chi_8 + \frac{\pi^2}{18}\chi_C + \frac{\pi^2}{6}\chi_\Delta , \tag{4.4}$$

for $\Delta_c>0$. For $\Delta_c<0$, the similar expression for the Fermi-liquid relation is obtained in which χ_Δ is replaced by $2\chi_\Delta$ in (4.4). It is seen that the Van Vleck susceptibility χ_V does not take part in this relation. Here, we write down the relation (4.4) for some limiting cases.

(a) $\Delta_c\to 0$: In the absence of the crystalline field, χ_7, χ_8 and χ_Δ are in proportion to the magnetic susceptibility $\tilde{\chi}_m$ for $\Delta_c=0$, that is, $\chi_7=(5/63)\tilde{\chi}_m$, $\chi_8=(26/63)\tilde{\chi}_m$ and $\chi_\Delta=(4/35)\tilde{\chi}_m$. Therefore the relation (4.4) is reduced to the result discussed in the article I.

(b) Coqblin-Schrieffer limit, $-\varepsilon_f>>\Delta$: The charge fluctuation of the localized electron is depressed in this limit, so that χ_C tends to zero in (4.4). In the absence of the crystalline field, this expression leads to

$$\gamma = (2\pi^2/21)\tilde{\chi}_m , \quad \tilde{\chi}_m = 35/(4\pi T_0) , \tag{4.5}$$

where T_0 is the Kondo temperature defined by (3.5b). On the other hand, for the large crystalline field:

$$\gamma = (6\pi^2/25)\chi_7 , \quad \chi_7 = 25/(36\pi T_e^{(7)}) , \quad \Delta_c>>T_0 , \tag{4.6}$$

$$\gamma = (9\pi^2/65)\chi_8 , \quad \chi_8 = 65/(18\pi T_e^{(8)}) , \quad -\Delta_c>>T_0 , \tag{4.7}$$

where $T_e^{(7)}$ $(T_e^{(8)})$ is the effective Kondo temperature for the $\Gamma_7(\Gamma_8)$ ground-state, (See Eqs.(4.10) and (4.11)).

4.2 T-linear specific heat

First, we discuss the properties of the T-linear specific heat in the Coqblin-Schrieffer limit. This limiting case of the degenerate Anderson model has been often used to understand the local properties of Ce (Yb) compounds, because many interesting phenomena of these compounds are considered to be caused by the virtual mixing of the different configurations [9-13]. Fortunately, the further analytical calculation is possible in this limiting case. In the absence of the magnetic field, the basic equations (3.7) and (3.8) can be solved with the aid of the standard Wiener-Hopf technique [14]. As a result, the coefficient of the T-linear specific heat is obtained in a closed form for $\Delta_c>0$ (Γ_7 ground state),

$$\gamma = \gamma_7 + \gamma_8 + \gamma_\Delta , \tag{4.8a}$$

$$T_0\gamma_7 = \frac{1}{24}\left(\frac{3}{\pi^3}\right)^{\frac{1}{2}}\Gamma\left(\frac{1}{6}\right)^3\left(\frac{\Delta_c}{T_0}\right)^2\left[1+\left(\frac{T_0}{a\Delta_c}\right)^3 f\left(4,\frac{\pi i}{2}\right)\right] , \tag{4.8b}$$

$$T_0\gamma_8 = \frac{3\pi^2}{8}\left(\frac{3}{2}\right)^{\frac{1}{2}} e^{-\frac{1}{4}}\Gamma\left(\frac{5}{4}\right)^{-1} f\left(4,\frac{\pi i}{4}\right)\frac{T_0}{\Delta_c} , \tag{4.8c}$$

$$T_0\gamma_\Delta = \frac{1}{12a\sigma_2(0)}\frac{T_0}{\Delta_c}\int_{-\infty}^{\infty}\exp(-i0^+\omega)f(4,\omega)d\omega \quad , \tag{4.8d}$$

where

$$f(n,\omega) = \frac{1}{G(\omega)}\frac{1}{2\pi i}\int_{-\infty}^{\infty}\frac{dx}{x-\omega}G(x)\,\frac{\sinh(nx)}{\sinh(6x)}\left(\frac{a\Delta_c}{T_0}\right)^{-6ix/\pi} \quad ,$$

$$G(\omega) = \frac{\sqrt{3}}{2}[3^6(-i\omega+0)/2^4e\pi]^{-\frac{i\omega}{\pi}}\Gamma(1-\frac{2i\omega}{\pi})\Gamma(1-\frac{4i\omega}{\pi})/\Gamma(1-\frac{i\omega}{\pi})\Gamma(1-\frac{6i\omega}{\pi}) \quad ,$$

and $\Gamma(x)$ represents the gamma function, $a=(3e/2)^{1/6}\Gamma(\frac{1}{6})/3\pi$ and $\sigma_2(k)=\rho_c^{(2)}(\Delta k+Q_2)/T_0$. Here $\gamma_7(\gamma_8)$ is the contribution to γ from the $\Gamma_7(\Gamma_8)$ state and γ_Δ is due to the excitation from the Γ_7 state to the Γ_8 state. On the other hand for $\Delta_c<0$, γ_7, γ_8 and γ_Δ are given by

$$T_0\gamma_7 = \frac{9}{4}\left(\frac{\pi^3}{2e}\right)^{\frac{1}{2}}\frac{T_0}{|\Delta_c|}\,f(2,\frac{\pi i}{2}) \quad , \tag{4.9a}$$

$$T_0\gamma_8 = \{\frac{\sqrt{\pi}}{16}\,6^{\frac{1}{4}}\,\Gamma(\frac{1}{6})^{\frac{3}{2}}/\Gamma(\frac{5}{4})\}\left(\frac{|\Delta_c|}{T_0}\right)^{\frac{1}{2}}[1+\left(\frac{T_0}{a|\Delta_c|}\right)^{\frac{3}{2}}f(2,\frac{\pi i}{4})] \quad , \tag{4.9b}$$

$$T_0\gamma_\Delta = \frac{1}{12a\,\sigma_4(0)}\frac{T_0}{|\Delta_c|}\int_{-\infty}^{\infty}\exp(-i0^+\omega)f(2,\omega)d\omega \quad . \tag{4.9c}$$

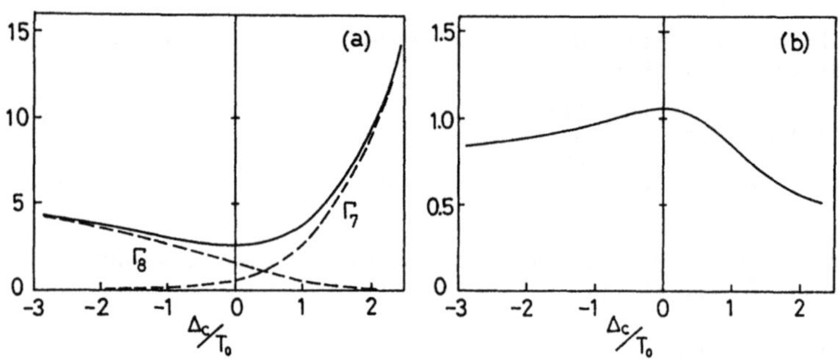

Fig.2. Universal curves as a function of Δ_c/T_0, for (a) $T_0\gamma$, (b) $R=\chi_m/\gamma$. Dashed curves in (a) represent the contribution from the Γ_7 (Γ_8) state.

In Fig.2-(a) these results for γ are shown as a function of Δ_c/T_0 [16]. It is seen that the value of γ is increased with increasing value of $|\Delta_c|$ and is dominated by the contribution from the $\Gamma_7(\Gamma_8)$ state for $\Delta_c>0$ ($\Delta_c<0$) even if $|\Delta_c|$ is not so large compared to T_0 ($|\Delta_c|\gtrsim 3T_0$). As for the magnetic susceptibility, this is not true, because the large contribution from the Van Vleck susceptibility remains for the rather large value of $|\Delta_c|/T_0$ [14]. The results for Wilson's ratio $R=\chi_m/\gamma$ are shown in Fig.2-(b). The value of R changes from $21/2\pi^2$ to its asymptotic value $25/6\pi^2$ ($65/9\pi^2$) for $\Delta_c>0$ ($\Delta_c<0$) with the increase of $|\Delta_c|$ (see Eqs.(4.7) and (4.8)). This dimensionless quantity has been often used to investigate the experimental results at low temperatures [9]. It should be noticed that the rather small crystalline field splitting affects the value of R considerably, especially for $\Delta_c>0$.

For the large crystalline field splitting, we define the effective Kondo temperature by $T_e=\pi(\nu_e-1)/6\gamma$ where the ground multiplet is assumed to be ν_e-fold degenerate. When $\Delta_c >> T_0$ ($\nu_e=2$, Γ_7 ground-state), we obtain from (4.8) the expression for T_e as [16]

$$T_e/T_0 = 4(\pi^5/3)^{\frac{1}{2}}\Gamma(1/6)^{-3}(\Delta_c/T_0)^{-2} \sim 0.234(\Delta_c/T_0)^{-2} \quad . \tag{4.10}$$

On the other hand, when $-\Delta_c >> T_0$ ($\nu_e=4$, Γ_8 ground-state) we obtain from (4.9)

$$T_e/T_0 = 8\sqrt{\pi}6^{-\frac{1}{4}}\Gamma(5/4)\Gamma(1/6)^{-\frac{3}{2}}(|\Delta_c|/T_0)^{-\frac{1}{2}} \sim 0.624(|\Delta_c|/T_0)^{-\frac{1}{2}} \quad . \tag{4.11}$$

These exact expressions for the Kondo temperature [19] are useful even if the value of $|\Delta_c|/T_0$ is not so large, as is understood from Eqs.(4.8) and (4.9) and also from Fig.2-(a).

In Fig.3 some numerical results are shown in the mixed valence regime in which $n_f=n_7+n_8$ is fractional ($0<n_f<1$).

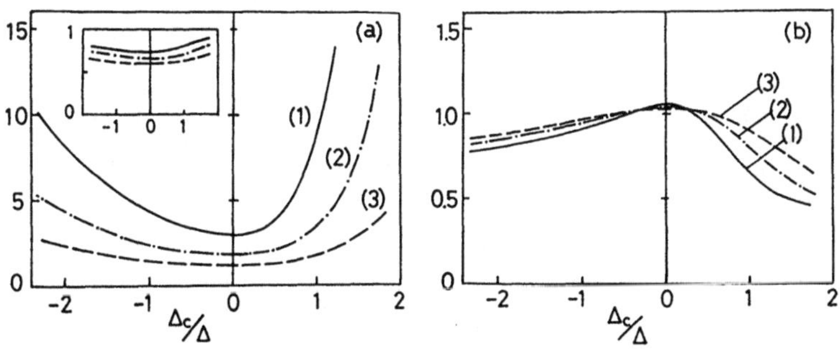

Fig.3. Results in the mixed valence regime: (a) $\Delta\cdot\gamma$, (b) $R=\chi_m/\gamma$ as a function of Δ_c/Δ: (1) $n_f(\Delta_c=0)=0.72$, (2) 0.65, (3) 0.59. The valence n_f is plotted vs Δ_c/Δ in the inset.

4.3 Temperature-dependent specific heat

The numerical calculation at finite temperatures is carried out by truncating the integral equations (3.2) at finite numbers n and converting the integral equations into a set of the matrix equations in which each pseudoenergy and its derivative are represented by 100 lattice points. These matrix equations are solved by the itration method. The specific heat obtained in the Coqblin-Schrieffer limit is shown in Figs.4 and 5 [16].

In Fig.4 the specific heat C_i is plotted as a function of the temperature for $\Delta_c>0$ (Γ_7 ground state). It is seen that the shape of the specific heat curve becomes asymmetric and has a shoulder structure at the low temperature side, when the value of Δ_c is gradually increased. This shoulder structure is caused by the low-lying excitation states within the Γ_7-doublet. The contribution to C_i from the Γ_7-doublet is also shown for $\Delta_c/T_0=2$ in Fig.4. When the value of Δ_c is further increased, the specific heat curve may have two peaks, as should be expected. In this case the behaviour at high temperatures is described approximately by the Schottky-formula

$$C_i = 2(\Delta_c/T)^2\exp(\Delta_c/T)[\exp(\Delta_c/T)+2]^{-2} \quad .$$

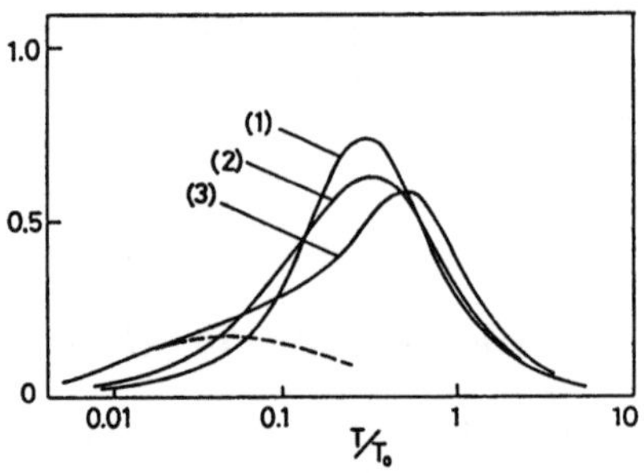

Fig.4. Plots of specific heat as a function of T/T_0 for the Γ_7 ground state: (1) $\Delta_c/T_0=0$, (2) 1, (3) 2. Dashed curve represents the contribution from Γ_7 state for $\Delta_c/T_0=2$.

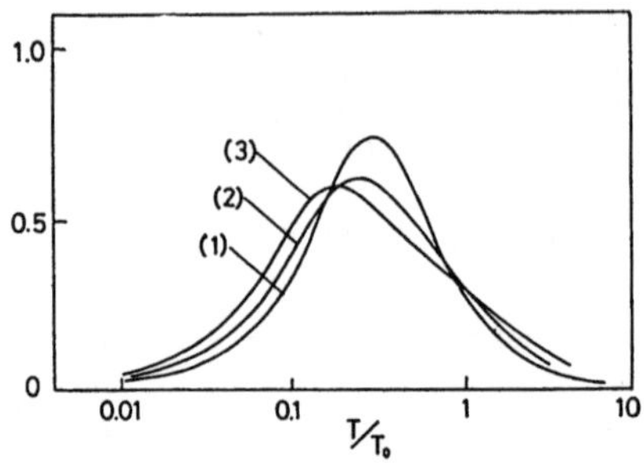

Fig.5. Plots of specific heat as a function of T/T_0 for the Γ_8 ground state: (1) $\Delta_c/T_0=0$, (2) -2, (3) -4.

On the other hand, the specific heat at low temperatures shows the universal Kondo-properties. They are described by the s-d exchange model (spin ½), in which the effect of the crystalline field does not appear explicitly but only through the effective Kondo temperature T_e defined by (4.10). The results for $\Delta_c<0$ (Γ_8 ground-state) are shown in Fig.5. As well as for $\Delta_c>0$. The curve C_i becomes asymmetric when the value of $-\Delta_c$ is increased. In this case, however, the shoulder structure is seen to appear gradually at the high temperature side, which is caused by the high-energy excitations from the Γ_8 ground-state to the Γ_7 excited states. These shoulder structures are considered to be characteristic of the case for the intermediate strength of the crystalline field ($|\Delta_c|\sim T_0$), though the shape of C_i depends on the sign of Δ_c. For $-\Delta_c>>T_0$, the curve of C_i at high temperatures is approximately given by $C_i = 2(\Delta_c/T)^2\exp(-\Delta_c/T)[2\exp(-\Delta_c/T)+1]^{-2}$ and the universal properties at low temperatures are described by the Coqblin-Schrieffer model ($\nu_e=4$) [4] with the effective Kondo temperature T_e defined by (4.11).

5. Summary

We have discussed the thermodynamic properties of the highly correlated degenerate Anderson model in the presence of the crystalline field and the spin-orbit coupling. The effect of these additional distrubance on the problem has been shown to be incorporated into the thermodynamic equations (3.2) through the value of W_j which can be calculated in the standard way (Eq.(4.1) for example). As a typical example, the properties of the specific heat has been investigated in detail for the case of Ce impurities in a cubic crystalline field. The coefficient of the T-linear specific heat has been shown to be dominated by the contribution from the ground multiplet even if the splitting of the crystalline field is not so large compared to T_0. Therefore the effective Kondo temperature defined by (4.10) and (4.11) is considered to be useful for the rather small strength of the crystalline field. The curve of the temperature-dependent specific heat has been shown to be asymmetric and have a shoulder structure when the crystalline field splitting grows. This shoulder structure is characteristic of the intermediate strength of the crystalline field ($|\Delta_c|\sim T_0$), but the shape of the curve depends on the sign of Δ_c.

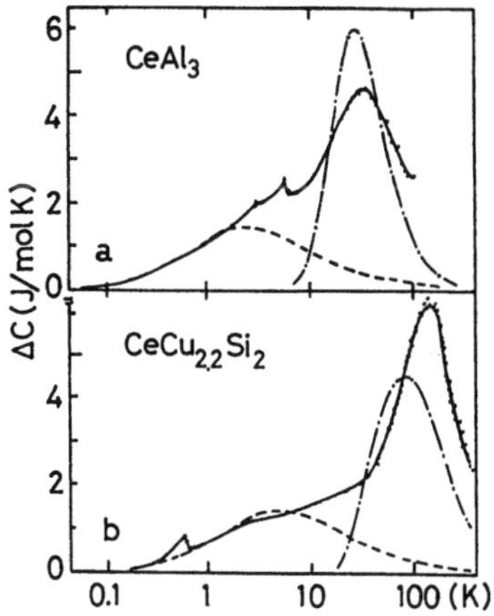

Fig.6. 4f-derived specific heat ΔC vs temperature on a logarithmic scale for polycrystalline samples of $CeA\ell_3$ (a) and $CeCu_{2.2}Si_2$ (b). Theoretical ΔC-results are included for comparison: (i) Bethe-ansatz for spin ½ (broken line); (ii) Schottky anomalies (dash-dotted line) calculated for three crystal-field doublets at 0K, 57K, 84K ($CeA\ell_3$), and at 0K, 140K, 364K ($CeCu_{2.2}Si_2$). Note that the anomalies in ΔC(T) of $CeA\ell_3$ near 3.5K and 6K must be ascribed to the onset of magnetic order in spurious precipitations of $CeA\ell_2$ and $Ce_3A\ell_{11}$ (after F. Steglich et al., ref.20).

The shoulder structure similar to the curve 3, in Fig.4 has been found experimentally for $CeCu_{2.2}Si_2$ and $CeA\ell_3$ [20] (See Fig.6). Although the crystalline field is not cubic but is tetragonal or hexagonal in these materials, the essential properties of the specific heat might be explained with the results discussed here, namely: in both materials the shoulder structure is caused by the low-lying excitation within the ground-doublet (Kondo effect) and the peak structure is caused by the direct excitation from the ground-doublet to the excited multiplets [20]. The detailed comparison with the experimental data will be published elsewhere.

References

1. A. Okiji and N. Kawakami: in this volume.
2. A.M. Tsvelick and P.B. Wiegmann: J. Phys. C15, 1707 (1982)
3. A.C. Hewson and J.W. Rasul: J. Phys. C16, 6799 (1983)
4. V.T. Rajan: Phys. Rev. Lett. 51, 308 (1983)
5. P. Schlottmann: Phys. Rev. Lett. 50, 1697 (1983)
6. N. Kawakami, S. Tokuono and A. Okiji: J. Phys. Soc. Jpn. 53, 51 (1984)
7. N. Kawakami and A. Okiji: Phys. Lett. 103A, 205 (1984)
8. P. Schlottmann: Z. Phys. B57, 23 (1984)
9. D.M. Newns and A.C. Hewson: J. Phys. F10, 2429 (1980)
10. I. Okada and K. Yosida: Prog. Theor. Phys. 49, 1483 (1973)
11. A. Ogawa and A. Yoshimori: Prog. Theor. Phys. 53, 315 (1975)
12. K. Yamada, K. Yosida and K. Hanzawa: Prog. Theor. Phys. 71, 450 (1984)
13. P. Nozieres and A. Blandin: J. Physique 41, 193 (1980)
14. P. Schlottmann: Z. Phys. B55, 293 (1984); Phys. Rev. B30, 1545 (1984)
15. N. Kawakami and A. Okiji: J. Phys. Soc. Jpn. 54, 685 (1985)
16. N. Kawakami and A. Okiji: to be published.
17. C.N. Yang: Phys. Rev. Lett. 19, 1312 (1967)
18. B. Sutherland: Phys. Rev. Lett. 20, 98 (1967)
19. If we rewrite Eqs.(4.12) and (4.13) with the band cut-off parameter, the results are in accordance with those obtained in refs.12 and 14.
20. F. Steglich, C.D. Bredl, W. Lieke, U. Rauchschwalbe and G. Sparn: Physica 126B, 82 (1984); F. Steglich: private communication.

Bethe-Ansatz Solution of the Degenerate Anderson Model: Applications to Ce and Yb Impurity Systems

*P. Schlottmann**

Institut für Festkörperforschung, Kernforschungsanlage Jülich
Postfach 1913, D-5170 Jülich, Fed. Rep. of Germany

We review the Bethe-ansatz solution of the orbitally degenerate Anderson model for a magnetic impurity in a metal. The groundstate properties of the integer and mixed-valent impurity are discussed in the presence of external potentials, e.g. magnetic and crystalline fields, and with spin-orbit coupling. The thermodynamic Bethe-ansatz equations are reviewed and their solution for Ce and Yb impurities are compared with experimental data. The Bethe-ansatz solution, the groundstate properties and the thermodynamics of a model for a mixed-valent impurity with two magnetic configurations are presented. This model is supposed to resemble some of the properties of Tm-ions.

I. Introduction

Intermediate valence systems are characterized by highly correlated f-states, which are neither band-like nor is their occupation a fixed integer. Two configurations, $(4f)^n$ and $(4f)^{n+1}$, have large probabilities of occupation, P_n and P_{n+1}, with the result that for thermodynamic quantities the system exhibits an average behavior corresponding to a nonintegral valence.

The origin of this behavior is the competing effects of three types of energies: (a) the strong electron-electron repulsion in the f-shell which makes it unfavorable to have more than two configurations; (b) the strong solid-state effects which completely delocalize the conduction-electron states; (c) the hybridization effects which, to a given degree of strength, mix the f- and conduction band states.

Mixed valent systems are usually described in terms of Anderson's model [1], which includes the above mentioned types of energies. We restrict ourselves here to impurity systems, i.e. isolated rare-earth ions in a metallic matrix, and discuss the applicability of impurity models to compounds in section V.3.

It is important for a realistic model to take the orbital degeneracy of the f-electrons, as well as the Hund's rules and the spin-orbit interaction, into account. The simplest case refers to mixed and integer-valent Ce ($j=5/2$) and Yb ($j=7/2$) impurities, where $n=0$ and the configurations $4f^0$ and $4f^1$ are considered (for Yb electrons are interchanged with holes).

During the last few years, considerable progress has been made towards the solution of the degenerate Anderson model within the 1/N ($N=2j+1$ being

*Heisenberg-fellow of the Deutsche Forschungsgemeinschaft; present address: Institut für Theoretische Physik, Universität Göttingen, W.-Germany

the degeneracy of the 4f-level) and the Bethe-ansatz approaches. We review the present status of the Bethe-ansatz solution in sections III-V.

In section VI, we consider a mixed valence impurity with two magnetic configurations coupled by s-wave conduction electrons via a hybridization matrix element. This model is expected to resemble some of the properties of Tm-impurities.

Since the Bethe-ansatz technique applied to the Kondo problem and the non-degenerate Anderson model (no orbital degeneracy) has been extensively reviewed by ANDREI, FURUYA and LOWENSTEIN [2], and TSVELICK and WIEGMANN [3], we restrict ourselves to discuss the two models mentioned above.

II. The Degenerate Anderson Model in the $U\to\infty$ Limit

Anderson's model [1] consists of localized highly correlated 4f-states of the impurity and extended conduction states of the host.

Due to the large Coulomb repulsion within 4f-shell only two configurations have a finite probability of occupation, which in the case of Ce-impurities are the $4f^0$- and $4f^1$-configurations. Due to the relatively large spin-orbit splitting of the $4f^1$ states (~ 0.3 eV), we consider only its Hund's rule ground-multiplet of total angular momentum $j=5/2$. All other states of the $4f^1$-configuration are neglected. The f-level of energy ε , measured from the Fermi level, is then either empty or singly occupied. We denote these states with bra and kets , $|0\rangle$ being the $4f^0$-singlet and $|m\rangle$ the $4f^1$-multiplet, where m ($|m|\leq j$) is the eigenvalue of the z-component of the total angular momentum. Note that the case of Yb impurities, where the two configurations are $4f^{13}$ and $4f^{14}$, can be obtained by interchanging electrons with holes and considering $j=7/2$. The effect of the excited spin-orbit multiplet on the groundstate properties is discussed in section IV.5.

The conduction states are expanded in partial waves at the impurity site and only the states with $\ell=3$ and total angular momentum j can couple to the f-states of the impurity. All other states can then be disregarded. The conduction states have then the quantum number $|\vec{k}|$, j and m . The dependence on $|\vec{k}|$ allows us to formulate the problem in one dimension in terms of forward moving particles. Hence, if the impurity is at $x=0$, $x<0$ represents incoming particles, while $x>0$ are the outgoing states.

A necessary condition for the integrability of the model is a linear dispersion.Since the important effects take place close to the Fermi level, a kinetic energy linearized around the Fermi level is not going to alter the physical properties of the model.

The f- and conduction states with the same symmetry hybridize, the matrix element being $V_{|k|}$. We neglect the $|k|$-dependence of the hybridization, i.e. we consider a contact potential. This approximation is necessary for the integrability of the system. It can be justified perturbatively from the leading contributions, which involve $|k|$-sums over $V^2_{|k|}$, such that an average matrix element can be taken.

The Hamiltonian is then given by [4]

$$H = \sum_m \left\{ \int dx\, c^+_m(x) \left(-i\frac{\partial}{\partial x}\right) c_m(x) + \varepsilon\, |m\rangle\langle m| + V \int dx\, \delta(x) \left(c^+_m(x)\, |0\rangle\langle m| + |m\rangle\langle 0|\, c_m(x)\right)\right\} \quad . \tag{II.1}$$

The first term is the linearized kinetic energy. Note that all electrons move to the right with the Fermi velocity (equated to unity). The distance between two arbitrary electrons is then a conserved quantity. The second term is the f-level energy and the last one the hybridization. Note that the number of electrons per spin component is conserved.

Depending on the f-level position with respect to the Fermi level, we may distinguish among the nonmagnetic situation, $\varepsilon>0$, the mixed-valent region, $\varepsilon\sim 0$, and the Kondo limit, $\varepsilon<0$. The crossover between these regions is smooth.

III. The Bethe-Ansatz Solution of Anderson's Model

In Bethe's hypothesis [5] the N-body wave-function is decomposed into linear combinations of products of two-body wave-functions. This hypothesis is exact for some completely integrable many-body systems in 1+1 dimensions, i.e. for systems having an infinite set of conserved (but in general unknown) currents. These systems usually involve zero-range interactions (δ-funcions) or nearest neighbor couplings on a lattice.

For our model, the N-particle wave function [3,4] is a linear superposition of (a) the N-electron Fermi sea with no f-electron, and (b) the (N-1)-electron Fermi sea with the impurity in the $4f^1$-state. In view of the special role of the two-body wave function, it is convenient to discuss first few-body wave functions, namely N=1,2,3,etc. and generalize the results then to arbitrary N (see [4]).

If we consider only one electron, the wave function is a plane wave to the right and to the left of the impurity. At the impurity site the electron has a resonant scattering with a phase-shift of $2\varphi_k$

$$\varphi_k = \text{arc tan}\left(\frac{1}{2} V^2 / (\varepsilon-k)\right) \qquad \text{(III.1)}$$

which relates the amplitudes to the right and to the left. The energy of the system is k .

Consider a system of two electrons with coordinates x_1 and x_2 and momenta k_1 and k_2 . Since all electrons move with the same velocity, the trajectory is along $x_1-x_2=\text{const.}$ When a particle crosses the impurity it resonates and its phase-shifts, according to (III.1). Hence the regions $x_1<x_2<0$ and $x_2<x_1<0$ determine completely the wave function for other arrangements of x_1 and x_2 with respect to the impurity (at the origin). Elsewhere, except on the line $x_1=x_2$, the wave function is a plane wave. There is an infinite Coulomb repulsion among the f-electrons implicit in our model, which acts at $x_1=x_2$. This interaction allows the electrons to exchange their momenta, such that the wave function for $x_1,x_2<0$ can be written as [4,6]

$$\psi_Q = \Sigma_P\, [Q,P]\, \exp[ik_{P1}x_{Q1} + ik_{P2}x_{Q2}] \quad , \qquad \text{(III.2)}$$

where $Q=(Q1,Q2)$ and $P=(P1,P2)$ are two permutations of the integers 1 and 2 . $[Q,P]$ are four coefficients arranged as a 2×2 matrix with columns denoted by ζ_P . Due to the continuity of the wave function and the discontinuity of its derivative at $x_1=x_2$ (due to the implicit 4f-repulsion) only two of the coefficients are independent. Hence the two columns ζ_P are related by

$$\zeta_{21} = Y_{12}^{12}\, \zeta_{12} \quad , \quad \text{where} \qquad \text{(III.3)}$$

$$Y_{12}^{ab} = \frac{iV^2}{k_1-k_2-iV^2}\,\hat{I} + \frac{k_1-k_2}{k_1-k_2-iV^2}\,\hat{P}_{ab} \qquad \text{(III.4)}$$

Here $\hat{I}$ is the unit matrix and $\hat{P}_{ab}$ interchanges the components a and b of ζ_P. The total energy of the system is k_1+k_2.

The arguments can be generalized to the N-particle problem [3,4,6]. Let the coordinates and momenta be x_i, k_i, $i=1,\ldots,N$. The sequence of coordinates is not changed with time since $x_i-x_j=\text{const}$ for all i and j. The electrons resonate when they pass the impurity site as in the one-body solution. It is then sufficient to consider the N! arrangements of $x_1,\ldots,x_N$ for all the electrons on the left of the impurity. There is a discontinuity of the derivative of the wave function at $x_i=x_j$ due to the infinite Coulomb repulsions. As a consequence, the momenta can be exchanged and the ansatz for the N-body wave function is then

$$\psi_Q = \Sigma_P\,[Q,P]\,\exp[ik_{P1}x_{Q1} + \ldots + ik_{PN}x_{QN}] \qquad \text{(III.5)}$$

where $P=(P1,P2,\ldots,PN)$ and $Q=(Q1,Q2,\ldots,QN)$ are two permutations of the integers $1,2,\ldots,N$. The coefficients $[Q,P]$ can be arranged as a $N!\times N!$ matrix with columns ζ_P related by

$$\zeta_P = Y_{ij}^{ab}\,\zeta_{P'} \quad . \qquad \text{(III.6)}$$

This follows from the two-body solution, since the matching of the wave function at the boundaries of the N! regions for different arrangements of coordinates only involves pairs of them. The total energy is $\Sigma_j k_j$.

It is important to verify the single-valuedness of the wave function, i.e. that the connection between two arbitrary points in the N-dimensional space of coordinates ($x_1,\ldots,x_N$) is independent of the path. This leads to product relations among the Y-operators (Yang-Baxter triangular relation [6]), which is equivalent to the factorizability condition of the S-matrix.

The next step consists of imposing periodic boundary conditions. The system is placed in a box of length L and one (arbitrary) coordinate x_j is shifted to x_j+L. During this shift the particle at x_j passes all other particles and also the impurity. Passing a particle means to change sector and involves a Y-operator. In this way, the periodic boundary conditions lead to operator eigenvalue equations, the eigenvalue being $\exp(ik_jL)$. Since there are N-particles, there are N such eigenvalue equations to be solved simultaneously [2-4,6]. This type of eigenvalue problem is common to a large class of models, e.g. the fermion gas with δ-function interaction, the Hubbard chain, the 1d-chiral Gross-Neveu model and the Kondo problem.

Our problem is related to the 1d-fermion gas with *attractive* δ-function interaction. For two spin components ($j=1/2$) the eigenvalue equations were solved by GAUDIN [7] and YANG [6] by means of a second (nested) Bethe-ansatz. SUTHERLAND [8] generalized their results to an arbitrary number of components by eliminating the components one by one with further nested Bethe-ansätze.

Let us denote with n_{j-m} the number of particles with spin-component m, N be the total number of electrons and define

$$M_i = \sum_{m=-j+i}^{j} n_{j+m} \quad , \quad M_0 = N \quad , \quad M_{2j+1} = 0 \quad , \qquad \text{(III.7)}$$

then $N \geq M_1 \geq \ldots \geq M_{2j} \geq 0$. In analogy to [8] one obtains the following set of coupled equations

$$\exp[i(\xi_\alpha^{(o)}L + 2\varphi_\alpha)] = \Pi_{\beta=1}^{M_1} e(\xi_\alpha^{(o)} - \xi_\alpha^{(1)}) \quad , \quad \alpha = 1, \ldots, N \qquad (III.8)$$

$$\Pi_{\beta=1}^{M_{\ell-1}} e(\xi_\alpha^{(\ell)}-\xi_\beta^{(\ell-1)}) \, \Pi_{\beta=1}^{M_{\ell+1}} e(\xi_\alpha^{(\ell)}-\xi_\beta^{(\ell+1)}) = -\Pi_{\beta=1}^{M_\ell} e[(\xi_\alpha^{(\ell)}-\xi_\beta^{(\ell)})/2] \quad ,$$
$$\alpha = 1, \ldots, M_\ell \ , \quad \ell = 1, \ldots, 2j \quad , \qquad (III.9)$$

where $\xi_\alpha^{(o)} \equiv k_\alpha$, L is the strength of the box, φ_α is the one-body phase shift and

$$e(x) \equiv (ix + \frac{1}{2} V^2) \ / \ (ix - \frac{1}{2} V^2) \quad . \qquad (III.10)$$

In summary, if the spin degeneracy is $2j+1$, there are $2j+1$-nested Bethe-ansätze, each of them involving a set of rapidities $\{\xi_\alpha^{(\ell)}\}$, $\ell=0,\ldots,2j$ and α being the running index. Solutions of the equations (III.8) and (III.9) are discussed in the following sections.

IV. Exact Properties at T=0

In this section we obtain the solution to (III.8) and (III.9) for the groundstate and discuss then the T=0-properties of the impurity.

IV.1 The groundstate equations

In the previous section we mentioned that the Bethe-ansatz solution of our impurity problem is related to a fermion gas with attractive δ-function interaction [7]. Attractive forces tend to bind the electrons. Bound electron complexes are described by complex (conjugated) rapidity values. Hence for the groundstate we have to search for complex conjugated solutions of the coupled nonlinear Bethe-ansatz equations. The structure and the construction of the bound states are discussed in [4,9].

The bound states correspond to spin component pairing in the electron gas. Due to Pauli's principle we can pair at most $2j+1$-electrons. There are then clusters (or states of paired electrons) of 1 up to $2j+1$-electrons. Since this pairing represents an energy reduction, the number of clusters of $2j+1$-electrons is maximum in the groundstate in the absence of external fields.

A complex of n-electrons ($n \leq 2j+1$) is characterized by one real $\xi^{(n-1)}$-value and in general complex $\xi^{(\ell)}$ values, $\ell < n-1$, which are related by

$$\xi_p^{(\ell)} = \xi^{(n-1)} + ipV^2 \quad , \qquad \ell \leq n-1$$
$$p = -\frac{1}{2}(n-\ell-1), \ -\frac{1}{2}(n-\ell-3), \ \ldots, \ \frac{1}{2}(n-\ell-1) \quad . \qquad (IV.1)$$

Hence, a complex of n electrons is completely determined by one real $\xi^{(n-1)}$ rapidity.

Inserting (IV.1) into (III.8) and (III.9) and logarithmizing the Bethe-ansatz equations one obtains [9]

$$(\ell+1)\ \xi_\alpha^{(\ell)}L = 2\pi\ I_\alpha^{(\ell)} - \left[\pi + 2\ \mathrm{arctg}\left((\xi_\alpha^{(\ell)}-\varepsilon)/(\ell+1)\ \frac{V^2}{2}\right)\right]$$
$$+ 2\ \Sigma_{q=0}^{2j}\ \Sigma_{P=0}^{P_{\ell q}}\ \Sigma_{\beta=1}^{N_q}\ \mathrm{arctg}\left((\xi_\alpha^{(\ell)}-\xi_\beta^{(q)})/(\ell+q-2p)\ \frac{V^2}{2}\right) \quad ,$$
$$\alpha = 1, \ldots N_\ell \ , \quad \ell = 0, \ldots, 2j \quad , \qquad (IV.2)$$

where $I_\alpha^{(\ell)}$ are sets of integers, which play the role of quantum numbers of the system and determine the ground and excited states, $p_{\ell q} = \min(\ell,q) - \delta_{\ell q}$, and

$$N_\ell = M_\ell - 2M_{\ell+1} + M_{\ell+2} = n_\ell - n_{\ell+1} \quad , \quad n_{2j+1} \equiv 0 \quad . \tag{IV.3}$$

Since the N_ℓ are all positive or zero the occupation numbers of the levels are labeled in decreasing order, $n_0 \geq n_1 \geq \dots \geq n_{2j}$, such that n_0 corresponds to the lowest lying level and n_{2j} is the less populated.

In the thermodynamic limit, the nonlinear equations for the rapidities, (IV.2), can be transformed into linear integral equations of the Wiener-Hopf type for density functions of the rapidities. We denote with $\sigma^{(\ell)}(\xi)$ the density function for $\xi^{(\ell)}$ and with $\tilde{\sigma}^{(\ell)}(\xi)$ the corresponding "hole" density function. In the groundstate $\sigma^{(\ell)}$ vanishes identically for $\xi>B_\ell$ and $\tilde{\sigma}^{(\ell)}$ for $\xi<B_\ell$, $\ell=0, \dots, 2j$. The parameters B_ℓ are discussed below; they depend on the external fields imposed on the impurity. $Q=B_{2j}$ plays the role of the Fermi level.

Assuming that the Fermi energy is much larger than the Zeeman energy and crystal field splittings, the system of integral equations for the groundstate becomes [9]

$$\tilde{\sigma}^{(\ell)}(\xi) + \sigma^{(\ell)}(\xi) + \sum_{q=0}^{2j+1} \int_{-\infty}^{B_q} d\xi' \, \sigma^{(q)}(\xi') \, K_{\ell q}(\xi-\xi') \tag{IV.4}$$

$$= \int_Q^\infty d\xi' \, \tilde{\sigma}^{(2j)}(\xi') \, F_\ell(\xi-\xi') + \frac{1}{L} F_{2j-\ell-1}(\xi-\varepsilon) \quad , \quad \ell = 0, \dots, 2j-1 \quad ,$$

where

$$F_\ell(\xi) = \frac{1}{(2j+1)V^2} \frac{\sin[\pi \frac{\ell+1}{2j+1}]}{\mathrm{Cosh} \frac{2\pi\xi}{(2j+1)V^2} + \cos[\pi \frac{\ell+1}{2j+1}]} \quad , \tag{IV.5}$$

and $K_{\ell q}(\xi)$ is the Fourier transform of $G_{\ell+1,q+1}(\omega) - \delta_{\ell,q}$

$$G_{\ell,q}(\omega) = \mathrm{Sinh}\left[\frac{V^2}{2} \omega \min(q,\ell)\right] \times \mathrm{Sinh}\left[\frac{V^2}{2} \omega \, (2j+1-\max(\ell,q))\right] \Big/ \mathrm{Sinh}\left[\frac{V^2}{2} \omega \, (2j+1)\right] \times \mathrm{Sinh}\left[\frac{V^2}{2} \omega\right] \quad . \tag{IV.6}$$

The integral equation obeyed by $\sigma^{(2j)}(\xi)$ is

$$\tilde{\sigma}^{(2j)}(\xi) + \sigma^{(2j)}(\xi) + \frac{1}{\pi} \sum_{q=0}^{2j} \sum_{p=0}^{q} \int_{-\infty}^{B_q} d\xi' \, \sigma^{(q)}(\xi') \, (2j+q-2p) \frac{V^2}{2} \Big/ \tag{IV.7}$$

$$\left\{(\xi-\xi')^2 + [(2j+q-2p) \frac{V^2}{2}]^2\right\} = \frac{\ell+1}{2\pi} + \frac{1}{\pi L} (2j+1) \frac{V^2}{2} \Big/ \left\{(\xi-\varepsilon)^2 + [(\ell+1) \frac{V^2}{2}]^2\right\}$$

The first term on the r.h.s. of (IV.4) represents the charge fluctuations in the system, while the second term is the Kondo driving term. From (IV.4) and (IV.7) it is clear that the density functions consist of an impurity and a host part, $\sigma^{(\ell)} = \sigma^{(\ell)}_{host} + \frac{1}{L} \sigma^{(\ell)}_{imp}$, where L is the length of the box. $\sigma^{(\ell)}_{host}(\xi)$ is then the density in the electron gas associated with all n_q for $q \leq \ell$ occupied and $q>\ell$ empty. The relative occupation of the levels

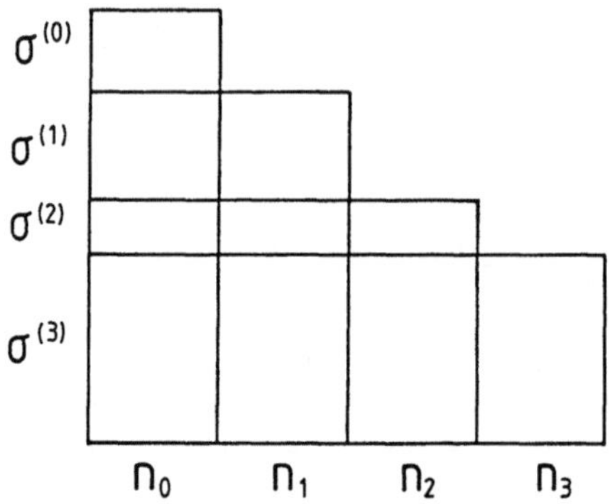

Fig. 1. Schematic representation of the occupation numbers for $j=3/2$ as a function of the ground-state density functions $\sigma^{(\ell)}(\xi)$. In the case of a pure Zeeman splitting we identify n_0, n_1, n_2 and n_3 with the occupations of the levels with $m=+3/2, +1/2, -1/2, -3/2$, respectively. The scheme holds for the impurity and for the host

(IV.3) is given in terms of the densities by

$$n_\ell - n_{\ell+1} = \int_{-\infty}^{B_\ell} d\xi\, \sigma_{imp}^{(\ell)}(\xi) \quad . \tag{IV.8}$$

The relation between the level occupations and the densities is schematically sketched in Fig. 1 for $j=3/2$ [10].

The degeneracy of the f-level is often lifted by "external" potentials, e.g. magnetic and crystal fields. Also the spin-orbit splitting can be included in this scheme. We diagonalize the ionic f-level Hamiltonian and new eigenstates (colors) replace the spin eigenstates. The number of electrons of each color is a conserved quantity and the above analysis for spin components remains unchanged when reinterpreted in terms of colors [11]. The B_ℓ are a set of (2j+1)-constants to be determined according to the external potentials imposed on the impurity. They determine the number of electrons of each color in the host and hence valence and relative population of the f-level states of the impurity through (IV.8). In the rest of this section we present a few examples that illustrate this discussion.

IV.2 Valence and susceptibility at $H=0$

If there is no f-level splitting all levels are equally occupied and according to (IV.8) $B_\ell=-\infty$ for $\ell=0, \ldots, 2j-1$. $B_{2j}\equiv Q$ represents the Fermi level and is finite. Hence the only non-vanishing density function is $\sigma^{(2j)}(\xi)$, given by (IV.7), which after fouriertransforming and some algebra reduces to (we drop the index $2j$)

$$\begin{aligned}\tilde{\sigma}(\omega) + \sigma(\omega)\, \exp[-jV^2|\omega|]\, \mathrm{Sinh}[\tfrac{2j+1}{2} V^2\omega] \,/\, \mathrm{Sinh}[\tfrac{V^2}{2}\omega] \\ = \frac{2j+1}{2}\,\delta(\omega) + \frac{1}{2\pi L}\exp[i(\varepsilon-Q)\omega - (2j+1)\tfrac{V^2}{2}|\omega|] \quad .\end{aligned} \tag{IV.9}$$

The first term of the r.h.s. describes the free electron gas, while the second one is the driving term of the impurity charge fluctuations. The solution of this Wiener-Hopf integral equation is straightforward and yields for the f-level occupancy [9,12,13]

$$n_f = \frac{1}{2} + \frac{i}{2\pi}\int_{-\infty}^{\infty}\frac{dx}{x}\,[-ix+0]^{ijx}\exp[ix\tilde{\varepsilon}-(j+\tfrac{1}{2})\pi\,|x|]\,\Gamma(1-i(j+\tfrac{1}{2})x)\,/\,\Gamma(1-\tfrac{i}{2}x) \;, \tag{IV.10}$$

where $\tilde{\varepsilon}$ is a dimensionless parameter related to the effective f-level position by

$$\tilde{\varepsilon} = (\pi/2\Gamma)\varepsilon + (j+\tfrac{1}{2})\ln(2j+1) + j\ln(D'/2e\Gamma) \quad . \qquad \text{(IV.11)}$$

Here D' is a cut-off energy for the electronic excitations which has been introduced a posteriori in the Bethe-ansatz scheme. Eq. (IV.10) yields a smooth monotonically decreasing f-level occupancy from 1 in the Kondo limit and asymptotically tending to 0 in the nonmagnetic region as shown in Fig. 2.

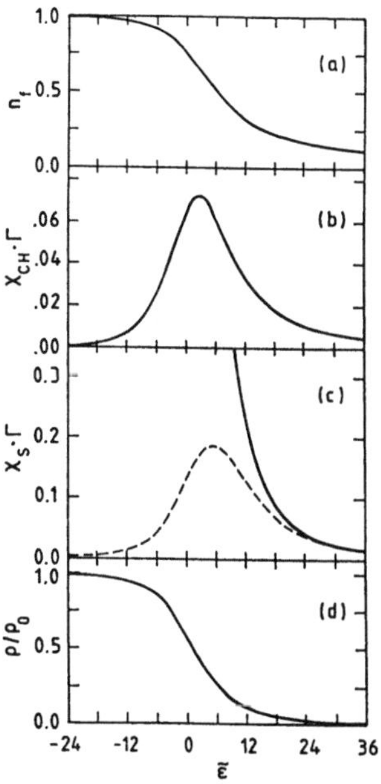

Fig. 2. (a) Valence, (b) charge susceptibility, (c) spin susceptibility, and (d) resistivity as a function of the invariant coupling $\tilde{\varepsilon}$. The resistivity is normalized to its value for a localized moment. The dashed line in (c) is the mixed-valent susceptibility, the full line is the total χ_s , and the difference between the full and dashed lines is the Kondo susceptibility. The Kondo part depends exponentially on $\tilde{\varepsilon}$

The charge susceptibility is obtained (IV.10) by differentiation with respect to ε and the resistivity is determined by the phase shift given by Friedel's sum rule [14]. The phase shift δ can be extracted from the Bethe-ansatz by considering the change of energy of the impurity if one electron is added to the Fermi sea [2,9], i.e.

$$2(2j+1)\delta = 2\pi\, \partial E_{imp}/\partial Q = 2\pi\, n_f \quad . \qquad \text{(IV.12)}$$

The resistivity and charge susceptibility are also shown in Fig. 2.

A magnetic field lifts the degeneracy of the $4f^1$ multiplet. According to (IV.8) all $B_\ell \neq -\infty$ in a magnetic field. The zero-field spin-susceptibility can be obtained by expansion in powers of the Zeeman energy, i.e. by assuming that $B_\ell \ll Q,\varepsilon$ for $\ell=0, \ldots, 2j-1$. Since the magnetic field affects $\tilde{\sigma}^{(2j)}$ only in higher order than linear, we can use the solution of (IV.9) in (IV.4). The r.h.s. of (IV.4) is then a known expression, from which we only need the asymptotic part for $\xi \to -\infty$. According to (IV.5) the asymptotic ξ-dependence is the same exponential for the impurity as for the host. Hence the impurity and host solutions of (IV.4) are proportional for small fields, the proportionality given by the r.h.s. of (IV.4). Since the host has a Pauli susceptibility, we obtain χ_s without explicitly solving the integral equations [9,12,13,15]

$$\chi_s\Gamma = \frac{j(j+1)}{6(2j+1)}\left[2\pi\,(j+\tfrac{1}{2})^{2j/(2j+1)}\,\exp[-\tilde{\varepsilon}/(j+\tfrac{1}{2})]\,/\,\Gamma(1+\frac{1}{2j+1}) - i\int_{-\infty}^{\infty}dx\right.$$
$$\left.\times\,[-ix+0]^{ijx}\,\exp[i\tilde{\varepsilon}x-(j+\tfrac{1}{2})\pi\,|x|]\,\Gamma(1-i(j+\tfrac{1}{2})x)\,/\,\Gamma(1-i\tfrac{x}{2})(x-2i/(2j+1))\right] \tag{IV.13}$$

χ_s consists of two terms, one due to spontaneous spin-flips of the localized f-electron on the Kondo energy scale and a second one due to the spin-fluctuations induced by the f-charge fluctuations. Note that there is a clear separation of the two energy scales in the groundstate. The susceptibility is always finite indicating a singlet groundstate. The Kondo term dominates for valences $n_f>1/2$, while for $n_f<1/2$ spin-fluctuations are mainly due to charge fluctuations. Note that in the expression for χ_s given in [9,12] a factor is missing in the Kondo term. The susceptibility is shown in Fig. 2.

RASUL and HEWSON [16] compared the above Bethe-ansatz results with the 1/N-perturbation expansion up to next-leading order and found excellent agreement for all f-level energies. They found a linear relation between the Bethe-ansatz cut-off, D', and the perturbative one, D, for a square density of states, (N=2j+1),

$$D' = \pi e\, N^{-N/(N-1)}\, D \quad . \tag{IV.14}$$

Note that, as discussed in [15], the leading 1/N-expressions do not agree with the Bethe-ansatz results in the Kondo regime.

IV.3 Coqblin-Schrieffer limit and solution for arbitrary fields

In a finite magnetic field the parameters B_ℓ are determined such that for small fields $n_\ell-n_{\ell+1}$ is proportional to H and independent of ℓ, $\ell\leq 2j-1$. All differences $B_\ell-B_{\ell'}$ are then constants independent of the field and only one B_ℓ is independent and parametrizes the magnetic field. From the symmetry of the integral equation we have that $B_\ell=B_{2j-\ell-1}$.

Since the magnetic field is usually much smaller than the hybridization width, Γ, the magnetization caused by the valence fluctuations (second term in (IV.13)) is always small and linear in the field [17]. The field, on the other hand, can be smaller or larger than the Kondo temperature, such that interesting features can be expected. The Kondo or Coqblin-Schrieffer limit is obtained by suppressing the charge fluctuations in the system, i.e. by taking $Q\to\infty$ in (IV.4). $\tilde{\sigma}^{(2j)}$ then decouples from all other $\sigma^{(q)}$ and the first term on the r.h.s. of (IV.4) vanishes [17].

The magnetization within the Coqblin-Schrieffer model has been obtained first by HEWSON et al. [18] by numerically solving the integral equations for various values of j. In order to obtain the magnetization only j (if j is an integer) or $j+1/2$ (if j is half-integer) integral equations need to be solved. The f-level occupation numbers and the magnetoresistance, however, require the complete solution, namely 2j integral equations obtained in [10,19]. The magnetization and magnetoresistance for $j=2$ are shown in Fig. 3. For $j=5/2$ the results can be found in 10. It has been noted by HEWSON et al. [18] that the magnetization for small fields behaves as $M/H = \chi_s + cH^2$, with $c<0$ for $j=1/2$, $c=0$ if $j=1$ and $c>0$ if $j\geq 3/2$.

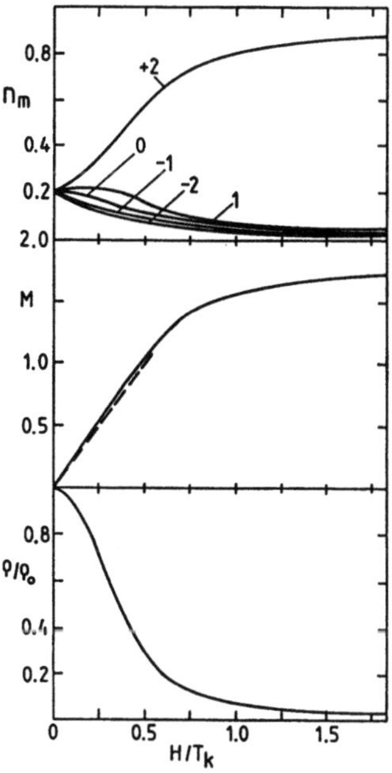

Fig. 3. Groundstate occupations of the f-levels, n_m, magnetization and magnetoresistance as a function of the magnetic field for the j=2 Coqblin-Schrieffer model [18,19]. T_K is the Kondo temperature defined via the zero-field susceptibility $\chi_s = j(j+1)/3T_K$ and ρ_0 is the resistivity in the absence of field. The index m denotes the spin component. The magnetization has been previously obtained by HEWSON, RASUL, and NEWNS [18]

IV.4 Effects of cubic and axial crystalline fields

In a cubic environment the j=5/2-multiplet of a Ce-ion splits into a Γ_7-doublet and a Γ_8-quartet. Only B_1 (if Γ_7 is the lower lying multiplet) and Q are finite, while B_0, B_2, B_3 and B_4 are $-\infty$ in the absence of a magnetic field. Eqs. (IV.4) and (IV.7) reduce to two coupled Wiener-Hopf integral equations, which have been solved numerically in [20]. The cubic crystal field splitting is parametrized by B_1 . For small crystal field splitting $B_1 \to -\infty$ such that the f-level occupations can be obtained in a similar way as χ_s in Sect. IV.2, yielding

$$\int d\xi\, \sigma^{(1)}_{imp}(\xi) \,/\, \int d\xi\, \sigma^{(1)}_{host}(\xi) = \chi_s^{imp} \,/\, \chi_s^{host} \quad , \qquad \text{(IV.15)}$$

where χ_s^{imp} is given by (IV.13), χ_s^{host} is the Pauli susceptibility and $\int d\xi \sigma^{(1)}_{host}(\xi)$ is the linear response of a free electron gas to the crystal field splitting. (IV.15) indicates that the susceptibility is independent of the mechanism that lifts the (2j+1)-fold degeneracy. We will use this in Sect. IV.6. The occupation numbers of the f-levels are obtained via (IV.8).

If the splitting is not small, the two coupled integral equations require a numerical solution [20]. We found that the valence grows with the splitting, the changes, however, are not very large. The relative occupations of the Γ_7- and Γ_8-levels strongly depend on the Kondo temperature of the j=5/2-sixtuplet. For a small Kondo temperature, i.e. a well localized f-electron, complete quenching of the orbital momentum can be obtained, such that the system behaves like a doublet.

The magnetic susceptibility can be obtained by applying a small magnetic field which can be treated as a perturbation. All B_ℓ are finite since all degeneracies are released but $B_o, B_2, B_3, B_4 << \varepsilon, B_1, Q$. Following the procedure of Sect. IV.1 we eliminate $\sigma^{(1)}(\xi)$ in (IV.4) and express it in terms of its hole function $\tilde{\sigma}^{(1)}(\xi)$. The feedback of the magnetic field on $\tilde{\sigma}^{(1)}(\xi)$ and $\tilde{\sigma}^{(5)}(\xi)$ is of higher order than the linear response and is neglected. Hence the equations for $\sigma^{(1)}$ and $\sigma^{(5)}$ decouple from the other four. The driving terms of the remaining four equations depend now on $\tilde{\sigma}^{(1)}$ and $\tilde{\sigma}^{(5)}$. The elimination of $\sigma^{(1)}$ has modified the kernel of the integral equations. The new kernel completely decouples $\sigma^{(o)}(\xi)$ from $\sigma^{(\ell)}(\xi)$, $\ell=2, 3,4$. The kernel for the $\sigma^{(o)}$ equation is the one of a spin 1/2 and corresponds to the Γ_7 . The kernel of the other three integral equations corresponds to a $j=3/2$ and is identified with the Γ_8 multiplet.

In a small magnetic field the solutions for the impurity and the host are again proportional. Since the susceptibilities in the free-electron gas are known, the impurity spin susceptibility can be obtained without actually solving the integral equations for $\ell=0,2,3$ and 4 . The susceptibility consists of three terms

$$\chi_s = \chi_{\Gamma_7} + \chi_{\Gamma_8} + \chi_{vv} \quad , \tag{IV.16}$$

the latter being the van Vleck susceptibility. For small crystal field one recovers the χ_s of the isotropic SU(6) impurity. With growing crystal field the multiplet is quenched and eventually the Γ_7 behaves like a j=1/2-impurity with a renormalized Kondo temperature

$$T_K \sim \varepsilon_F(\varepsilon_F/b_4)^2 \exp[(\pi/2\Gamma)(\varepsilon-\varepsilon_F)] \quad . \tag{IV.17}$$

Note that this renormalization typically represents 4 orders of magnitude [20].

In the Coqblin-Schrieffer limit the charge fluctuations are suppressed and the integral equations for $\sigma^{(1)}$ and $\sigma^{(5)}$ decouple. The remaining Wiener-Hopf equation for $\sigma^{(1)}$ can be solved analytically, such that explicit results can be obtained for that case [21].

The case of a j=5/2 multiplet in an axial crystalline field is discussed in [22] for the Coqblin-Schrieffer limit.

IV.5 Influence of spin-orbit coupling

We considered the 14-fold $4f^1$-multiplet split by spin-orbit coupling into a j=5/2- and a j=7/2-multiplet [23]. If j=5/2 is the lower-lying multiplet, i.e. the case of Ce, then the B_5 and Q are finite, while all other B_ℓ are $-\infty$, $5\neq\ell\leq 12$, and we have two coupled Wiener-Hopf integral equations which have been solved numerically. The valence, the occupations of the j=5/2 and 7/2 multiplets and the magnetic susceptibility have been discussed as a function of the spin-orbit splitting. The calculation procedure is similar to the one used to solve the impurity in a crystal field [20-22]. An explicit solution can be obtained in the Coqblin-Schrieffer limit [23].

The importance of the j=7/2-multiplet in the groundstate of Ce is a function of the f-level position [23]. In the Kondo regime the excited multiplet is irrelevant ($T_K << \xi_{so}$), while it is of crucial importance for the weakly magnetic impurity. The valence is much more strongly affected than the susceptibility. In general, the average number of f-electrons is underestimated if the excited j=7/2 levels are disregarded, i.e. in the $\xi_{so} \to \infty$ limit.

The true valence is then smaller than the one obtained with the j=5/2 multiplet only. Similarly to (IV.17) the Kondo temperature of the j=5/2 multiplet is renormalized by the spin-orbit splitting, as noted previously by GUNNARSSON and SCHÖNHAMMER [24] by means of the 1/N approximation.

IV.6 Universality of the Coqblin-Schrieffer limit

The Coqblin-Schrieffer limit of the model is obtained by suppressing all charge fluctuations, i.e. by taking $Q \to \infty$ (see Sect. IV.3). In this limit, the properties of the model have a universal dependence on the magnetic field and temperature, i.e. there is only one characteristic energy scale in the system, namely the Kondo temperature. The universality is usually characterized by the so-called Wilson-number [2], which relates the high and low temperature Kondo scales of the zero-field susceptibility.

Since the Wilson-number is defined for the *linear response* to a magnetic field it is independent of the mechanism lifting the degeneracy (see Sect. IV.4). In particular we may choose to split the (2j+1)-fold multiplet into two arbitrary submultiplets [25]. This involves only one Wiener-Hopf integral equation and can be solved analytically. Defining T_0^* such that

$$\chi(T=0) = C/T_0^* \qquad \text{(IV.18)}$$

C being the Curie constant, and T_K^* such that the $\ell n^{-2} T/T_K^*$ term of the high temperature series vanishes, we were able to prove [25] HEWSON and RASUL's conjecture [26]

$$T_K^*/T_0^* = \beta\gamma \exp[-\tfrac{3}{2} (1+\tfrac{1}{N})] \;/\; \Gamma(1+\tfrac{1}{N}) \qquad \text{(IV.19)}$$

Here $N=2j+1$, $\ell n\, \gamma$ is Euler's constant and $\beta=1.93890$. Note that in [25] the definition of T_0^* differs by a factor of 2 . Eq. (IV.19) correctly reproduces the Wilson number for N=2 [2].

V. Thermodynamic Properties of the Anderson Model

In this section we review the thermodynamic Bethe-ansatz equations, discuss their solution and compare our results to the experiments.

V.1 Thermodynamic Bethe-ansatz equations

The thermodynamic Bethe-ansatz equations of the model (II.1) have been independently derived by several authors [27-30]. The excitations of the system, given by the solutions of (III.8) and (III.9), are needed for this purpose. The excited states are characterized by rapidities and can be classified according to

(i) real charge rapidities, belonging to the set $\{\xi_\alpha^{(o)}\}$, which correspond to unpaired propagating electrons,

(ii) complex spin and charge rapidities, which correspond to bound states of electrons with different spin components (colors), and

(iii) strings of complex spin rapidities, which correspond to bound spin states.

The classes (i) and (ii) are already present in the groundstate and have been discussed in Sect. IV.1. In class (iii) there is a set of strings of complex spin rapidities $\{\xi_\alpha^{(\ell)}\}$, $\ell=1,\ldots,2j$. A string of length n is given by

$$\xi_{\alpha n}^{(\ell)\mu} = \Lambda_{\alpha n}^{(\ell)} + i\mu \frac{V^2}{2} \quad , \quad \mu = -(n-1), -(n-3), \ldots, (n-1) \quad , \qquad (V.1)$$

where $\Lambda_{\alpha n}^{(\ell)}$ is a real parameter and α is the running index within the set.

The above rapidities are inserted into (III.8) and (III.9) and the resulting coupled equations for the real $\{\xi_\alpha^{(\ell)}\}$ and $\{\Lambda_{\alpha n}^{(\ell)}\}$ are logarithmized. In this way quantum numbers like the $I_\alpha^{(\ell)}$ in (IV.2) are introduced. We introduce density functions for each class of excitations with index ℓ for the length of the string, i.e. $\rho^{(\ell)}(\xi)$ for $\xi^{(\ell)}$ and $\sigma_n^{(\ell)}(\xi)$ for $\Lambda_n^{(\ell)}$. Similarly we define the corresponding "hole"-functions. "Particle" and "hole" density functions are, of course, not independent, but linearly related in the thermodynamic limit. It is convenient to introduce the functions ε_ℓ and η_n^ℓ

$$\rho_h^{(\ell)}/\rho^{(\ell)} = \exp(\varepsilon_\ell/T) \quad , \quad \sigma_{n,h}^{(\ell)}/\sigma_n^{(\ell)} = \eta_n^\ell \quad . \qquad (V.2)$$

The internal energy and the distribution entropy can be expressed as functionals of the particle and hole densities. The conservation of the number of electrons of each spin-component (color) is included in the free energy functional by means of Lagrange multipliers. The Lagrange multipliers play the role of the chemical potential, crystal and magnetic fields, spin-orbit splitting, etc. Finally, the free energy functional is minimized with respect to the densities and the thermodynamic Bethe-ansatz equations are obtained [27-30]

$$\varepsilon_{2j} = \xi - A + T\,\Sigma_{q=0}^{2j-1} \int d\xi'\, \ell n(1+e^{-\varepsilon_q/T})\, F_q(\xi-\xi') + T\int d\xi'\, \ell n(1+e^{\varepsilon_{2j}/T})$$
$$\int \frac{d\omega}{2\pi}\, \exp[i(\xi-\xi')\omega - \frac{V^2}{2}|\omega|]\, \mathrm{Sinh}(jV^2\omega) \,/\, \mathrm{Sinh}\left((2j+1)\frac{V^2}{2}\omega\right) \qquad (V.3)$$

$$\ell n(1+e^{\varepsilon_\ell/T}) = \int d\xi'\, \ell n(1+e^{\varepsilon_{2j}/T})\, F_\ell(\xi-\xi') + \Sigma_{q=0}^{2j-1}\int d\xi' \Big\{ \ell n(1+e^{-\varepsilon_q/T})$$
$$\int \frac{d\omega}{2\pi}\, e^{i(\xi-\xi')\omega}\, 2\,\mathrm{Cosh}(\frac{V^2}{2}\omega)\, G(\omega)_{\ell+1,q+1} - \ell n(1+\eta_1^{q+1}) \int \frac{d\omega}{2\pi}\, e^{i(\xi-\xi')\omega}\, G(\omega)_{\ell+1,q+1} \Big\}$$
$$(V.4)$$

where F_ℓ and $G_{\ell,q}$ are given by (IV.5) and (IV.6) and A is the chemical potential. There are further 2j-infinite sets of nonlinear integral equations for the string excitations

$$\ell n\, \eta_n^\ell = \int d\xi' \Big\{ \ell n\left[(1+\eta_{n+1}^\ell)(1+\eta_{n-1}^\ell)/(1+(\eta_n^{\ell+1})^{-1})(1+(\eta_n^{\ell-1})^{-1})\right]$$
$$+ \delta_{n,1}\, \ell n(1+e^{-\varepsilon_{\ell-1}/T}) \Big\} \,/\, [2V^2\, \mathrm{Cosh}(\pi(\xi-\xi')/V^2)] \quad . \qquad (V.5)$$

The asymptotic boundary conditions for large n are given by the crystal and magnetic fields via the Lagrange multipliers [30]. For a pure Zeeman splitting we have

$$\lim_{n\to\infty} \frac{1}{n}\, \ell n\, \eta_n^\ell = H/T \quad , \quad \ell = 1, \ldots, 2j \quad , \qquad (V.6)$$

and the impurity free energy is given by

$$F_{imp} = -\frac{T}{\pi}\, \Sigma_{\ell=0}^{2j} \int d\xi\, \ell n(1+e^{-\varepsilon_\ell/T})\, (\ell+1)\frac{V^2}{2} \,/\, \left[(\xi-\varepsilon)^2 + \left((\ell+1)\frac{V^2}{2}\right)^2\right] \quad . \qquad (V.7)$$

In [30] we have shown that the above equations contain the Coqblin-Schrieffer limit and the groundstate equations as special cases. We have also extracted the low temperature Fermi liquid behavior.

V.2 Numerical solution of the equations and results

Eqs. (V.3)-(V.7) have been solved numerically [31-33] and the valence, entropy, charge and spin susceptibility and the specific heat have been obtained. The procedure is described in [31]. We obtained analytically the behavior of the thermodynamic density functions for large string indices, as well as a complete analytic solution for energies far from the Fermi level, and solved the rest numerically. The results [31-33] are summarized below.

At low temperatures, the number of f-electrons depends strongly on the f-level position. At high temperatures, it asymptotically approaches the value $n_f = (2j+1)/(2j+2)$. The temperature dependence is largest in the weak magnetic case. In the Kondo regime, there is a strong variation only at low T on the scale of the Kondo temperature.

The entropy asymptotically approaches $\ell n(2j+2)$ at high T and vanishes at $T=0$ since the groundstate is a singlet. The low T behavior strongly depends on the f-level position. In the Kondo regime, the initial slope is large, of the order of $1/T_K$, and it bends to a smaller slope when it reaches $\ell n(2j+1)$, i.e. when the spin degrees of freedom are released since the thermal bath overcomes the Kondo temperature. A small high T increase of the entropy is due to the charge excitations. In the quasi-nonmagnetic regime, the increase of S with T is slower and spin and charge degrees of freedom are mixed and no longer separated by two different energy scales.

The charge susceptibility as a function of $\varepsilon-\varepsilon_F$ has a resonance shape at all temperatures. The resonance energy and the width are temperature dependent. The width also increases with j . With increasing temperature, the position of the maximum is shifted to higher values of $\varepsilon-\varepsilon_F$. At low T the width is slightly reduced with T , but for $T>\Gamma$ the resonance broadens and becomes very wide at high T . The maxima value of the resonance grows when the width decreases and vice versa.

The spin susceptibility has a maximum as a function of T . The maximum is shifted to lower T when the f-level is lowered. As expected for a Fermi liquid χ depends as $a+bT^2$ at low T . For $N\geq 4$, b is always positive, while for $N=3$ it is positive in the mixed valence regime and vanishes asymptotically in the Kondo limit. For $N=2$, b changes sign approximately at $\varepsilon-\varepsilon_F \sim 0$, being negative in the Kondo regime. Hence a negative b uniquely characterizes a system of being spin $1/2$ and close to a Kondo compensation. At high T the free ion Curie susceptibility is asymptotically approached.

The specific heat shows one peak that shifts toward lower energies when the f-level position is lowered. At low T , C depends as $\gamma T+\beta T^3$, as expected for a Fermi liquid. For $N\geq 4$, β is always positive, while for $N=3$ it is positive in the mixed valence regime and vanishes asymptotically in the Kondo region. For $N=2$, β changes sign, being negative in the Kondo regime. This behavior is the same as χ_s , such that a negative β also characterizes a spin 1/2 system close to a Kondo screening. At high T , the free ion specific heat is asymptotically reached.

The thermodynamic properties of Ce in a magnetic field are discussed in [34].

V.3 Comparison with experiment

We have compared our results with experiments for mixed valent Ce impurities in Th [32], for the Kondo compounds YbCuAl [32] and $CeSn_3$ [35] and found good agreement between theory and experiment.

In order to compare our results with compounds, we considered the compounds as a dense impurity system with two main additional features [32]: (a) the chemical potential is renormalized when a finite fraction of f-electrons is promoted into the conduction band [36], and (b) conduction electron mediated interactions between rare earth sites. In a Kondo compound, however, the number of f-electrons is almost constant and the amplitude of the RKKY-interaction among sites is small if j is sufficiently large (1/N-expansion).

An isolated isotropic impurity obeys the following low temperature Fermi-liquid relation [15,30]

$$(2j+1)\gamma = \frac{\pi^2}{3}\left\{2j\frac{\chi_s}{C} + \chi_{ch}\right\} \quad , \qquad (V.8)$$

where γ is the coefficient linear in T of the specific heat, χ_{ch} is the charge susceptibility and $C=j(j+1)/3$ is the Curie constant. If j is sufficiently large the main contribution to the r.h.s. of (V.8) comes from χ_s. In a compound we expect the RKKY-interaction to contribute to order $1/(2j+1)$ on the r.h.s. In Fig. 4 [34], we plotted γ vs χ_s for several Yb-compounds ($j=7/2$). The straight line corresponds to (V.8) by neglecting χ_{ch} and RKKY-interactions. All compounds are close or below the straight line. The cases $YbAl_2$, $YbCu_2$ and $YbZn_2$ have a small susceptibility and are probably near to being divalent. Then χ_{ch} in (V.8) is important and can no longer be neglected. The plot justifies the applications of impurity

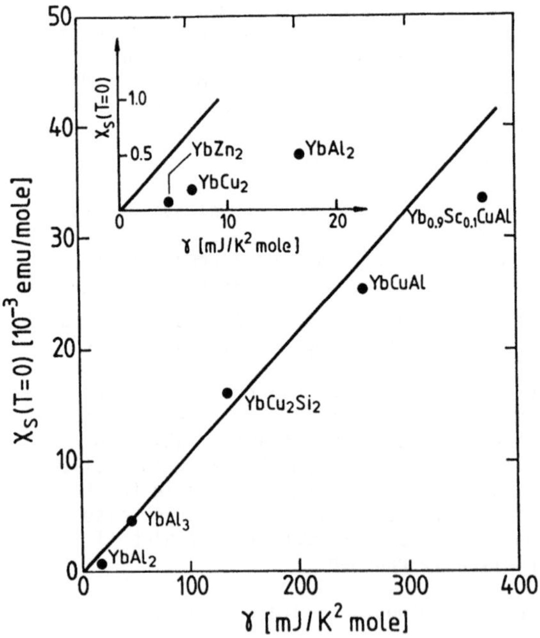

Fig. 4. Susceptibility vs linear specific heat coefficient for various Yb-compounds. The straight line is the impurity relation, (V.8), where χ_{ch} has been neglected. The data are from [38] and [39]

models to Yb-compounds with a large susceptibility, i.e. for systems being close to trivalence. A similar figure has been given by NEWNS and HEWSON [37]. Note that the susceptibility of $YbCu_2Si_2$ is anisotropic due to crystal field effects. Similar arguments hold for Ce-compounds.

In [32] we found good agreement between the susceptibility and specific heat data for YbCuAl and our impurity theory for $T_K \sim 100$ K. Also the low T magnetization as a function of field is consistent [26] and the calculated valence is in reasonable agreement with the one obtained from L_{III}-edge measurements. In [35] we successfully compared the theory with susceptibility and specific heat measurements for $CeSn_3$.

VI. Mixed-Valent Impurity With Two Magnetic Configurations

The mixed-valence model described in Sect. II involves one nonmagnetic (singlet) configuration, $4f^0$, and one magnetic (multiplet) configuration, $4f^1$. In this section we discuss a model with two magnetic configurations, e.g. $4f^n$ and $4f^{n+1}$, with angular momenta J_1 and J_2. We assume that the conduction states only carry an angular momentum of 1/2 such that $J_1 = J_2 \pm 1/2$. The Hamiltonian is given by [40,41]

$$H = \sum_\sigma \int dx\, c_\sigma^+(x) \left(-i \frac{\partial}{\partial x}\right) c_\sigma(x) + \varepsilon \sum_{M_1} |J_1M_1\rangle\langle J_1M_1| \tag{VI.1}$$
$$+ V \sum_{\sigma M_1 M_2} (J_2M_2\ \tfrac{1}{2}\ \sigma|J_2\ \tfrac{1}{2}\ J_1M_1) \int dx\, \delta(x)\, [c_\sigma^+(x)\, |J_2M_2\rangle\langle J_1M_1| + \text{h.c.}] ,$$

where notations are similar as in Sect. II and the Clebsch-Gordan coefficient selects $M_1=M_2+\sigma$. Note that if $J_2=0$ we recover (II.1) for $j=1/2$. Also the case $J_1=J_2-1/2$ can be mapped onto $J_1=J_2+1/2$ by adequately redefining J_2, exchanging electrons with holes and rescaling the hybridization. In the integer-valent case the model reduces to the Kondo exchange model for an arbitrary spin [42]. The case $J_2=1/2$ and $J_1=1$ has been independently solved by PROETTO, ALIGIA and BALSEIRO [43].

Experiments for dilute Tm alloys seem to indicate a magnetic groundstate [44,45]. For mixed-valent Tm-ions the Hund's rule groundstates of the two configurations $4f^{12}$ and $4f^{13}$ are both magnetic ($J=6$ and $J=7/2$, respectively). Our model is supposed to resemble some of the properties of Tm-ions.

VI.1 Bethe-ansatz solution

The construction of the Bethe-ansatz is analogous to Sect. III. We first discuss the one- and two-particle problems. Due to the $(2J_2+1)$ internal degrees of freedom of the $4f^n$-configuration, the scattering of the impurity is not just given by a change in phase but by the following scattering matrix [41]

$$S^{\sigma\sigma'}_{M_2M_2'}(k) = \delta_{\sigma\sigma'}\delta_{M_2M_2'} - (\sigma M_2|M_2+\sigma)(\sigma' M_2'|M_2'+\sigma') \frac{iV^2}{k-\varepsilon+iV^2/2} P^{\sigma\sigma'}_{M_2M_2'} \tag{VI.2}$$

where σM_2 ($\sigma' M_2'$) are the spin indices before (after) scattering, the symbols are short-hand notations for the Clebsch-Gordan coefficients and

$$P^{\sigma\sigma'}_{M_2M_2'} = \delta_{\sigma,\sigma'}\delta_{M_2M_2'} + \delta_{\sigma',-\sigma}\delta_{M_2',M_2+2\sigma} \quad . \tag{VI.3}$$

The two-body wave function is similar to the one in Sect. III, except that (assuming $J_1=J_2+1/2$) V^2 in (III.4) is to be replaced by $c=V^2/(2J_2+1)$. Also the generalization to the N-body wave function is straightforward, however, the single-valuedness of the wave function is to be verified with respect to the impurity scattering matrix (VI.2) as well.

Imposing periodic boundary conditions and following the procedure of [42] we obtain for N-M electrons with spin-up and M electrons with spin-down by means of a second Bethe-ansatz [41]

$$\exp[i(k_jL+2\varphi_j)] = \Pi_{\beta=1}^{M} e(k_j-\Lambda_\beta) \quad , \qquad j = 1, \ldots, N \tag{VI.4}$$

$$((\Lambda_\alpha-\varepsilon-iJ_2c)/(\Lambda_\alpha-\varepsilon+iJ_2c))\, \Pi_{j=1}^{N} e(\Lambda_\alpha-k_j) = -\Pi_{\beta=1}^{M} e[(\Lambda_\alpha-\Lambda_\beta)/2] \qquad \alpha=1,\ldots,M \tag{VI.5}$$

Here L is the length of the box, φ_j is given by (III.1) and $e(x)$ is given by (III.10) with V^2 replaced by c .

These results were also obtained by PROETTO et al. [43] for $J_2=1/2$ and $J_1=1$.

VI.2 Groundstate properties

The groundstate equations are obtained by following the scheme of Sect. IV.1. Since $c>0$ there is an effective attractive interaction between electrons of opposite spin. Hence, there exist complex conjugated k_j as solutions of (IV.4), related to the real Λ_α values by

$$k_\alpha^\pm = \Lambda_\alpha \pm i\,\frac{c}{2} \quad , \qquad \alpha = 1, \ldots, M \tag{VI.6}$$

which correspond to spin-paired electrons. There are also N-2M real k_j values corresponding to unpaired charge modes.

Inserting (VI.6) into (VI.4) and (VI.5) and logarithmizing we obtain the thermodynamic limit [41]

$$\rho_h(k) + \rho(k) + \frac{1}{\pi}\int_{-\infty}^{Q} d\Lambda'\, \sigma(\Lambda')\, \frac{c/2}{(\Lambda'-k)^2+(c/2)^2} = \frac{1}{2\pi} + \frac{1}{\pi L}\,\frac{V^2/2}{(k-\varepsilon)^2+(V^2/2)^2} \tag{VI.7}$$

and

$$\sigma_h(\Lambda) + \sigma(\Lambda) + \frac{1}{\pi}\int_{-\infty}^{Q} d\Lambda'\, \sigma(\Lambda')\, \frac{c}{(\Lambda-\Lambda')^2+c^2} + \frac{1}{\pi}\int_{-\infty}^{B} dk\, \rho(k)\, \frac{c/2}{(\Lambda-k)^2+(c/2)^2} =$$

$$= \frac{1}{\pi} + \frac{1}{\pi L}\,\frac{c(J_2+1)}{(\Lambda-\varepsilon)^2+[c(J_2+1)]^2} \tag{VI.8}$$

where $\rho(k)$ and $\sigma(\Lambda)$ are the distribution density functions for the k- and Λ-rapidities. Note that $\rho(k)\equiv 0$ for $k>B$ and $\sigma(\Lambda)=0$ for $\Lambda>Q$ and the "hole"-functions, $\rho_h(k)$ and $\sigma_h(\Lambda)$, vanish identically for $k<B$ and $\Lambda<Q$, respectively. The integration limits B and Q are determined from the magnetization and the number of particles

$$N/L = 2\int_{-\infty}^{Q} d\Lambda\, \sigma(\Lambda) + \int_{-\infty}^{B} dk\, \rho(k) \quad , \tag{VI.9a}$$

$$S_z/L = \frac{1}{2}\int_{-\infty}^{B} dk\, \rho(k) + J_2/L \quad . \tag{VI.9b}$$

Since the Fermi energy, ε_F, is much larger than the Zeeman energy, H, we have that $Q \gg B$.

In zero-field $B=-\infty$ and $\rho \equiv 0$, such that (VI.8) reduces to a single Wiener-Hopf equation for σ. The solution yields the f-level occupation [41]

$$n_f = n + \frac{1}{2} + (i/\sqrt{\pi}) \int \frac{dx}{2\pi} \frac{1}{x} \Gamma(\tfrac{1}{2}-ix) \left[\frac{-ix+0}{e}\right]^{ix} \exp[i(\varepsilon-Q) \frac{2\pi}{c} x - 2\pi(J_2+1)|x|] \tag{VI.10}$$

There are two contributions to the magnetization: One arises from spin-fluctuations caused by the charge fluctuations, the other one is more important and due to the Kondo effect. The latter contribution is just the magnetization of the Kondo exchange model [2,3,42] for a spin J_1.

VI.3 Thermodynamics

In analogy to Sect. V.1 we classify solutions of (VI.4) and (VI.5) according to

(i) (N-2M') real charge rapidities, which correspond to unpaired propagating electrons;

(ii) 2M' complex rapidities, which correspond to paired electron states, $k^{\pm} = \Lambda'_\alpha \pm ic/2$, where Λ'_α is a real spin rapidity, $\alpha=1,\ldots,M'$; and

(iii) M_n strings of complex spin rapidities of length n, $n=1,\ldots,$, which correspond to bound spin states, $\Lambda^{\mu}_{\alpha,n} = \Lambda_{\alpha,n} + i\mu c/2$, $\mu=-(n-1),-(n-3),\ldots,(n-1)$, $\Lambda_{\alpha,n}$ being a real parameter.

The integers M' and M_n satisfy the relation

$$M' + \Sigma_{n=1}^{\infty} n M_n = M \quad . \tag{VI.11}$$

The above rapidities are inserted into (VI.4) and (VI.5) and the resulting equations are logarithmized. The usual distribution functions for the rapidities are defined: $\rho(k)$, $\sigma'(\Lambda)$, and $\sigma_n(\Lambda)$, as well as the corresponding "hole" distribution functions. Particle and hole functions are not independent, but linearly coupled by integral equations in the thermodynamic limit. It is convenient to introduce the following functions

$$\rho_h/\rho = \exp(\varepsilon_0/T) \ , \quad \sigma'_h/\sigma' = \exp(\varepsilon_1/T) \ , \quad \sigma_{n,h}/\sigma_n = \eta_n \quad . \tag{VI.12}$$

The distribution functions ρ, σ' and σ_n are actually determined by minimizing the free energy functional, subject to the conservation of the number of particles and the total spin [41]. We obtain (V.3)-(V.6) specialized for $j=1/2$ with V^2 replaced by c. The impurity-free energy is given by [41]

$$F_{imp} = (\varepsilon-A) - \frac{T}{2} \int d\xi \, \ell n(1+\exp(\varepsilon_1/T)) \, G_{2J_2+1}(\varepsilon-\xi) - \frac{T}{2} \int d\xi \, \ell n(1+\eta_{2J_2}) \, G_0(\varepsilon-\xi) \tag{VI.13}$$

where

$$G_\ell(\xi) = \int \frac{d\omega}{2\pi} \exp[i\xi\omega - \ell \frac{c}{2} |\omega|] \, / \, \mathrm{Cosh} \, (\frac{c}{2} \omega) \quad . \tag{VI.14}$$

From these equations we have recovered [41] the groundstate equations of Sect. VI.2, the Kondo limit corresponding to the exchange model for an arbitrary spin J_1 [46] and the high-temperature limit (perturbation expansion).

VII. Concluding Remarks

In summary we reported the present status of the Bethe-ansatz solution of the degenerate Anderson model in the $U\to\infty$ limit with excluded multiple f-occupation. In Sect. II-V we considered the cases of Ce and Yb, i.e. one singlet and one magnetic configuration, while in Sect. VI the case of two magnetic configurations was discussed. The latter is supposed to resemble some properties of Tm-ions. The N-body wave function for the models factorizes and the general solution was constructed by nested Bethe-ansätze.

The model for Ce and Yb shows several crossovers which have been extracted from the Bethe-ansatz solution: as a function of the f-level position the system goes from a nonmagnetic situation, through the mixed-valent into the Kondo region; the Kondo state can be broken up by a sufficiently strong magnetic field and the system becomes asymptotically free; if the orbital angular momentum is quenched by a strong crystal field, the symmetry is lowered and the Kondo temperature is drastically modified by the crystal field; and as a function of temperature the system gradually goes from a singlet state at $T=0$ to a manifold of $(2j+2)$ degrees of freedom.

For the model with two magnetic configurations we briefly discussed the groundstate properties and the thermodynamic Bethe-ansatz equations. If $J_2\neq 0\neq J_1$ the groundstate is always magnetic. The model also contains the Kondo exchange model for an arbitrary spin as its integer-valent limit. For a comparison with experiments for Tm-systems generalizations of the model including an orbital momentum channel are needed.

References

1 P.W. Anderson, Phys. Rev. 124, 41 (1961)
2 N. Andrei, K. Furuya and J.H. Lowenstein, Rev. Mod. Phys. 55, 331 (1983)
3 A.M. Tsvelick and P.B. Wiegmann, Adv. Phys. 32, 453 (1983)
4 P. Schlottmann, Z. Phys. B 49, 109 (1982)
5 H.A. Bethe, Z. Phys. 71, 205 (1931)
6 C.N. Yang, Phys. Rev. Lett. 19, 1312 (1967)
7 M. Gaudin, Phys. Lett. 24A, 55 (1967); Thesis, Paris (1967)
8 B. Sutherland, Phys. Rev. Lett. 20, 98 (1968)
9 P. Schlottmann, Z. Phys. B 51, 49 (1983)
10 P. Schlottmann, J. Mag. Mag. Mat., in press, invited paper, ICVF, Köln 1984
11 P. Schlottmann, Phys. Rev. B 30, 1454 (1984)
12 P. Schlottmann, Phys. Rev. Lett. 50, 1967 (1983)
13 E. Ogievetski, A.M. Tsvelick and P.B. Wiegmann, J. Phys. C 16, L797 (1983)
14 D.C. Langreth, Phys. Rev. 150, 516 (1966)
15 P. Schlottmann, Phys. Lett. 100A, 509 (1984)
16 J.W. Rasul and A.C. Hewson, J. Phys. C 17, 3337 (1984)
17 P. Schlottmann, Z. Phys. B 51, 223 (1983)
18 A.C. Hewson, J.W. Rasul and D.M. Newns, Phys. Lett. 93A, 311 (1983)
19 P. Schlottmann, unpublished
20 P. Schlottmann, Phys. Rev. B 30, 1454 (1984)
21 P. Schlottmann, Z. Phys. B 55, 213 (1984)
22 P. Schlottmann, Proc. ICCF 5, Sendai, Japan (1985)
23 P. Schlottmann, J. Phys. C 18, (1985)
24 O. Gunnarsson and K. Schönhammer, Phys. Rev. B 28, 4315 (1983)
25 P. Schlottmann, J. Phys. C 17, L267 (1984)
26 A.C. Hewson and J.W. Rasul, J. Phys. C 16, 6799 (1983)

27 P. Schlottmann, Z. Phys. B 52, 127 (1983)
28 A.M. Tsvelick, J. Phys. C 17, 2299 (1984)
29 N. Kawakami, S. Tokuono and A. Okiji, J. Phys. Soc. Japan 53, 51 (1984)
30 P. Schlottmann, Z. Phys. B 54, 207 (1984)
31 P. Schlottmann, Z. Phys. B 56, 127 (1984)
32 P. Schlottmann, Z. Phys. B 57, 23 (1984)
33 N. Kawakami and A. Okiji, Phys. Lett. 103A, 205 (1984)
34 P. Schlottmann, Proc. ICCF 5, Sendai, Japan (1985)
35 P. Schlottmann, J. Appl. Physics, in print
36 A.C. Hewson, J. Phys. C 10, 4973 (1977)
37 D.M. Newns and A.C. Hewson, in Valence Fluctuations in Solids, eds. L.M. Falicov, W. Hanke and M.B. Maple, North-Holland (Amsterdam, 1981), p. 27
38 J.C.P. Klaasse, F.R. de Boer and P.F. de Châtel, Physica 106B, 178 (1981)
39 B.C. Sales and R. Viswanathan, J. Low Temp. Phys. 23, 449 (1976)
40 P. Schlottmann, in Valence Instabilities, eds. P. Wachter and H. Boppart, North Holland (Amsterdam, 1982), p. 471
41 P. Schlottmann, Z. Phys. B 59, xxx (1985)
42 V.A. Fateev and P.B. Wiegmann, Phys. Lett. 81A, 179 (1981)
43 C.R. Proetto, A.A. Aligia and C.A. Balseiro, Phys. Lett. 107A, 93 (1985); and preprints
44 A. Berger, P. Haen, F. Holtzberg, F. Lapierre, J.M. Mignot, T. Penney, O. Peña and R. Tournier, J. Phys. (Paris) 40, C5-364 (1979)
45 E. Holland-Moritz, J. Mag. Mag. Mat. 38, 253 (1983)
46 V.A. Fateev and P.B. Wiegmann, Phys. Rev. Lett. 46, 356 (1981)

Part III

Dynamical Properties of Single Impurity

Kondo Effect versus Crystal Field

S. Maekawa, S. Kashiba, S. Takahashi, and M. Tachiki

The Research Institute for Iron, Steel and Other Metals
Tohoku University, Katahira, Sendai 980, Japan

In Ce compounds and alloys, the scattering of conduction electrons by Ce^{3+} ions is accompanied by transition between the crystal field energy levels. We present a theory for the magnetic and transport properties which takes into account the Kondo effect and the crystal field self-consistently. Through the analyses of the magnetic susceptibility, the neutron quasi-elastic and inelastic scattering spectra, and the resistivity, we propose the crystal field level-scheme in $CeCu_2Si_2$.

1. Introduction

The purpose of this paper is to study the magnetic and transport properties in Ce compounds and alloys by taking into account the Kondo effect and the crystal field self-consistently. The atomic ground state of a Ce^{3+} ion has total angular momentum J=5/2, in which the orbital angular momentum (L=3) is antiparallel to the spin (S=1/2) because of the strong spin-orbit interaction. The 6-fold degeneracy of the ground state is partly removed by the crystal field. The exchange scattering of conduction electrons by Ce ions is strongly enhanced by the Kondo effect. The scattering is accompanied by transition between the crystal field energy levels in Ce ions and induces the shift and broadening of the levels. Therefore, the Kondo scattering and the crystal field are correlated with each other and are determined self-consistently [1-3].

We will use the Coqblin-Schrieffer exchange model [4]. By extending Abrikosov's pseudofermion method [5] to the 4f states, the dynamical as well as static susceptibilities are obtained as functions of the self-energy of the pseudofermions. The self-energy is solved in the self-consistent ladder approximation [1,2] in Sec. 2. In Sec. 3, we calculate the susceptibility, its anisotropy, and the neutron scattering spectra in $CeCu_2Si_2$. By comparing the theoretical results with the experimental data, the crystal field level scheme in this compound is derived. In Sec. 4, we formulate the electrical resistivity, and examine the structure in the resistivity vs. temperature curve and the anisotropy. Finally, the summary and discussion is given in Sec. 5. In this paper we take units of $\hbar=k_B=1$.

2. Magnetic Susceptibility

The exchange interaction between the angular momentum $\vec{J}$ and conduction electrons is described by the Coqblin-Schrieffer model [4]:

$$H = \sum_m (E_m + \lambda) f_m^\dagger f_m + \sum_{k,m} \xi_k c_{km}^\dagger c_{km} - J \sum_{k,k'} \sum_{m,m'} f_{m'}^\dagger f_m c_{km}^\dagger c_{k'm'} \, , \tag{2.1}$$

where f_m is the annihilation operator of 4f electron in the eigenstate m and E_m is the crystal field energy. Here, only the ground multiplet with J=5/2 is considered. In (2.1), the fictitious chemical potential ($-\lambda$) has been introduced in the 4f states in order to retain the condition, $n_f \equiv \sum_m f_m^{\dagger} f_m = 1$, in a Ce^{3+} ion; λ will be taken to be infinity later. c_{km} is the operator of conduction electron with wave number $|\vec{k}|$ and m. We assume that the energy of conduction electron ξ_k measured from the Fermi energy is a function of $|\vec{k}|$. J is the exchange interaction constant ($J<0$). As seen in (2.1), the exchange scattering of conduction electrons induces transition between the crystal field levels of Ce^{3+} ions. The Kondo scattering results in the renormalization of the exchange interaction. Therefore, the crystal field level scheme may be modified by the Kondo scattering.

It has been derived that the thermal average of a physical quantity A, $\langle A \rangle$, is given by the corresponding average $\langle A \rangle_\lambda$ and $\langle n_f \rangle_\lambda$ with the fictitious chemical potential ($-\lambda$) as [2,5-9]

$$\langle A \rangle = \lim_{\lambda \to \infty} (\langle A \rangle_\lambda / \langle n_f \rangle_\lambda) \ . \tag{2.2}$$

Here, the physical quantity $\langle A \rangle$ may be a function of frequency. Therefore, our main task is to solve various quantities in the fictitious space with λ, in which the usual field theoretical method can be used.

Let us examine the perturbation expansion of the thermodynamic Green's fuction of 4f electrons, $G_{fm}(i\omega_n)$, in the space with the chemical potential ($-\lambda$). Since we are interested in the limit of large λ, it turns out that the processes with a 4f electron only contribute to the physical quantities and the conduction electrons in the intermediate processes are not renormalized. We calculate the Green's function of 4f electron in the self-consistent ladder approximation [1,2]. The corresponding Dyson equation is expressed in Fig. 1, where the single and double dashed lines denote the bare and renormalized 4f Green's functions, respectively. The solid lines denote the Green's functions of conduction electron. The wavy and double wavy lines are the bare (J) and renormalized vertices, respectively. Various self-consistent approximations similar to ours have been studied in the Anderson impurity model [10,11]. Some of them are discussed in this volume by several authors [11]. In the sense of the diagrammatic method, our approximation can be thought to be a version of the above approximations in the Coqblin-Schrieffer model.

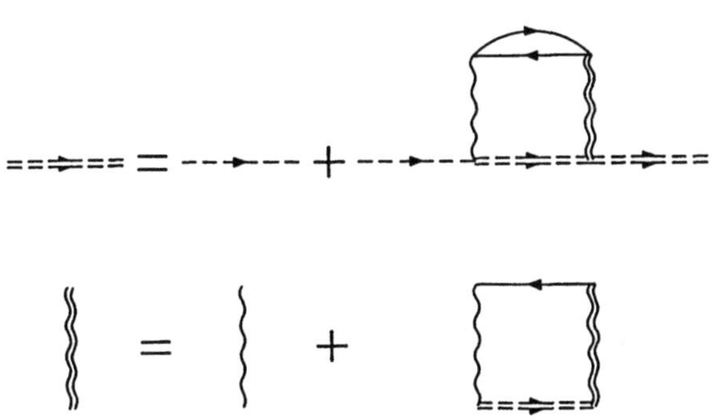

Fig. 1 Dyson equation for the Green's function of 4f electron in the space with λ in the self-consistent ladder approximation [1,2].

From Fig. 1, we obtain the self-energy of 4f electron as

$$\Sigma(i\omega_n) = - J\rho \int_{-D}^{D} d\xi_1 f(-\xi_1)\{1-[1-J\rho\sum_m \int_{-D}^{D} d\xi_2 f(\xi_2) G_{fm}(i\omega_n+\xi_2-\xi_1)]^{-1}\} , \quad (2.3)$$

where

$$G_{fm}(i\omega_n) = [i\omega_n - E_m - \lambda - \Sigma(i\omega_n)]^{-1} = \int_{-\infty}^{\infty} dx \frac{\rho_m(x)}{i\omega_n - x} . \quad (2.4)$$

Here, 2D is the conduction band width, ρ is the density of states of conduction electrons per spin, $f(\xi)$ is the Fermi distribution function, and $\rho_m(x)$ is the spectral function defined by

$$\rho_m(x) = - \frac{1}{\pi} Im\, G_{fm}(i\omega_n \to x + io) , \quad (2.5)$$

$$\int_{-\infty}^{\infty} dx \rho_m(x) = 1 . \quad (2.6)$$

Since the self-energy and the spectral function have the significant contributions only around $\omega=\lambda$ in the limit of large λ, it is convenient to redefine them around $\omega=\lambda$ as

$$\bar{\Sigma}(i\omega_n) \equiv \Sigma(i\omega_n + \lambda) , \quad (2.7)$$

$$\bar{\rho}_m(x) \equiv \rho_m(x + \lambda) . \quad (2.8)$$

Then, (2.3) is rewritten as

$$\bar{\Sigma}(i\omega_n) = - J\rho \int_{-D}^{D} d\xi_1 \frac{f(-\xi_1)\bar{\Pi}(i\omega_n - \xi_1)}{1 + \bar{\Pi}(i\omega_n - \xi_1)} , \quad (2.9)$$

$$\bar{\Pi}(i\omega_n - \xi_1) = - J\rho\sum_m \int_{-D}^{D} d\xi_2 \frac{f(\xi_2)}{i\omega_n + \xi_2 - \xi_1 - E_m - \bar{\Sigma}(i\omega_n + \xi_2 - \xi_1)} . \quad (2.10)$$

The μ component of magnetic moment operator in a Ce^{3+} ion is expressed as

$$M_\mu = g_J\mu_B \sum_{m,m'} \langle m|J_\mu|m'\rangle f^\dagger_m f_{m'} , \quad (2.11)$$

where g_J is the Lande g-factor (g_J=6/7) and μ_B the Bohr magneton. The diagrammatic equation for the dynamical susceptibility is shown in Fig. 2. In the self-consistent ladder approximation, however, we find that the vertex correction (Λ) in Fig. 2 vanishes when the exchange interaction in (2.1) is used,because of the conservation relation of the quantum number m. Then, the imaginary part of the $\mu\nu$ component of the susceptibility in the space with λ is obtained as

$$Im\, \chi^{(\lambda)}_{\mu\nu}(\omega) = \pi(g_J\mu_B)^2 \sum_{m_1 m_2} \langle m_1|J_\mu|m_2\rangle\langle m_2|J_\nu|m_1\rangle$$

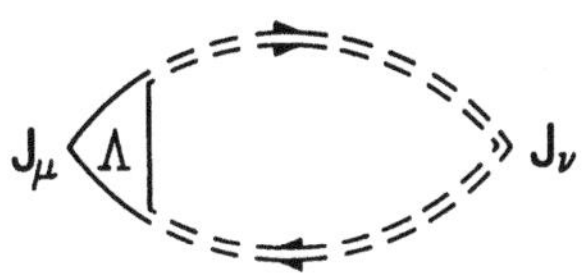

Fig. 2 Equation for the dynamical susceptibility in the space with λ in the self-consistent ladder approximation [1,2].

$$\times \int_{-\infty}^{\infty} d\omega'[f(\omega' + \lambda) - f(\omega' + \omega + \lambda)]\bar{\rho}_{m_1}(\omega)\bar{\rho}_{m_2}(\omega' + \omega) \ , \qquad (2.12)$$

Using the relation (2.2), we have the imaginary part of the real susceptibility,

$$Im\ \chi_{\mu\nu}(\omega) = \lim_{\lambda\to\infty} Im\ \chi^{(\lambda)}_{\mu\nu}(\omega)/\langle n_f\rangle_\lambda \ , \qquad (2.13)$$

where the denominator is given by

$$\langle n_f\rangle_\lambda = \sum_m \int_{-\infty}^{\infty} dx f(x + \lambda)\bar{\rho}_m(x) \ . \qquad (2.14)$$

The neutron scattering intensity is expressed as

$$\frac{d^2\sigma}{d\Omega d\omega} \propto \sum_{\mu,\nu} (\delta_{\mu\nu} - \frac{q_\mu q_\nu}{q^2}) \frac{Im\ \chi_{\mu\nu}(\omega)}{(1 - e^{-\beta\omega})} \ . \qquad (2.15)$$

where $\vec{q}$ is the scattering wave vector. The static susceptibility is also obtained from (2.13) using the Kramers-Kronig relation as

$$\chi_{\mu\nu}(0) = \frac{1}{\pi} P\int_{-\infty}^{\infty} d\omega \frac{Im\ \chi_{\mu\nu}(\omega)}{\omega} \ . \qquad (2.16)$$

It was shown in Ref. 3 that the static susceptibility calculated in the self-consistent ladder approximation agrees quantitatively with the exact Bethe-Ansatz results in the wide temperature-range. The neutron scattering spectra and the static susceptibilities have been calculated in $Ce_xLa_{1-x}Al_2$, CeB_6, and $CeCu_2Si_2$, and compared with the experimental results in Ref. 3. In the next Section, we examine $CeCu_2Si_2$ in the more detail.

3. Magnetic Properties in $CeCu_2Si_2$

The magnetic anisotropy in non-cubic crystals is a measure of the crystal field. The anisotropy of the susceptibility in the tetragonal crystal $CeCu_2Si_2$ has been observed by several groups [12-14]. In this compound, the c-axis is the easy axis of magnetization. This crystal shows the superconductivity below about 0.6 K. The neutron quasi-elastic and inelastic scattering experiments have been done by Horn et al. [15] in poly-crystalline $CeCu_2Si_2$. They observed a clear inelastic peak at the energy ω=364 K. We write the wave functions of the 4f electron as

$$\begin{cases} |0\pm\rangle = a|\pm5/2\rangle + b|\mp3/2\rangle \ , \\ |1\pm\rangle = |\pm1/2\rangle \ , \\ |2\pm\rangle = b|\pm5/2\rangle - a|\mp3/2\rangle \ , \end{cases} \qquad (3.1)$$

where a and b are parameters.

We calculate the susceptibility in $CeCu_2Si_2$ in the following way: One of the crystal field splitting parameters is taken to be 364 K. The other value, a, and the exchange interaction constant $J\rho$ are taken to be adjustable parameters. The μ component of the spin-polarization of the conduction electrons $<s_\mu>$ is taken into account in the mean field approximation, i.e., $<s_\mu>=J\rho<J_\mu>/7$ where $<J_\mu>$ is the polarization of a Ce^{3+} ion. Then, we have the $\mu\mu$ component of the susceptibility per Ce^{3+} ion [3],

$$\chi_{\mu\mu} = [1 + \frac{2}{7}(1 + \frac{1}{g_J})J\rho]\chi^f_{\mu\mu}(0) , \tag{3.2}$$

where $\chi^f_{\mu\mu}(0)$ is the $\mu\mu$ component of the susceptibility of a Ce^{3+} ion. In Fig. 3, the theoretical results of χ^{-1}_{zz} and χ^{-1}_{xx} (solid lines) in $CeCu_2Si_2$ are compared with the experimental data by Onuki et al. [14] where the z direction is parallel to the c-axis. In the calculation, we have chosen that the doublet $|0\pm>$ is the ground states, $|1\pm>$ the first excited states with the excitation energy 300 K, $|2\pm>$ the second excited states with 364 K, a=0.87, and b=0.49. We also chose the values $J\rho$=-0.0385 and D=10^4K. As seen in Fig. 3, the theoretical results agree well with the experimental data. Recently, Bredl et al. [16] have observed the specific heat in $CeCu_2Si_2$ and concluded that the specific heat anomaly may be described by two closely spaced excited doublets at 360 K. Their conclusion is consistent with ours. Although the neutron scattering experiments [15] posed that the first excited doublet has the energy ω=140 K, the level scheme results in much smaller anisotropy of the susceptibility. This fact was also pointed out by Onuki et al. [14].

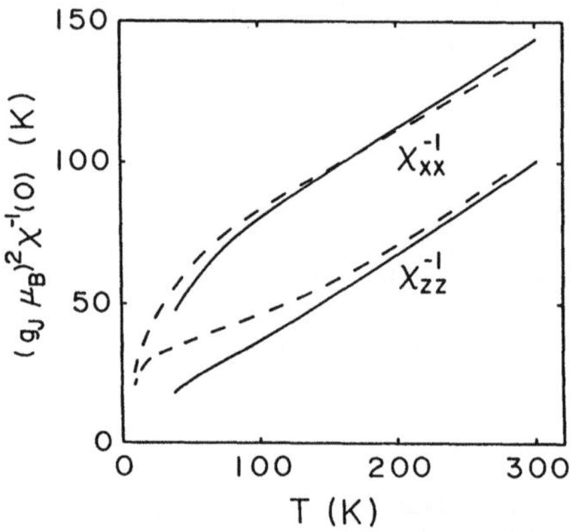

Fig. 3 Theoretical results of the inverse susceptibility per Ce ion (solid lines) are compared with the experimental data (dashed lines) by Onuki et al. [14] in $CeCu_2Si_2$. The z direction is taken to be parallel to the c-axis.

We have also calculated the quasi-elastic line width and the neutron inelastic scattering intensity by using our level scheme. The results are shown in Figs. 4 and 5, respectively. The theoretical results are compatible with the experimental data by Horn et al. [15]. In particular, we note in Fig. 5 that there exists strong scattering intensity between the quasi-elastic and inelastic lines (ω=100-200 K), since the lines are broad because of the Kondo effect. In the next section, the electrical resistivity vs. temperature curve in $CeCu_2Si_2$ is calculated.

4. Electrical Resistivity

Cornut and Coqblin [17] have studied the influence of the crystal field on the resistivity in the third-order perturbation theory in terms of $J\rho$,

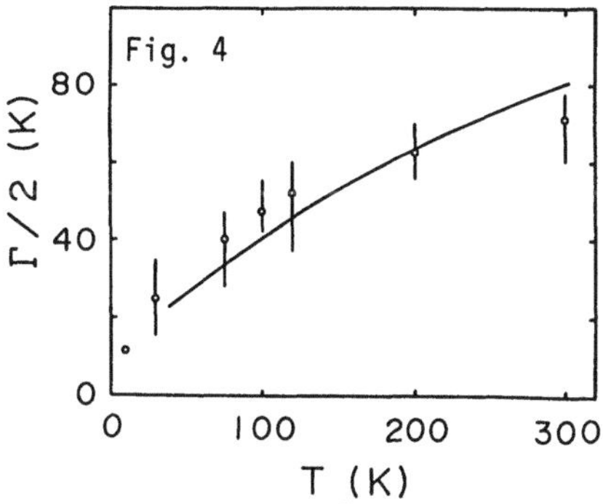

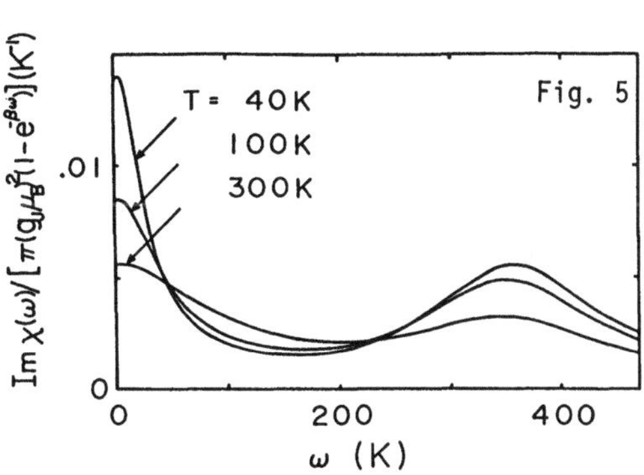

Fig. 4 Theoretical results of the half-width of the quasi-elastic line (solid line) are compared with the experimental results by Horn et al. [15] in $CeCu_2Si_2$.

Fig. 5 Theoretical results of the neutron scattering intensity in $CeCu_2Si_2$.

and observed the peak structure in the resistivity vs. temperature curve: When temperature is much higher than the crystal field splitting, all the crystal field levels contribute to the Kondo resistivity, so that the large resistivity is obtained. When temperature becomes much lower than the splitting, only the ground state will mainly contribute to the resistivity. Therefore, a peak of the resistivity may appear at a certain temperature [18]. However, as seen in the previous section, the Kondo scattering modifies the crystal field level scheme, so that the resistivity will be influenced by the modification, which was not included in the third-order perturbation theory. In this section, we apply the self-consistent ladder approximation to the calculation of the resistivity.

We expand the T-matrix of the conduction electrons in the power of the exchange interaction $J\rho$ in the fictitious space with the chemical potential $(-\lambda)$, and retain the ladder diagrams. The diagrammatic equation of the T-matrix is shown in Fig. 6, where the renormalized Green's functions of 4f electron are used. Here, we note that the T-matrix does not have the off-diagonal components, i.e., $T_{m,m'}=T_m\delta_{m,m'}$, since the exchange interaction is assumed to be independent of m.

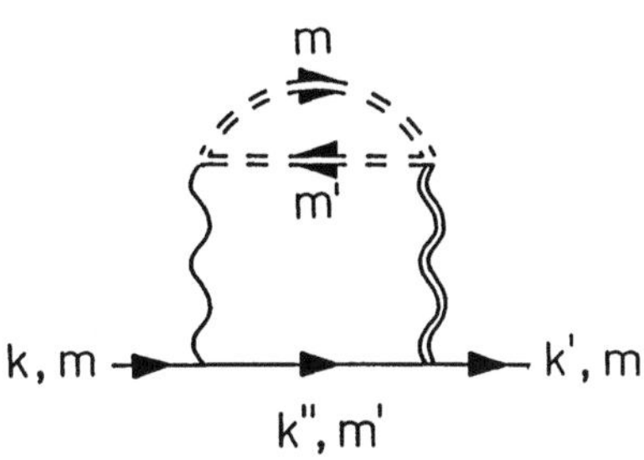

Fig. 6 Diagram for the T-matrix of conduction electron in the space with λ in the self-consistent ladder approximation.

It is easy to see that the T-matrix is written as

$$T_m^{(\lambda)}(i\omega_n) = -J T \sum_{\omega_1} \frac{G_{fm}(i\omega_1)\bar{\Pi}(i\omega_1 - i\omega_n)}{1 + \bar{\Pi}(i\omega_1 - i\omega_n)} , \qquad (4.1)$$

where $\overline{\Pi}$ is given in (2.10). Then, using the relation (2.2), we have the real T-matrix,

$$T_m(\omega) = \lim_{\lambda\to\infty} T_m^{(\lambda)}(i\omega_n \to \omega + io)/\langle n_f\rangle_\lambda \ . \tag{4.2}$$

In the numerical calculation, it is useful to introduce the following spectral function $\overline{\eta}(x)$;

$$\frac{\overline{\Pi}(i\omega_n)}{1+\overline{\Pi}(i\omega_n)} = \int_{-\infty}^{\infty} dx \frac{\overline{\eta}(x)}{i\omega_n - x} \quad , \tag{4.3}$$

where $\overline{\eta}(x)$ has the significant contribution around $x\sim 0$ in the limit of large λ.

Inserting (2.4) and (4.3) into (4.1), and using (2.14) and (4.2), we have the imaginary part of the T-matix,

$$Im\ T_m(\omega) = \frac{\pi J(1+e^{\beta\omega})\int_{-\infty}^{\infty} dx\overline{\rho}_m(x)\overline{\eta}(x-\omega)e^{-\beta x}}{\sum_{m'}\int_{-\infty}^{\infty} dx\overline{\rho}_{m'}(x)e^{-\beta x}} \ . \tag{4.4}$$

When the transition matrix from the state $|\vec{k}\sigma\rangle$ with $\vec{k}$ and σ being wave-vector and spin, respectively, to $|\vec{k}'\sigma'\rangle$ is written as $\langle\vec{k}'\sigma'|T|\vec{k}\sigma\rangle$, the transport relaxation time with $\vec{k}$ and σ is given by

$$\tau_{tr\sigma}^{-1}(\vec{k}) = 2\pi c_i \sum_{\sigma'}\int \frac{d^3k'}{(2\pi)^3}\delta(\xi_k - \xi_{k'})|\langle\vec{k}\,\sigma|T|\vec{k}'\sigma'\rangle|^2(1-\cos\theta_{kk'}) \ , \tag{4.5}$$

where c_i is the number density of Ce ions and $\theta_{k,k'}$ is the angle between the wave-vectors $\vec{k}$ and $\vec{k}'$. Since the term with $\cos\theta_{k,k'}$ does not contribute to $\tau_{tr\sigma}(\vec{k})$ in our system, we obtain

$$\tau_{tr\sigma}^{-1}(\vec{k}) = -2c_i\sum_m|\langle\vec{k}\ \sigma|km\rangle|^2\ Im\ T_m(\xi_k) \ , \tag{4.6}$$

where we used the optical theorem,

$$Im\ T_m(\omega) = -\pi\rho|T_m(\omega)|^2 \ . \tag{4.7}$$

Inserting (4.6) into the expression of the conductivity tensor;

$$\sigma_{\mu\nu} = -\frac{e^2}{m^2}\rho\sum_\sigma\int d\xi_k\int\frac{d\Omega_k}{4\pi}k_\mu k_\nu\tau_{tr\sigma}(\vec{k}\,)\frac{\partial f(\xi_k)}{\partial\xi_k} \ , \tag{4.8}$$

with Ω_k being the solid angle of $\vec{k}$, we find

$$\sigma_{\mu\nu} = -\frac{2e^2}{m^2}\rho\int d\xi_k\int\frac{d\Omega_k}{4\pi}k_\mu k_\nu\frac{\partial f(\xi_k)}{\partial\xi_k}[\tau_{im}^{-1}-2c_i\sum_m|\langle\vec{k}\uparrow|km\rangle|^2 Im\ T_m(\xi_k)]^{-1}, \tag{4.9}$$

where τ_{im} is the relaxation time due to impurity potentials. In (4.9), we assumed $\tau_{tr\sigma}(\vec{k})=\tau_{tr-\sigma}(\vec{k})$.

Let us first examine the resistivity in the second order of $J\rho$. Using the spectral function, $\bar{\rho}_m(x)=\delta(x-E_m)$, we have

$$Im\ T_m(\omega) = -\pi J^2\rho(1+e^{\beta\omega})\sum_{m'} f(\omega+E_{m'}-E_m)\langle n_m\rangle_0 \,, \tag{4.10}$$

$$\langle n_m\rangle_0 = e^{-\beta E_m}/\sum_{m'} e^{-\beta E_{m'}} \quad . \tag{4.11}$$

with f(x) being the Fermi distribution function. Introducing (4.10) into (4.6), we find the conductivity tensor in the second order of $J\rho$. In this case, since the Kondo effect is not taken into account, the resistivity decreases with decreasing temperature [19]. We note that the anisotropy appears due to the crystal field through the matrix element $|\langle\vec{k}\vec{\sigma}|km\rangle|^2$ in (4.6). It is obvious that such anisotropy also exists in the Kondo resistivity. The Kondo effect modifies both the spectral densities of $\bar{\rho}_m(x)$ and $\bar{\eta}(x-\omega)$ in (4.4), so that the effect is included in $T_m(\omega)$ and is independent of the matrix element $|\langle k\sigma|km\rangle|^2$ in (4.6).

The resistivity vs. temperature curve has a broad peak around 100 K in $CeCu_2Si_2$ [12-14,20]. To understand the origin of the peak, we calculate the resistivity in the following model systems: The excited quartet is separated from the ground doublet by Δ_0. We neglect, for simplicity, the anisotropy of the resistivity and take the matrix element $|\langle\vec{k}\sigma|km\rangle|^2$ to be constant in (4.9). The theoretical results with various choices of Δ_0 are shown in Fig. 7, where we took $J\rho$=-0.0315 and τ_{im}^{-1}=0. As seen in Fig. 7, a broad peak in the resistivity vs. temperature curve appears around 100 K when the crystal field splitting Δ_0 is 300 K. This is partly because the crystal field levels are broad due to the Kondo effect. Such broadening effect of the levels has not been included in the third-order perturbation theory [17]. The theoretical resistivity with Δ_0=300 K agrees with the experimental results, and is consistent with our level scheme of $CeCu_2Si_2$ proposed in the previous section. The detailed study of the resistivity and its anisotropy will be given in a separate paper.

5. Summary and Discussion

We have studied the magnetic properties and the electrical resistivity in Ce compounds and alloys, taking into account the Kondo effect and the crystal

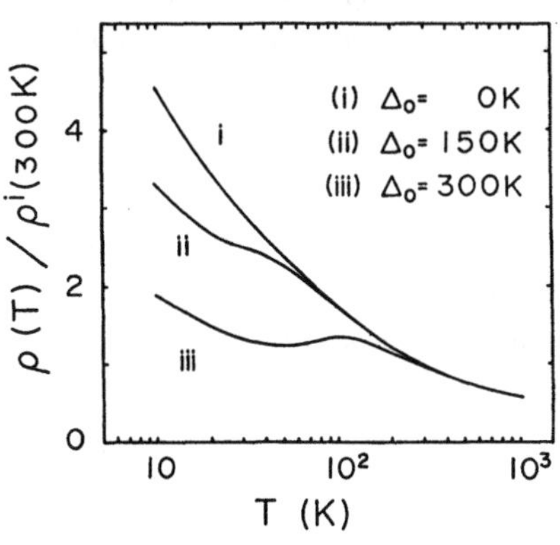

Fig. 7 Theoretical results of the resistivity $\rho(T)$ in the systems with an excited quartet separated from the ground doublet by Δ_0. ρ^i(300K) indicates the resistivity at 300K with Δ_0=0K.

field self-consistently. In Ce systems, the scattering of conduction electrons by Ce^{3+} ions is accompanied by transition between the crystal field energy levels. Therefore, various interesting phenomena are seen in Ce systems.

The crystal field determines the anisotropy of the susceptibility in non-cubic crystals. We have applied the theory to the tetragonal crystal $CeCu_2Si_2$. Through the analyses of the susceptibility, the neutron quasi-elastic and inelastic scattering spectra, and the electrical resistivity, we have proposed the crystal field level scheme in $CeCu_2Si_2$ which agrees with the specific heat study [16].

In this paper, we have concentrated on the single-ion Kondo effect. In Ce systems, the single-ion effect seems to overcome the intersite interaction even in the systems with high Ce concentration. Therefore, we can study various aspects of the Kondo effect including the dynamics. The theory presented in this paper explains both static and dynamic properties.

Acknowledgements

We are indebted to Professor F. Steglich for valuable discussions about $CeCu_2Si_2$. This work has been supported by the Grant-in-Aid from Ministry of Education, Science and Culture.

References

1. S. Maekawa, S. Takahashi, S. Kashiba, and M. Tachiki, J. Appl. Phys. 57, 3169 (1985).
2. S. Maekawa, S. Takahashi, S. Kashiba, and M. Tachiki, J. Phys. Soc. Jpn, 54, 1955 (1985).
3. S. Maekawa, S. Kashiba, S. Takahashi, and M. Tachiki, Porc. of ICCF-5 (to be published in J. Magn. Magn. Mat.).
4. B. Coqblin and J. R. Schrieffer, Phys. Rev. 185, 847 (1969).
5. A. A. Abrikosov, Physics 2, 5 (1965).
6. A. Zawadowski and P. Fazekas, Z. Physik 226, 235 (1969).
7. P. Fulde, L. L. Hirst, and A. Luther, Z. Physik 230, 155 (1970).
8. J. L. Black, K. Vlader, and A. Zawadowski, Phys. Rev. B26, 1559 (1983).
9. P. Coleman, Phys. Rev. B29, 3035 (1984).
10. See, for example, D. R. Hamann, Phys. Rev. 186, 549 (1969) and references therein.
11. Sections by Y. Kuramoto, J. W. Wilkins, and P. Coleman in this volume.
12. W. Assmus, M. Herrmann, U. Rauchschwalbe, S. Riegel, W. Lieke, H. Spille, S. Horn, G. Weber, F. Steglich, and G. Cordier, Phys. Rev. Lett. 52, 469 (1984).
13. B. Batlogg, J. P. Remeika, A. S. Cooper, and Z. Fisk, J. Appl. Phys. 55, 2001 (1984).

14. Y. Onuki, T. Hirai, T. Komatsubara, S. Takayanagi, A. Sumiyama, A. Furukawa, Y. Oda, and H. Nagano, Porc. of ICCF-5 (to be published in J. Magn, Magn. Mat.).
15. S. Horn, E. Holland-Moritz, M. Loewenhaupt, F. Steglich, H. Scheuer, A. Benoit, and J. Flouquet, Phys. Rev. B23, 3171 (1981).
16. C. D. Bredl, W. Lieke, R. Schefzyk, M. Lang, U. Rauchwalbe, F. Steglich, S. Riegel, R. Felten, G. Weber, J. C. P. Klaasse, J. Arts, and F.R. de Boer, to appear in J. Magn. Magn. Mat. 47-48 (1985).
17. B. Cornut and B. Coqblin, Phys. Rev. B5, 4541 (1972).
18. K. Yamada, K. Yosida, and K. Hanzawa, Prog. Theor. Phys. 71, 450 (1984).
19. P. Fulde, Handbook on the Physics and Chemistry of Rare Earths, ed. K. A. Gschneider, Jr. and L. Eyring (North Holland, 1979) Chapter 17.
20. D. Muller, S. Hussain, E. Cattaneo, H. Schneider, W. Schlabitz, and D. Wohlleben, Valence Instabilities, ed. P. Wachter and H. Boppart (North-Holland, 1982) p. 463.

The Frequency-Dependent Susceptibility in the Anderson Model

O. Gunnarsson
Max-Planck-Institut für Festkörperforschung
D-7000 Stuttgart, Fed. Rep. of Germany

K. Schönhammer
Institut für Theoretische Physik, Universität Göttingen
D-3400 Göttingen, Fed. Rep. of Germany

Abstract: We calculate the dynamic susceptibility $\chi(\omega)$ in the Anderson impurity model, using a $1/N_f$ expansion, where N_f is the degeneracy of the f-level. For $N_f\to\infty$ the Shiba Fermi liquid relation is satisfied. In this limit Im $\chi(\omega)/\omega$ has a very steep rise at $\omega=T_K$, the Kondo temperature. An interpretation for this rise is given. Crystal-field and spin-orbit splittings have also been included.

1. Introduction

The dynamic spin susceptibility $\chi(\omega)$, which is related to the inelastic neutron scattering cross-section, gives important information about the low-energy excitations of the system. In this paper we apply a method [1], developed for electron spectroscopies, to the calculation of $\chi(\omega)$ in the Anderson model of a Ce compound. This method uses the suggestion [2] that $1/N_f$ may be treated as a small parameter, where N_f is the 4f level degeneracy. KURAMOTO and KOJIMA [3] and BICKERS et al [4] have calculated $\chi(\omega)$ using somewhat different $1/N_f$ methods. $\chi(0)$ was calculated by RASUL and HEWSON [5], who in particular studied the accuracy of the $1/N_f$ expansion, by comparing with the exact Bethe Ansatz results. In these calculations the f-f Coulomb interaction U was assumed to be infinite. BICKERS et al [4] considered spin-orbit splitting. Here we also treat crystal-field and spin-orbit splitting and allow U to be finite.

We consider a generalized Anderson impurity model

$$H = \sum_{\nu=1}^{N_f} \{\int_{-B}^{B} d\varepsilon\, \varepsilon\, \phi_{\varepsilon\nu}^{+}\phi_{\varepsilon\nu} + \varepsilon_{f\nu}\phi_{\nu}^{+}\phi_{\nu} + \int_{-B}^{B} d\varepsilon\, [V(\varepsilon)\phi_{\nu}^{+}\phi_{\varepsilon\nu} + \text{H.C.}]\} + U \sum_{\nu<\nu'} n_{\nu}n_{\nu'} \; . \tag{1}$$

The first term describes the conduction states with a band width 2B, the second term the f-states and the third term the hopping between these states. The hopping matrix element $V(\varepsilon)$ is assumed to be independent of ν. The fourth term includes the Coulomb interaction between the f-electrons. If $\varepsilon_{f\nu}$ is independent of ν, this model describes a spin with $j=(N_f-1)/2$. We also consider the cases when the f-level is split into a j=5/2 and a j=7/2 level by the spin-orbit splitting, and when the j=5/2 level is further split into a Γ_7 doublet and a Γ_8 quartet by a cubic crystal-field. In particular, we discuss the case of a large degeneracy without crystal-field or spin-orbit splitting for U=∞. In this limit the calculation fulfills the Shiba Fermi liquid relation [6]. Im $\chi(\omega)/\omega$ has a sharp increase of the order N_f at $\omega=T_K$, where T_K is the Kondo temperature. A physical interpretation of this behaviour is given. We also briefly discuss the cases when a crystal-field splitting is introduced or U is finite. The method is presented in Sec. 2, the result for a large N_f is given in Sec. 3 and the effects of crystal-field splitting are discussed in Sec. 4.

2. Method

The dynamic longitudinal susceptibility is written as

$$\chi(z) = \langle\phi|S_z(z+H-E_0)^{-1}S_z|\phi\rangle - \langle\phi|S_z(z-H+E_0)^{-1}S_z|\phi\rangle \quad (2.1)$$

where $|\phi\rangle$ is the ground-state and E_0 is the corresponding energy. For a spin j the operator S_z is

$$S_z = g \sum_{m=-j}^{j} m\, \phi_m^+\phi_m \quad (2.2)$$

where g is the Lande g-factor. For a spin-orbit split level there are two terms of the type (2.2) corresponding to j=5/2 and j=7/2 and a cross term. In the presence of a cubic crystal field we replace the j=5/2 operator by [7]

$$g_1\{\frac{5}{6}(\phi_2^+\phi_2-\phi_1^+\phi_1) + \frac{11}{6}(\phi_3^+\phi_3-\phi_4^+\phi_4) + \frac{1}{2}(\phi_5^+\phi_5-\phi_6^+\phi_6) + \frac{\sqrt{20}}{3}(\phi_3^+\phi_1-\phi_4^+\phi_2) + \frac{\sqrt{20}}{3}(\phi_1^+\phi_3-\phi_2^+\phi_4)\} \quad (2.3)$$

where the indices 1 and 2 refer to the Γ_7 and 3-6 to the Γ_8 representation.

We can now use the same methods as were applied in [1,8] to the calculation of electron spectra. Thus, we introduce basis states $\{|i\rangle\}$ and calculate the ground-state variationally

$$|\phi\rangle = \sum C_i\, |i\rangle \quad (2.4)$$

The choice of $\{|i\rangle\}$ was discussed extensively in [1] and [8]. To invert the operator $(z+H-E_0)$ we introduce a basis set $\{|\mu\rangle\}$, which is obtained from $S_z|i\rangle$ or $HS_z|i\rangle$. We have earlier calculated the resolvent operator using a folding technique [1] and a time-dependent method [8]. Here we use a moment (LANCZOS) method [9] to obtain $\langle\gamma|z+H-E_0|\mu\rangle$ in a tridiagonal form, after which the matrix can easily be inverted.

As an illustration we consider the case of no spin-orbit or crystal-field splitting with U=∞. We introduce the state

$$|\,0\,\rangle \quad (2.5)$$

where the f-level is empty and the conduction levels below the Fermi energy, ε_F, are filled. This singlet state couples to

$$|\varepsilon\rangle = 1/\sqrt{N_f} \sum_m \phi_m^+ \phi_{\varepsilon m} |0\rangle \quad . \tag{2.6}$$

These states [10] give a contribution to the energy of the order $(1/N_f)^0$. The states of order $(1/N_f)$ are

$$|E\varepsilon\rangle = 1/\sqrt{N_f} \sum_m \phi_{Em}^+ \phi_{\varepsilon m} |0\rangle \tag{2.7}$$

and

$$|E\varepsilon\varepsilon'\rangle = 1/\sqrt{N_f(N_f-1)} \sum_{m \neq m'} \phi_{Em}^+ \phi_{\varepsilon m} \phi_{m'}^+ \phi_{\varepsilon' m'} |0\rangle \tag{2.8}$$

where $\varepsilon < \varepsilon_F$ and $E > \varepsilon_F$. States with $m=m'$ give a contribution of the order $(1/N_f)^2$. To calculate the resolvent operator, we introduce "magnetic" basis states

$$|\varepsilon S_z\rangle = 1/\sqrt{C} \sum_m m \, \phi_m^+ \phi_{\varepsilon m} |0\rangle \tag{2.9}$$

where $C = j(j+1)(2j+1)/3$. These states were obtained from $S_z|\varepsilon\rangle$, while $S_z|0\rangle = 0$. The states $|\varepsilon S_z\rangle$ couple to

$$|E\varepsilon S_z\rangle = 1/\sqrt{C} \sum_m m \, \phi_{Em}^+ \phi_{\varepsilon m} |0\rangle \tag{2.10}$$

via H. Two additional sets of states are obtained from $H|E\varepsilon S_z\rangle$ and $S_z|E\varepsilon\varepsilon'\rangle$. Since these states are not orthogonal, we introduce orthogonal linear combinations

$$|E\varepsilon\varepsilon' 1 S_z\rangle = 1/\sqrt{2(N_f-2)C} \sum_{m \neq m'} (m+m') \phi_{Em}^+ \phi_{\varepsilon m} \phi_{m'}^+ \phi_{\varepsilon' m'} |0\rangle \tag{2.11}$$

and

$$|E\varepsilon\varepsilon' 2 S_z\rangle = 1/\sqrt{2N_f C} \sum_{m \neq m'} (m-m') \phi_{Em}^+ \phi_{\varepsilon m} \phi_{m'}^+ \phi_{\varepsilon' m'} |0\rangle . \tag{2.12}$$

The calculation of the resolvent operator requires the inversion of a matrix. With the states (2.9-12) this inversion can be performed analytically and we obtain

$$\chi(z) = \frac{1}{3} j(j+1) g^2 (F(z) + F(-z)) \tag{2.13}$$

where

$$F(z) = \int d\varepsilon \, a(\varepsilon)^2 g(\varepsilon+z) + \int dE \int d\varepsilon \int d\varepsilon' \frac{d(E,\varepsilon,\varepsilon')^2}{z+\Delta E-E-\varepsilon_f+\varepsilon+\varepsilon'} \, . \tag{2.14}$$

Here $a(\varepsilon)$ and $d(E,\varepsilon,\varepsilon')$ are the coefficients of $|\varepsilon\rangle$ and $|E\varepsilon\varepsilon'\rangle$, respectively, in the ground-state and $\Delta E = E_0 - \langle 0|H|0\rangle$. The function $g(z)$ is given by

$$g(z) = (z+\Delta E-\varepsilon_f-\mu(z))^{-1} \tag{2.15}$$

where

$$\mu(z) = \int_0^B dE \, V(E)^2 (z+\Delta E-E-(N_f-1)\Gamma(-z-\Delta E+\varepsilon_f+E))^{-1} \tag{2.16}$$

$$\equiv \int dE \, V(E)^2 \, h(z+E)$$

and

$$\Gamma(z) = \int_{-B}^0 d\varepsilon \, V(\varepsilon)^2 (z-\varepsilon)^{-1} \tag{2.17}$$

The susceptibility (2.13) is correct to order $(1/N_f)$.

3. The large degeneracy case

We now consider $N_f\to\infty$. The result (2.13-2.17) with an accuracy $(1/N_f)$ is needed even for $N_f\to\infty$, since $\lim_{\omega\to 0}$ Im $\chi(\omega)/\omega$ is one order smaller in $(1/N_f)$ than the maximum value of Im $\chi(\omega)/\omega$. For simplicity we assume that $\pi V(\varepsilon)^2\equiv\Delta$ is a constant. First we study the SHIBA relation [6] between $\lim_{\omega\to 0}$ Im $\chi(\omega)/\omega$ and $\chi(0)$. The function $g(z)$ (Eq. (2.15)) has a pole at z_p

$$z_p + \Delta E - \varepsilon_f - \mu(z_p) = 0 \qquad (3.1)$$

with the weight ω_p. For $N_f\to\infty$ we find that $z_p\to 0$ and $\omega_p\to 0$, so that the pole need not be considered in this limit. Even for a finite N_f there is, however, no pole in $\chi(z)$ since $g(z)$ appears in an integral (2.14). The function $h(z)$, defined in Eq. (2.16) has a pole at z_1 with the weight ω_1. As $N_f\to\infty$ $z_1\to 0$, while ω_1 stays finite

$$\omega_1 \to 1 - n_f \qquad (3.2)$$

where n_f is the f-occupancy. It therefore follows that

$$\text{Im}\ \mu\ (\omega-i0) = (1-n_f)\Delta \qquad z_1<\omega<\varepsilon_f-\Delta E+\alpha\Delta \qquad (3.3)$$

where α is of the order one. The real part of $\mu(\omega-i0)$ is of the order $1/N_f$ unless $\omega-z_1 < B\exp(-N_f)$, since $\mu(\omega)$ has a logarithmic singularity at $\omega=z_1$. For $z_1+B\exp(-N_f)<<\omega<\varepsilon_f-\Delta E+\alpha\Delta$ we then obtain

$$\text{Im}\ g(\omega-i0) = \frac{(1-n_f)\Delta}{(\omega+\Delta E-\varepsilon_f)^2+(1-n_f)^2\Delta^2} \ . \qquad (3.4)$$

Since $z_1\to 0$ as $N_f\to\infty$, this relation is true for any arbitrary small ω in the limit $N_f\to\infty$. We then have

$$\lim_{\omega\to 0} \frac{1}{\omega}\ \text{Im}\ \chi(\omega) = \frac{1}{3}\ j(j+1)g^2(1-n_f)^2\ \frac{N_f\Delta^2}{\pi(\Delta E-\varepsilon_f)^4} \ . \qquad (3.5)$$

Furthermore

$$\text{Re}\ \chi(0) = \frac{1}{3}\ j(j+1)\ g^2(1-n_f) \qquad \frac{N_f\Delta}{\pi(\Delta E-\varepsilon_f)^2} \qquad (3.6)$$

which leads to SHIBA's result [5]

$$\lim_{\omega\to 0} \frac{1}{\omega}\ \text{Im}\ \chi(\omega) = \frac{3\pi\ \chi(0)^2}{j(j+1)(2j+1)g^2} \ . \qquad (3.7)$$

Fig. 1 shows results for Im $\chi(\omega)/\omega$ for $N_f=500$. In this case $z_p=0.0008$ and the corresponding pole weight is negligible ($<10^{-4}$). The pole of $h(z)$ is located at $z_1=0.0008$ and determines the size of the gap in Fig. 1. The Kondo temperature is $T_K=\Delta E-\varepsilon_f=0.04$. At this energy Im $\chi(\omega)/\omega$ has a very steep rise. For $\omega>T_K$ we obtain

$$\frac{1}{\omega}\ \text{Im}\ \chi(\omega) = \frac{1}{3}\ j(j+1)g^2(1-n_f)N_f\Delta\ \frac{1}{\omega^3} \qquad (3.8)$$

and this quantity is a factor N_f/n_f larger for $\omega = T_K$ than for $\omega=0$. To understand this large difference, we calculate the low-lying magnetic states in the space (2.9-2.12). To lowest order in $1/N_f$ there are states of the type

$$1/\sqrt{C} \sum_m m \; \psi^+_{Em}\psi_{\varepsilon m}|\phi>^{(1)} \tag{3.9}$$

where $|\phi>^{(1)}$ is the lowest order ground-state (obtained in the space (2.5-6)) with the energy $\Delta E^{(1)}+<0|H|0>$. Since $\Delta E^{(1)}\to\Delta E$ for $N_f\to\infty$, the lowest excitation energy of (3.9) goes to zero and the unphysical gap at $\omega=0$ is closed. The full eigenstates of this type give a contribution of the order $1/N_f$. Another set of approximate eigenstates are given by

$$'|\varepsilon \; S_z> \tag{3.10}$$

in Eq. (2.9). These states have the excitation energy

$$\varepsilon_f - \Delta E - \varepsilon \;=\; T_K - \varepsilon \tag{3.11}$$

and they couple to the ground-state via S_z with a strength $(1/N_f)^0$. This explains why Im $\chi(\omega)/\omega$ increases at $\omega=T_k$ by a factor of the order N_f. The states (3.9) and (3.10) are quite different, although they are both triplet states. The state (3.9) is locally very similar to the ground-state, and it only differs from this by the presence of an electron-hole pair excitation. The state (3.10), on the other hand, is a non-symmetric combination of $\psi^+_m\psi_{\varepsilon m}|0>$, while the ground-state contains a symmetric combination of these states.

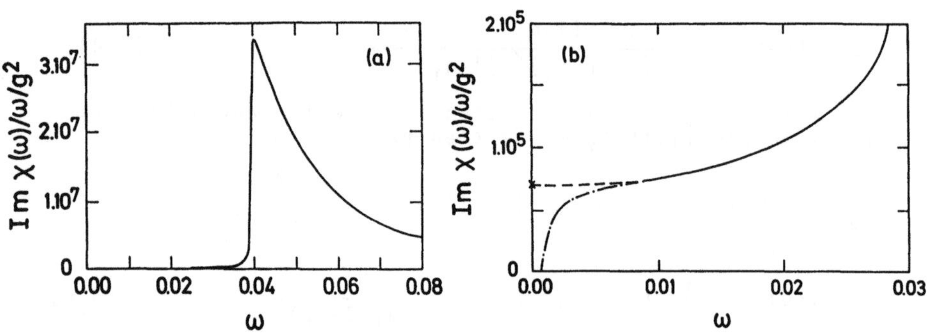

Fig. 1 Im $\chi(\omega)/\omega$ as a function of ω. Observe that the vertical scales in Figs. 1a and 1b differ by a factor 200. The cross in Fig. 1b marks the exact Shiba result for $\omega=0$. The parameters are $N_f\Delta=1$, $\varepsilon_f=-1$, B=1 and $N_f=500$.

4. Crystal-Field Splitting

The generalization of the formalism in Sec. 2 to include crystal-field splitting is fairly straightforward. In addition to (2.5-12) we have included states with two conduction electrons and two conduction holes and states with two conduction electrons, one f-electron and three conduction holes. This is sufficient to obtain results accurate to order $(1/N\;)^2$. To be

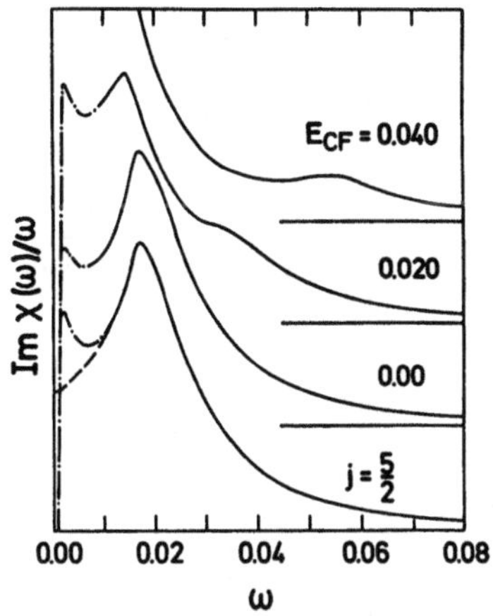

Fig. 2 Im $\chi(\omega)/\omega$ as a function of ω for different values of E_{CF}. In the lowest curve the j=7/2 level is neglected. The cross shows the Shiba result and the broken curve an extrapolation to this result. The parameters are $14\Delta=0.4$ eV, B=6 eV and $\varepsilon_f=-0.49$ eV (three uppermost curves) and $\varepsilon_f=-0.26$ eV (lowest curve).

able to obtain an analytical expression for $\chi(\omega)$, we have, however, in most calculations, neglected some contributions of order $(1/N\ )^2$.

We assume that the Γ_7 level is an energy E_{CF} above Γ_8 and that the spin-orbit splitting is $E_{SO}=0.25$ eV. In Fig.2 we show Im $\chi(\omega)/\omega$ for different values of E_{CF}. To relate to the previous section, we also show at the bottom of Fig. 2 results for a j=5/2 spin without crystal-field splitting and neglecting the j=7/2 level. The threshold-like behaviour in Fig. 1 survives, to some extent, in Fig. 2 as a peak (at $\omega\sim0.017$ eV in the bottom curve). The weight at smaller ω is, however, greatly enhanced compared with Fig. 1. In the large N_f limit the peak occurs at

$$T_K = B\exp\left(\pi(\varepsilon_f-T_K)/N_f\Delta\right) \tag{4.1}$$

while for $N_f=6$ this formula of order $(1/N_f)^0$ only gives a semi-quantitative estimate (0.030 eV) of the peak position. The SHIBA result [6] is marked by a cross, and an extrapolation to this value is shown. For small values of ω there is a spurious peak which deviates from this extrapolation. Similar results were obtained by KURAMOTO and KOJIMA [3], except that they have low-energy anomalies of a somewhat different type.

In the second lowest curve in Fig. 2 ($E_{CF}=0$) the j=7/2 level is also included. The Kondo temperature is then in the large N_f limit [1,11]

$$T_K = \left(1+B/(T_K+E_{SO})\right)^{N_{f2}/N_{f1}} B\exp\left(\pi(\varepsilon_f-T_K)/N_{f1}\Delta\right) \tag{4.2}$$

where N_{f1}=6 and N_{f2}=8. For $E_{SO}+T_K<B$ the inclusion of the j=7/2 level leads to a very large enhancement of T_K. To keep the energy scale unchanged, we have therefore changed ε_f from -0.260 to -0.490. Although (4.2) is only correct to order $(1/N_f)^0$ it gives an accurate prediction (-0.487) of the ε_f needed to obtain the same energy scale. The two lower curves in Fig. 2 have a very similar shape. This approximate universality was first observed by BICKERS et al.[4].

In the two uppermost curves in Fig. 2 we have used a non-zero E_{CF}. The generalization of Eq. (4.2) to this case is straightforward. It shows that T_K is reduced as E_{CF} is increased, and it gives a reasonable prediction of the change of the energy scale. For E_{CF}=0.04 eV there is a clearly visible inelastic peak at $\omega\sim0.055$ eV and for E_{CF}=0.02 eV there is a shoulder at $\omega\sim0.035$ eV. For $N_f\to\infty$ the theory gives an inelastic peak at $\omega=T_K+E_{CF}$, i.e. the two peaks are separated by E_{CF}. The energy required to produce this excitation is the sum of the crystal-field splitting and the energy it costs to break up the Kondo singlet.

In these calculations we have used a constant $V(\varepsilon)^2$. In most mixed valence Ce compounds $V(\varepsilon)^2$ increases as ε moved below ε_F. These conduction states far below ε_F influence T_K and for these more realistic $V(\varepsilon)^2$ one should use a more negative ε_f than considered here. We have considered the case $U<\infty$ and allowed an energy-dependent $V(\varepsilon)^2$ [12]. Using the parameters deduced earlier [13] for $CePd_3$ from electron spectroscopies, we obtain an energy-scale for Im $\chi(\omega)/\omega$ in fair agreement with the measured quasi-elastic line width [12].

References

1. O. Gunnarsson and K. Schönhammer, Phys.Rev.Lett. 50, 604 (1983); Phys.Rev. B28, 4315 (1983)
2. T.V. Ramakrishnan in: Valence fluctuations in Solids (Eds. L.M. Falicov, W. Hanke and M.P. Maple, North Holland, Amsterdam, 1981), p 13; P.W. Anderson, ibid p. 451.
3. Y. Kuramoto and H. Kojima, Z.Phys. 57, 95 (1984)
4. N.E. Bickers, D.L. Cox and J.W. Wilkins, Phys.Rev.Lett. 54, 230 (1985)
5. J.W. Rasul and A.C. Hewson, J.Phys.C16, L933 (1983); 17, 2555 (1984); 17, 3337 (1984)
6. H. Shiba, Prog. Theor. Phys. 54, 967 (1975)
7. K.R. Lea, M.J.M. Leask and W.P. Wolf, J.Phys.Chem.Solids 23, 1381 (1962)
8. O. Gunnarsson and K. Schönhammer, Phys.Rev. (in press)
9. R. Haydock, Solid State Physics 35, 215 (1980)
10. The states (2.5-6) were first introduced for N_f=2 by C.M. Varma and Y. Yafet, Phys.Rev. B13, 2950 (1976)

11. See also A. Ogawa and A. Yoshimori, Prog. Theor. Phys. 53, 315 (1975) and Ref. 4.

12. O. Gunnarsson and K. Schönhammer, J.Magn.Magn.Mater. (to be published)

13. O. Gunnarsson, K. Schönhammer, J.C. Fuggle, F.U. Hillebrecht, J.-M. Esteva, R.C. Karnatak and B. Hillebrand, Phys.Rev. B28, 7330 (1983)

Part IV

Spectroscopy Properties

Ground-State and Spectroscopic Properties of the Finite U Anderson Model

O. Gunnarsson
Max-Planck-Institut für Festkörperforschung
D-7000 Stuttgart, Fed. Rep. of Germany

K. Schönhammer
Institut für Theoretische Physik, Universität Göttingen
D-3400 Göttingen, Fed. Rep. of Germany

Abstract: We study the Anderson impurity model at T=0 taking into account double occupancy of the f-level. Particular emphasis is given to the limit when the f-level degeneracy N_f is large. In this limit there is an important asymmetry between the f^0 and f^2 configurations. We define a Kondo temperature δ_u for $N_f \to \infty$ and relate δ_u to the susceptibility and the f^0 weight. δ_u depends on the band width in a nontrivial way. It is shown that the valence photoemission spectrum can change substantially as double occupancy is taken into account.

1. Introduction

Ce mixed valence compounds are often described by the Anderson model[1]. Traditionally it was assumed that the f^0 and f^1 configurations are almost degenerate, and that the f^2 configurations have substantially higher energies. The f-f Coulomb interaction U has therefore usually been assumed to be infinite, so that the f^2 configurations can be neglected. There is now, however, much evidence [2-6] that the lowest f^0 and f^1 configurations are separated by a few eV and that the f^1-f^2 separation is of the same order of magnitude. The aim of this paper is therefore to study the Anderson impurity model at T=0 including double occupancy of the f-level. The thermodynamic properties of this model have been studied extensively for $N_f=2$ [7], where N_f is the "f-level" degeneracy. As N_f is increased, many qualitatively new features appear [8-13], and we therefore focus on a large N_f. Using the idea [10] that $1/N_f$ can be treated as a small parameter, methods have been developed for calculating thermodynamic properties for $U=\infty$ [11, 8, 12-14] and for $U<\infty$ [8, 9]. In [8,9] we developed a $1/N_f$ method for calculating spectroscopic properties for T=0 and $U<\infty$. Alternative methods [13, 14] for obtaining spectra were later developed for T>0 and $U=\infty$.

We consider the Anderson impurity model

$$H = \sum_{\nu=1}^{N_f} \left\{ \int_{-B}^{B} d\varepsilon\, \varepsilon\, \psi^+_{\nu\varepsilon}\psi_{\nu\varepsilon} + \varepsilon_f \psi^+_\nu \psi_\nu + \int V(\varepsilon)[\psi^+_\nu \psi_{\varepsilon\nu} + \text{H.C.}]\right\} + U \sum_{\nu<\mu} n_\nu n_\mu \quad (1.1)$$

where the first term describes the conduction states with a band width 2B, the second the f-level and the third the hopping between these states. The last term includes the Coulomb interaction U. We have introduced a combined spin and orbital index ν, transformed the conduction states into this representation and absorbed the conduction density of states in $V(\varepsilon)^2$ [8]. We set the Fermi energy $\varepsilon_F=0$. In the large N_f limit we keep [8]

$$\tilde{V}(\varepsilon) = \sqrt{N_f}\, V(\varepsilon) \tag{1.2}$$

constant, since many properties of interest then converge for $N_f \to \infty$.

If $2\varepsilon_f+U=0$ and $N_f=2$, the f^0 amd f^2 weights are equal due to electron-hole symmetry. For a large N_f we find, however, that the f^2 weight can be much larger than the f^0 weight, since the f^0-f^1 and f^1-f^2 couplings are very different for a large N_f. We define a Kondo temperature δ_u and show that it is related to ε_f, U, $V(\varepsilon)$ and B in an interesting way. δ_u can depend very strongly on U. Relations for the f^0 weight and the static, T=0 susceptibility in terms of δ_u are given. Finally we show that the valence photoemission spectrum can be strongly affected if finite values of U are allowed. In Sec. 2 we present our methods, and the ground-state and spectroscopic properties are discussed in Secs. 3 and 4, respectively.

2. Method

We use a variational approach and introduce basis states. To order $(1/N_f)^0$ we need

$$|0\rangle, \tag{2.1}$$

where the f-level is empty and the conduction states below ε_F are occupied,

$$|\varepsilon\rangle = \frac{1}{\sqrt{N_f}} \sum_{\nu} \psi_\nu^+ \psi_{\varepsilon\nu} |0\rangle \tag{2.2}$$

and

$$|\varepsilon\varepsilon'\rangle = \frac{1}{\sqrt{N_f(N_f-1)}} \sum_{\nu\neq\nu'} \psi_\nu^+ \psi_{\varepsilon\nu} \psi_{\nu'}^+ \psi_{\varepsilon'\nu'} |0\rangle \;. \tag{2.3}$$

The state (2.2) and (2.3) correspond to f^1 and f^2 configurations, respectively. The states (2.1) and (2.2) were introduced by VARMA and YAFET [15] for $N_f=2$. In [9] we discussed the states needed to obtain the accuracy $(1/N_f)^2$. The ground-state $|\phi\rangle$ is written as a linear combination of these basis states $\{|i\rangle\}$

$$|\phi\rangle = \sum c_i |i\rangle \tag{2.4}$$

and the ground-state energy E_0 is minimized [8].

In Table 1 we show results for $\Delta E \equiv E_0 - \langle 0|H|0\rangle$ and the f^0 and f^2 weights for two values of $\tilde{\Delta} \equiv N_f \Delta \equiv \pi N_f\, V(0)^2$ for $N_f=14$. For $V(\varepsilon)^2$ we used the energy-dependence

$$V(\varepsilon)^2 = V(0)^2 \sqrt{1-(\varepsilon/B)^2} \;. \tag{2.5}$$

The first row ($(1/N_f)^0$) in Table 1 shows results obtained using the states (2.1-3) and the two following rows results which are of higher order in $(1/N_f)$ [16]. In some cases the

$(1/N_f)^0$ results are not satisfactorily converged, while this is true for the $(1/N_f)$ results in all the cases we have considered. The method for calculating the valence spectrum has been described in Ref. 9.

Table 1. $\Delta E-\varepsilon_f$ and the f^0 and f^2 weights for two values of the hybridization $\tilde{\Delta}$ and N_f=14. Basis sets of different orders in $(1/N_f)$ have been considered. All energies are in eV.

Basis set	$\tilde{\Delta}$ = 0.42			$\tilde{\Delta}$ = 1.17		
	$\Delta E-\varepsilon_f$	$\omega(f^0)$	$\omega(f^2)$	$\Delta E-\varepsilon_f$	$\omega(f^0)$	$\omega(f^2)$
$(1/N_f)^0$	-0.125	0.000	0.029	-0.361	0.028	0.076
$(1/N_f)^1$	-0.138	0.003	0.030	-0.404	0.034	0.080
$(1/N_f)^2$	-0.138	0.003	0.030	-0.407	0.035	0.080

3. Ground-State Properties for a Large Degeneracy

In the discussion of the ground-state properties it is useful to define a Kondo temperature, δ_u. We do this in terms of "magnetic" states

$$|\varepsilon S_z\rangle = 1/\sqrt{C}\sum_{\nu} \nu\, \psi_\nu^+\psi_{\varepsilon\nu}\, |0\rangle \tag{3.1}$$

$$|\varepsilon\varepsilon' S_z 1\rangle = 1/\sqrt{2(N_f-2)C}\sum_{\nu\neq\nu'}(\nu+\nu')\psi_\nu^+\psi_{\varepsilon\nu}\psi_{\nu'}^+\psi_{\varepsilon'\nu'}|0\rangle \tag{3.2}$$

$$|\varepsilon\varepsilon' S_z 2\rangle = 1/\sqrt{2N_f C}\sum_{\nu\neq\nu'}(\nu-\nu')\psi_\nu^+\psi_{\varepsilon\nu}\psi_{\nu'}^+\psi_{\varepsilon'\nu'}|0\rangle \tag{3.3}$$

where ν runs from $-(N_f-1)/2$ to $(N_f-1)/2$ and $C=N_f(N_f^2-1)/12$. In contrast to the states (2.2-3) these states are not symmetric in ν. The lowest state in the space (3.1-3) has the energy $\varepsilon_f+\Delta E_{mag}+\langle 0|H|0\rangle$, where

$$\Delta E_{mag} = N_f\, \Gamma\, (\Delta E_{mag}-\varepsilon_f-U) \tag{3.4}$$

and

$$\Gamma(z) = \int_{-B}^{0} \frac{V(\varepsilon)^2}{z-\varepsilon}\, d\varepsilon\ . \tag{3.5}$$

In the limit $N_f\to\infty$ we define δ_u as

$$\Delta E = \varepsilon_f + \Delta E_{mag} - \delta_u \tag{3.6}$$

where $\Delta E=E_0-\langle 0|H|0\rangle$ and E_0 is the ground-state energy. We show below that with this definition, δ_u is directly related to the f^0 weight, the static susceptibility $\chi(0)$ and the valence photoemission spectrum. Elsewhere [17] we have shown that δ_u gives the position of a peak in Im $\chi(\omega)$ for $N_f\to\infty$.

We first consider the case when U>>B, where 2B is the band-width, since ground-state properties can be obtained analytically in this limit. We then show numerically that many of these results are quite accurate even for U<B. The ground-state in the space (2.1-3) is given by

$$|\phi\rangle = A\left(|0\rangle + \int_{-B}^{0} d\varepsilon\; a(\varepsilon)|\varepsilon\rangle + \int_{-B}^{0} d\varepsilon \int_{-B}^{\varepsilon} d\varepsilon'\; b(\varepsilon,\varepsilon')|\varepsilon\varepsilon'\rangle\right). \tag{3.7}$$

Using the Schrödinger equation we obtain

$$\Delta E = \int \tilde{V}(\varepsilon)\, a(\varepsilon)\, d\varepsilon \tag{3.8}$$

and

$$\left[\varepsilon_f - \varepsilon - \Delta E - N_f\, \tilde{\Gamma}(2\varepsilon_f + U - \varepsilon - \Delta E)\right] a(\varepsilon) + \tilde{V}(\varepsilon)\left[1 - \int d\varepsilon' \frac{\tilde{V}(\varepsilon')a(\varepsilon')}{2\varepsilon_f + U - \Delta E - \varepsilon - \varepsilon'}\right] = 0 \tag{3.9}$$

where we have used that

$$b(\varepsilon,\varepsilon') = \frac{\tilde{V}(\varepsilon)a(\varepsilon') + \tilde{V}(\varepsilon)a(\varepsilon')}{2\varepsilon_f + U - \Delta E - \varepsilon - \varepsilon'} . \tag{3.10}$$

At some places we have replaced N_f-1 by N_f. For U>>B we make the approximation [9]

$$\int d\varepsilon' \frac{\tilde{V}(\varepsilon')a(\varepsilon')}{2\varepsilon_f + U - \Delta E - \varepsilon - \varepsilon'} \approx \frac{1}{2\varepsilon_f + U - \Delta E}\int d\varepsilon'\; \tilde{V}(\varepsilon')a(\varepsilon')$$
$$= \frac{\Delta E}{2\varepsilon_f + U - \Delta E} . \tag{3.11}$$

We can now solve for $a(\varepsilon)$ and obtain an implicit equation for ΔE.

$$\Delta E \approx -\left(1 - \frac{\Delta E}{2\varepsilon_f + U - \Delta E}\right) N_f\, \tilde{\Gamma}\,(\varepsilon_f - \Delta E - E_{mag}) . \tag{3.12}$$

For a constant $\pi N_f\, V(\varepsilon)^2 \equiv \tilde{\Delta}$ this gives

$$\delta_u = B\exp\left\{\frac{\pi}{\tilde{\Delta}}\,\varepsilon_0(U)\right\} \quad \text{where} \tag{3.13}$$

$$\varepsilon_0(U) = \left[\frac{1}{\varepsilon_f + \Delta E_{mag} - \delta_u} - \frac{1}{\varepsilon_f + U - \Delta E_{mag} + \delta_u}\right]^{-1} . \tag{3.14}$$

For $-\varepsilon_f, U+\varepsilon_f >> B, \tilde{\Delta}$ we can neglect ΔE_{mag} and δ_u compared with $-\varepsilon_f$ and ε_f+U. The results (3.13-14) then reduce to a form which can also be obtained by performing a Schrieffer-Wolf transformation and by using the expression [18] for the Kondo temperature of the Coqblin-Schrieffer model for $N_f\to\infty$. If we instead assume $U>>B, -\varepsilon_f$ and $\tilde{\Delta}$ we obtain

$$\varepsilon_0(U) = \varepsilon_f - \delta_u - \frac{B\tilde{\Delta}/\pi - (\varepsilon_f - \delta_u)^2}{\varepsilon_f - \delta_u + U} . \tag{3.15}$$

For U<B this formula is not valid and δ_u is calculated numerically from (3.6).

To lowest order in 1/U, $a(\varepsilon)$ is given by

$$a(\varepsilon) = \frac{V(\varepsilon)}{\varepsilon - \delta_u}\left(1 - \frac{\Delta E}{U + 2\varepsilon_f - \Delta E}\right) . \tag{3.16}$$

We now consider the spin-fluctuation limit, where the f^0 weight $\omega(f^0)$ is much smaller than 1. Using (3.7), (3.10) and (3.16) we find to lowest order in 1/U

$$\omega(f^0) = A^2 = \left[1 + \frac{\tilde{\Delta}}{\pi\delta_u}\left(1 - 2\frac{\varepsilon_f}{\varepsilon_f+U}\right)\right]^{-1} \tag{3.17}$$

$$\omega(f^2) = \frac{\tilde{\Delta}B}{\pi}\left(\frac{1}{U+\varepsilon_f}\right)^2 \tag{3.18}$$

and

$$\chi(0) = \frac{1}{12}\,(N_f^2-1)\,\frac{1-\omega(f^0)}{\delta_u}\,. \tag{3.19}$$

For $U=\infty$ these results (for $N_f\to\infty$) have been derived [8] under the assumption $\delta_u \ll B$.

Numerically we find that the range of validity of (3.17) and (3.19) is broader than the derivation suggests.

In Fig. 1 we show $\chi(0)$ as a function of U for two values of ε_f. In both cases Eq. (3.19) is essentially correct to within the plotting accuracy. For the parameters in Fig. 1a Eq. (3.17) for $\omega(f^0)$ is correct to within 2% for U>5 and for the parameters in Fig. 1b it is correct to within 8% for U>B=24.

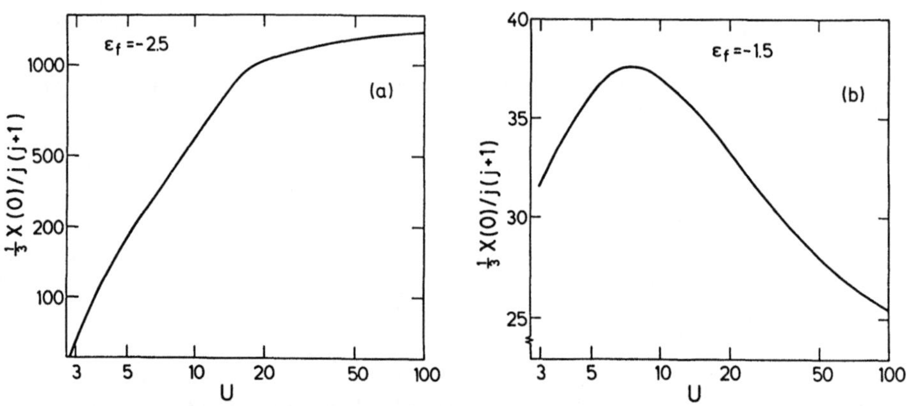

Fig. 1 The static susceptibility at T=0 as a function of U for ε_f=-2.5 (a) and ε_f=-1.6 (b). We have used the parameters $\tilde{\Delta}$=0.75 and B=24, and we assumed that $\tilde{V}(\varepsilon)$ is a constant. All energies are in eV. Observe that a logarithmic scale was used in (a).

Equations (3.12-13) show that δ_u has an interesting dependence on U. If $B\tilde{\Delta}/\pi > (\varepsilon_f-\delta_u)^2$, a reduction of U leads to a decrease of δ_u and an increase of $\chi(0)$, while the opposite happens for $B\tilde{\Delta}/\pi < (\varepsilon_f-\delta_u)^2$. This is illustrated in Fig. 1. For ε_f=-2.5, $\chi(0)$ is reduced by more than one order of magnitude, while for ε_f=-1.5 $\chi(0)$ initially increases as U is reduced. For U~8, far outside the range of validity of Eq. (3.12-13), the χ curve for ε_f=-1.5 has a maximum, and a further reduction of U leads to a reduction of $\chi(0)$ also for ε_f=-1.5.

Equation (3.17) shows that in the spin-fluctuation limit $\omega(f^0)$ is proportional to δ_u, which decreases exponentially as Δ is decreased (see Eq. (3.14)). This is in contrast to $\omega(f^2)$ (Eq. (3.18)) which decreases linearly with Δ. The f^2 weight can therefore be larger than the f^0 weight, even if the f^2 configurations have a higher energy than the f^0 configuration. In the $N_f \to \infty$ limit there is therefore a very important asymmetry between the coupling to the f^0 states and to the f^2 states. The results in Table 1 illustrate that these effects are also present to a large extent for $U \sim B$ and $N_f=14$. For $N_f\Delta=0.42$, for instance, the f^2 weight is 10 times larger than the f^0 weight, although the lowest f^0 and f^2 configurations are degenerate.

The f^2-weight is approximately the same for the groundstate (Eq. (3.18)) and for the lowest "magnetic" states in the space (3.1-3.3). The relatively strong coupling between the f^1 and f^2 configurations therefore has a rather small influence on attempts to line up the f-spin along an external field. The susceptibility can therefore be large even if the f^2 weight is not small. In the space (3.1-3.3) there is, however, no coupling to f^0 states, and the corresponding energy is lost when the spin is aligned. The susceptibility (3.19) therefore is closely related to the f^0 weight via δ_u.

It is interesting to apply scaling ideas to Eq. (1.1) in the limit $N_f \to \infty$, since in this limit we can compare with an exact numerical solution. Generalizing poor man's scaling [19-21] to arbitrary N_f one obtains

$$\frac{d\varepsilon_f}{dB} = \frac{\Delta}{\pi}\left(\frac{N_f-1}{B+\varepsilon_f+U} - \frac{N_f}{B} + \frac{1}{B-\varepsilon_f}\right) \tag{3.20}$$

$$\frac{dU}{dB} = \left(\frac{N_f}{B} - \frac{2(N_f-1)}{B+U+\varepsilon_f}\right)\frac{\Delta}{\pi} \tag{3.21}$$

$$\frac{d\Delta\tilde{E}}{dB} = \frac{N_f\Delta}{B} \tag{3.22}$$

where $\Delta\tilde{E}$ is a c-number in the scaled Hamiltonian. In deriving (3.20-22) we have assumed that f^3 and higher configurations are suppressed and that $\Delta(\varepsilon)\equiv\Delta$ is constant. In the denominators on the right-hand side we have assumed that $\Delta E \approx \varepsilon_f$. The band edges are at B and -B, respectively. For $N_f=2$ ε_f and U are not scaled if $B \gg U+\varepsilon_f$, $-\varepsilon_f$ [21], while for $N_f>2$ U is scaled also in this limit.

In Table 2 we show results for a case where scaling is started at B=10000. Scaling B from 10000 to 30 leads to an increase of U from 5 to 6.8 but has a small effect on ε_f. The physical properties ΔE, δ_u, $\omega(f^0)$, $\omega(f^2)$ and χ are almost unchanged, as they should be, since B is much larger than all other quantities in the problem. As B is further reduced, the scaled Hamiltonian starts to give different properties than the unscaled one. For instance, for $B \sim U$ there is a difference of about 15% in δ_u, $\omega(f^0)$ and χ and about 40% in $\omega(f^2)$. Thus poor man's scaling breaks down before the Hamiltonian has scaled to something which is appreciably simpler to solve. It is interesting that for $U=\infty$ and $N_f=\infty$ scaling is valid for $B \gg \delta_u$ [8], while here we find a much more restrictive requirement: $B >> U+\varepsilon_f$. To understand this, we note that for $U=\infty$ the conduction holes are within a few δ_u's below ε_F. For a finite U there are, however, holes on an energy-scale $2\varepsilon_f+U-\Delta E$ below ε_F, where $2\varepsilon_f+U-\Delta E \approx U+\varepsilon_f+\Delta E_{mag}$ in the spin-fluctuation limit. This can be seen from Eq. (3.10).

Table 2. The results for ε_f and U obtained from the scaling equations (3.20-22) and the results obtained from the scaled Hamiltonian for ΔE, δ_u, $\omega(f^0)$, $\omega(f^2)$ and $\chi(0)$. We have used $\tilde{\Delta}=0.75$ and assumed that $V(\varepsilon)^2$ is a constant. All energies are eV.

B	ε_f	U	ΔE	δ_u	$\omega(f_0)$	$\omega(f_2)$	$\frac{\chi(0)}{\frac{1}{3}j(j+1)}$
10000	-2.500	5.000	-4.942	0.0179	0.0265	0.0623	54.9
1000	-2.499	5.731	-4.942	0.0178	0.0265	0.0620	55.0
100	-2.489	6.443	-4.946	0.0176	0.0262	0.0597	55.7
30	-2.460	6.770	-4.949	0.0172	0.0255	0.0544	57.2
10	-2.387	6.973	-4.954	0.0160	0.0239	0.0435	61.4
5	-2.299	7.018	-4.955	0.0146	0.0221	0.0337	67.4
3	-2.209	7.001	-4.956	0.0132	0.0203	0.0261	74.7
1	-1.951	6.837	-4.954	0.0096	0.0157	0.0126	103.4

4. Valence photoemission spectrum

The need for including f^2 configurations is obvious for most spectroscopies. In Bremsstrahlung isochromat spectroscopy an f-electron is added to the system and the probability of finding an f^2 configuration is usually large. In the core spectroscopies, the core hole attraction pulls f^2 configurations below the f^1 configurations and the f^2 configurations are essential. In valence photoemission an f-electron is removed and the importance of f^2 configurations is less obvious. In Fig. 2 we show that the valence photoemission spectrum is nevertheless strongly influenced by the f^2 configurations. The reason for this was discussed in Ref. 9.

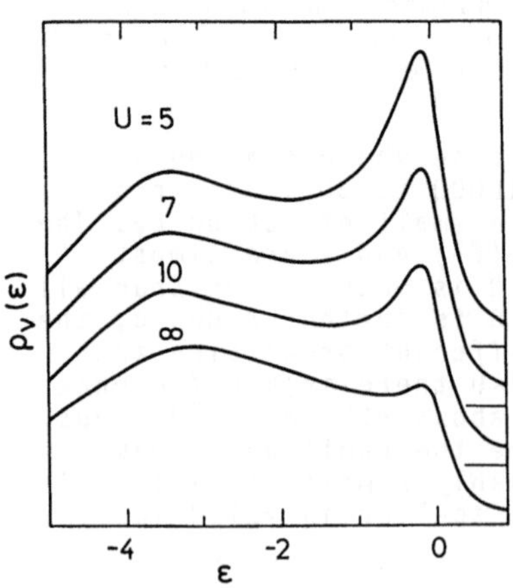

Fig. 2 The valence photoemission spectrum for different values of U. The $V(\varepsilon)^2$ has the energy-dependence (2.5), B=6 and $\tilde{\Delta}=1.7$. All energies are in eV.

For ε close to $\varepsilon_F=0$, the spectrum $\rho_v(\varepsilon)$ is given by [9]

$$\rho_v(\varepsilon) = A^4 \left\{a(\varepsilon)\left[1 - \int_{-B}^{0} d\varepsilon' \frac{a(\varepsilon')\,\tilde{V}(\varepsilon')}{2\varepsilon_f+U-\Delta E-\varepsilon-\varepsilon'}\right] - \tilde{V}(\varepsilon)\int_{-B}^{0} d\varepsilon' \frac{a(\varepsilon')^2}{2\varepsilon_f+U-\Delta E-\varepsilon-\varepsilon'}\right\}^2 . \tag{4.1}$$

The first term, proportional to $a(\varepsilon)$, gives a contribution which decays on the energy scale δ_u. The last term gives a qualitative new contribution for a finite U, as it decays on a much larger energy scale $2\varepsilon_f+U-\Delta E$ which is typically several eV. In the spin-fluctuation limit this term dominates. The variation of $V(\varepsilon)$ over this energy scale can be very substantial. A detailed knowledge of $V(\varepsilon)$ is therefore desirable, since $V(\varepsilon)$ can lead to additional structure in the spectrum.

The Kondo-like peak in Fig. 2 at $\varepsilon\to0$ is a genuine many-body effect due to the large Coulomb interaction U. The model can, however, also give rise to interesting structures of a more one-particle-like nature. This is illustrated by the results in Fig. 3, for CeP, CeAs and CeBi [22]. These results were obtained using a singlet ground-state. Forcing the impurity model into a magnetic state has a small effect on the spectrum. The similarity between the U=6 and U=∞ curve shows that the last term in Eq. (4.1) gives a small contribution. We also find that the first two terms in (4.1) are small. The spectra in Fig. 3 can be understood in terms of a non-degenerate model with non-interacting electrons, provided that we replace $V(\varepsilon)$ by $\sqrt{N_f}\,V(\varepsilon)$ for $\varepsilon<\varepsilon_F=0$ and by 0 otherwise [9]. The f-level is close to the peak of the conduction density of states and it interacts strongly with these states ($N_f\Delta_{av}$=0.7 eV). This hybridization leads to two peaks in the f-spectrum. The weights of these peaks depend sensitively on ε_f. Hybridization effects for the pnictides have been discussed earlier by SAKAI et al. [23].

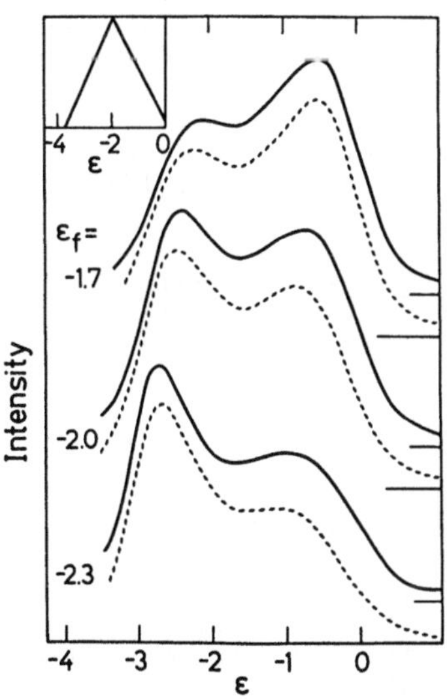

Fig.3 The valence photoemission spectrum for different values of ε_f. The shape of $V(\varepsilon)^2$ is shown in the insert and may represent the pnictides. The full curves show results for U=6 and the dashed curves for U=∞. The average value $N_f\Delta_{av}$ of $\pi N_f V(\varepsilon)^2$ over the occupied part of the conduction band is 0.7. All energies are in eV.

5. Concluding remarks

We have defined a Kondo temperature δ_u for a large N_f and showed that with this definition a number of ground-state and dynamic properties can be described in terms of δ_u. We find that δ_u can be strongly renormalized from its $U=\infty$ limit. We have related the static susceptibility $\chi(0)$ and a structure in Im $\chi(\omega)/\omega$ to δ_u in a way which does not depend on U explicitly. On the other hand we find that both the total energy and the f^0 weight have such an explicit U dependence. Furthermore the f^2 weight goes as $1/U^2$ for U>>B . There are two energy scales, δ_u and $2\varepsilon_f+U-\Delta E$, for the weight at small energies in the valence photoemission spectrum. The inclusion of these effects has improved the agreement between theory and experiment.

References

1. P.W. Anderson, Phys.Rev. 124, 41 (1961).

2. B. Johansson, Philos.Mag. 30, 469 (1974); J.Phys. F 4; L169 (1974).

3. J.F. Herbst, R.E. Watson and J.W. Wilkins, Phys.Rev. B13, 1439 (1976);B17, 3089 (1978).

4. J.K. Lang, Y. Baer and P.A. Cox, Phys.Rev.Lett. 42, 74 (1979).

5. J.W. Allen, S.-J. Oh, I. Lindau, J.M. Lawrence, L.I. Johansson annd S.B. Hagström, Phys.Rev.Lett. 46 1100 (1981); M. Croft, J.H. Weaver, D.J. Peterman and A. Franciosi, Phys.Rev. Lett. 46, 1104(1981).

6. J.C. Fuggle, F.U. Hillebrecht, J.-M. Esteva, R.C. Karnatak, O. Gunnarsson and K. Schönhammer, Phys.Rev. B27, 4637 (1983); J.C. Fuggle, F.U. Hillebrecht, Z. Zolnierek, R. Lässer, Ch. Freiburg, O. Gunnarsson and K. Schönhammer, Phys.Rev. B27, 7330 (1983); O. Gunnarsson, K. Schönhammer, J.C. Fuggle, F.U. Hillebrecht, J.-M. Esteva, R.C. Karnatak and B. Hillebrand, Phys.Rev. B28, 7330 (1983); F.U. Hillebrecht, J.C. Fuggle, G.A. Sawatzky, M. Campagna, O. Gunnarsson and K. Schönhammer, Phys. Rev. B30, 1777 (1984).

7. H.R. Krishna-murthy, J.W. Wilkins and K.G. Wilson, Phys.Rev. B21, 1003 (1980); B21, 1044 (1980); N. Kawakami and A. Okiji, Phys.Lett. 86A, 483 (1981); J. Phys.Soc. Jpn. 51, 1145 (1982).

8. O. Gunnarsson and K. Schönhammer, Phys.Rev.Lett. 50, 604 (1983); Phys.Rev. B28, 4315 (1983).

9. O. Gunnarsson and K. Schönhammer, Phys.Rev. B31, xxxx (1985).

10. T.V. Ramakrishnan, in Valence Fluctuations (Eds.: L.M. Falicov, W. Hanke and M.P Maple, North-Holland, Amsterdam, 1981),p.13; P.W. Anderson, ibid, p. 451.

11. T.V. Ramakrishnan and K. Sur, Phys.Rev. B26, 1798 (1982).

12. N. Read and D.M. Newns, J.Phys. C16, 3273 (1983; 16, L1055 (1983); F.C. Zhang and T. K. Lee, Phys.Rev. B28,33 (1983); J.W. Rasul and A.C. Hewson, J.Phys. C16, L933 (1983); 17, 2555 (1984); 17, 3337 (1984).

13. Y. Kuramoto, Z.Phys. B53, 37 (1983).

14. P. Coleman, Phys.Rev. B29, 3035 (1984); N. Bickers, D.L. Cox and J.W. Wilkins, Phys.Rev.Lett. 54, 230 (1985).

15. C.M. Varma and Y. Yafet, Phys.Rev. B13, 2950 (1976).

16. The f^2 states of order $(1/N_f)^2$ have not been included for numerical reasons.

17. O. Gunnarsson and K. Schönhammer, these proceedings and J.Magn.Magn.Mater. (to be published).

18. N. Read and D.M. Newns, J.Phys. C16, 3273 (1983).

19. P.W. Anderson, J.Phys. C3, 2436 (1970).

20. J.H. Jefferson, J.Phys. C10, 3589 (1977).

21. F.D.M. Haldane, Phys.Rev.Lett. 40, 416 (1978).

22. A. Franciosi, J.H. Weaver, N. Mårtensson and M. Croft, Phys.Rev. B24, 3651 (1981).

23. O. Sakai, H. Takahashi, M. Takeshige and T. Kasuya, Solid State Commun. 52, 997 (1984).

PES and BIS Spectra in Ce-Compounds

M. Takeshige, O. Sakai, and T. Kasuya

Department of Physics, Tohoku University, Sendai 980, Japan

c-f hybridization matrices for Ce-monopnictides and Ce-metals are calculated based on the detailed band calculation. They have a broad peak at about 0.1 Ry below the Fermi level. PES and BIS are calculated by using the hybridization matrices and also by including the d-f Coulomb interaction. Calculated PES shows double peak structure. The intensity ratio and the location of the peaks show good agreement with experiments. The deeper peak is due to the bonding effect of the c-f hybridization. The shallower peak in Ce-monopnictides is mainly ascribed to the anti-bonding effect, while the peak in Ce-metals is composed of the anti-bonding peak and those due to the Fermi edge effects. The d-f Coulomb interaction reduces the magnitude of the hybridization and broadens the width of the peaks.

1. Introduction

Experimental and theoretical works on the photoemission spectroscopy (PES) and the bremsstrahlung isochromat spectroscopy (BIS) have been extensively made to study the f-state in Ce-compounds [1]. In many compounds, such as the Ce-metals [2,3] and the Ce-monopnictides [4], PES shows double peak structure ascribed to the f-state. It is now generally accepted that this structure originates from the hybridization between the band electrons and the f-electron. The deeper peak is due to the bonding effect of the c-f hybridization [5,6]. On the other hand, we have three possible processes which can cause the peak structure near the Fermi level: the Kondo scattering [7], the relaxation of the band electrons to the spin-orbit excited states [8] and the anti-bonding effect [6,9]. There are ambiguities to conclude what is the main mechanism in actual materials because most calculations are based on the simplified model for the electronic band structure, and perhaps the main mechanism may be different in different materials.

In a previous paper, we have calculated the hybridization matrix for CeSb based on the detailed band calculation, and shown that it has a broad peak at about 0.1 Ry below the Fermi level [6]. The calculated hybridization causes actually double peak structure. The shallow peak at about 0.02 Ry below the Fermi level originates mainly from the anti-bonding effect. The peak structures due to the Fermi edge effect are small,because CeSb is a semi-metallic compound. On the other hand, in metallic compounds the Fermi edge effect is expected to become important,as well as the anti-bonding effect. Therefore, we choose Ce-metals and Ce-monopnictides as typical materials,and study photoexcitation spectra by using the hybridization matrices obtained from the detailed band calculations.

In the previous paper, we have used a model in which the d-f Coulomb interaction is not explicitly considered. When the f-electron is excited by the photon, the induced f-hole is screened not only by p-electron through the p-f mixing but also by the d-electrons through the intra-atomic d-f Coulomb interaction. In this paper we report the results of this effect on PES [10]. Before, we also neglected the d-f mixing matrix because it is smaller than p-f mixing. However, for Ce-monopnictides with smaller anions, particularly, for CeN and CeP, it can not be neglected. Therefore the d-f mixing is also included here. We use the terminology c-f hybridization in this context.

At first, we determine various bonding parameters of LCAO band scheme by fitting the bands to those obtained for La-compounds by the APW method. Next, we calculate the c-f hybridization matrices by using the p-f and d-f bonding parameters of the LCAO model, and then study the photoexcitation spectra based on them. In this paper, we are mainly concerned with the determination of the band parameters. Recent results of the calculation for the photoexcitation spectra are only briefly discussed, since details are given in a separate paper [11]. The calculation of the band structure of CeSb, and comparison of it with experimental results, is also shown in a separate paper [12].

2. Determination of Bonding Parameters

We use the following model Hamiltonian,

$$H = H_1 + H_2 , \tag{1}$$

where H_1 is the single electron term written as follows,

$$H_1 = \sum_{\lambda\mathbf{k}} \varepsilon_{\lambda\mathbf{k}} c^+_{\lambda\mathbf{k}} c_{\lambda\mathbf{k}} + \sum_{\mathbf{i}\mathbf{i}'} \sum_{\eta\eta'} \varepsilon_{\eta\eta'}(\mathbf{i}-\mathbf{i}') f^+_{\mathbf{i}\eta} f_{\mathbf{i}'\eta'}$$
$$+ \sum_{\lambda\mathbf{k}\mathbf{i}\eta} \{ v_\eta(\lambda\mathbf{k}) e^{i\mathbf{k}\mathbf{i}} f^+_{\mathbf{i}\eta} c_{\lambda\mathbf{k}} + \text{h.c.} \} . \tag{2}$$

Here $c^+_{\lambda\mathbf{k}}$ is the creation operator of usual band state with p- and d-characters, λ and $\mathbf{k}$ are the band suffix and wave number vector, respectively. $f^+_{\mathbf{i}\eta}$ is the creation operator of f-state at site $\mathbf{i}$, symmetry $\eta = (j\Gamma\nu)$, where j stands for $j = 5/2$ and $j = 7/2$, Γ denotes the irreducible representation of the cubic group, and ν the component of it. The quantity $\varepsilon_{\lambda\mathbf{k}}$ is the band energy and $\varepsilon_{\eta\eta'}(\mathbf{h})$ is the bonding integral between f-states. In the third term $v_\eta(\lambda\mathbf{k})$ is the c-f hybridization matrix element. The term H_2 in (1) represents the electron-electron interactions,

$$H_2 = \frac{u_{ff}}{2} \sum_{\mathbf{i}} \sum_{\eta(\neq)\eta'} f^+_{\mathbf{i}\eta} f_{\mathbf{i}\eta} f^+_{\mathbf{i}\eta'} f_{\mathbf{i}\eta'} + u_{df} \sum_{\mathbf{i}} \sum_{\zeta\eta} d^+_{\mathbf{i}\zeta} d_{\mathbf{i}\zeta} (f^+_{\mathbf{i}\eta} f_{\mathbf{i}\eta} - 1) . \tag{3}$$

The first is the f-f Coulomb interaction and the second is the d-f Coulomb interaction. The quantities u_{ff} and u_{df} are the interaction constants, and ζ denotes the symmetry of the d-orbit.

The single electron term H_1 is determined by the usual band theory. In the present paper, we use LCAO band scheme for La-monopnictides, and LCAO-OPW scheme for La-metals.

2a. Ce-monopnictides

At first we determine the parameters of the LCAO band scheme by fitting it to the band calculation made for LaSb by Hasegawa [13] based on the APW method. We use 5d and 4f orbits of La, and 5s and 5p orbits of Sb. The two-center bonding parameters for pairs farther than the nearest neighbour are neglected.

In the fitting process, we assume that both the two center parameters between f-orbits and the cubic crystal-field splitting of f-states can be neglected, i.e. we set $\varepsilon_{\eta\eta'}(\mathbf{h}) = \varepsilon_f \delta_{\eta\eta'} \delta_{\mathbf{h}0}$. The 4f energy level ε_f is chosen to be 0.4142 Ry equal to that of the Γ_{25} state. The d-d, d-s, d-p and p-p bonding parameters are determined by the similar method to that described before [14].

By using the energy-shift of the Γ_{15} state at 0.4707 Ry from ε_f and the energy of the lower Γ_{15} state, 0.2600 Ry, we estimate the p-f hybridization matrix element at Γ-point, $2(pf\sigma)-\sqrt{6}(pf\pi)$ to be 0.09334 Ry. The energy of the fictitious unhybridized Γ_{15} state of p-band is estimated to be 0.3165 Ry, which is consistent with the band calculation for YSb [13]. Combining the result obtained from Γ_{15} states and the energies at X_4' and X_5', we estimated as $(pf\sigma)$ = 0.02686 Ry and $(pf\pi)$ = -0.01617 Ry. However, there are ambiguities in these values because it is impossible to perfectly reproduce the X_4' and X_5' energy levels by the present LCAO model. We need to include atomic orbits of higher energy, bonding parameters for farther pairs, and/or non-orthogonality effects.

Recently, TAKAHASHI and KASUYA have shown that the crystal-field splitting due to the p-f hybridization is very sensitive to the ratio $(pf\pi)/(pf\sigma)$, and estimated it to be -0.7 from the analysis of experiments [15]. Present parameters have ratio near -0.7. Inversely, if we assume the ratio to be -0.7 and use the data from Γ_{15} states, $(pf\sigma)$ and $(pf\pi)$ are estimated as 0.02513 and -0.01759 Ry, respectively. From now on we use the latter method. The d-f bonding parameters are determined from the energies of Δ_2 and Σ_2 states, but there are many ambiguities in these parameters.

The energy bands calculated by using the parameters given in Table 1 rather well reproduce the APW bands. However, calculated density of states has a double peak structure separated by 0.05 Ry in the valence band region, in contrast with a broad single peak of valence band in APW calculation. This difference is mainly ascribed to the discrepancy of the Σ_4 branch, which gives the higher energy sub-peak in LCAO. In later calculations, we modify the bonding parameters to improve the fitting along the Σ axis as described in the caption of Table 1. The LCAO bands, calculated by using the modified

Table. 1 Values of the parameters used in LCAO band calculations for La-monopnictides. To improve the fitting along the Σ axis, we modify the $(dp\sigma)$ and include the p-p bonding parameters of the 2nd nearest pnictogen-pnictogen neighbors. The new parameters are $(dp\sigma)$ = 0.0803, 0.0726 and 0.0789 Ry ; $(pp\sigma)$2nd = 0.0128, 0.0128 and 0.0106 Ry ; $(pp\pi)$2nd = -0.0064, -0.0064 and -0.0053 Ry for LaP, LaAs and LaSb, respectively.

	LaP	LaAs	LaSb	(Ry)
ε(5d-t2g)	0.4547	0.4448	0.3795	
ε(5d-eg)	0.4961	0.4774	0.4031	
ε(4f)	0.5140	0.4942	0.4142	
ε(s)	-0.3990	-0.4480	-0.3274	
ε(p)	0.2459	0.2448	0.2288	
$(dd\sigma)$	-0.0638	-0.0577	-0.0454	
$(dd\pi)$	0.0175	0.0188	0.0164	
$(dd\delta)$	0.0027	0.0007	-0.0002	
$(pp\sigma)$	0.0183	0.0213	0.0280	
$(pp\pi)$	-0.0020	-0.0023	-0.0030	
$(ds\sigma)$	-0.1636	-0.1602	-0.1376	
$(dp\sigma)$	0.0502	0.0558	0.0563	
$(dp\pi)$	-0.0619	-0.0582	-0.0480	
$(fs\sigma)$	0.0	0.0	0.0	
$(fp\sigma)$	0.0265	0.0266	0.0251	
$(fp\pi)$	-0.0186	-0.0186	-0.0176	
$(fd\sigma)$	0.0158	0.0161	0.0121	
$(fd\pi)$	-0.0092	-0.0073	-0.0052	
$(fd\delta)$	0.0069	0.0037	-0.0004	

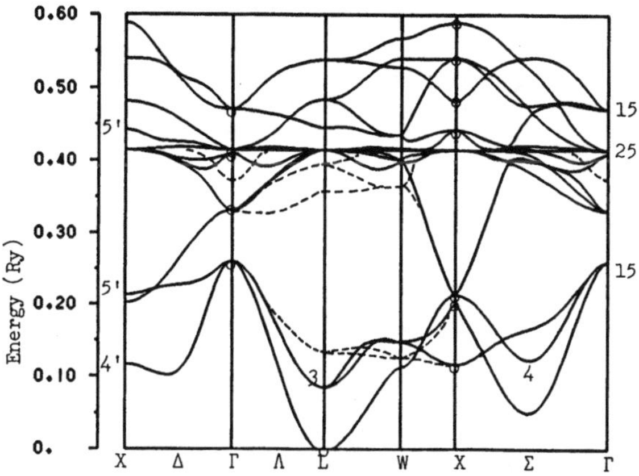

Fig. 1 LCAO bands for LaSb calculated by using the modified parameters to improve the fitting along the Σ axis. Small circles indicate the points used as the data, and dashed lines are the branches which can not be well reproduced by this LCAO scheme. Fermi energy = 0.253 Ry. If we use the parameters in Table 1, the L3 branch is well reproduced but Σ4 branch rises up.

parameters to improve the fitting along the Σ axis, are shown in Fig. 1. Small circles in Fig. 1 indicate the points used as the data, and dashed lines the branches which can not be well reproduced by the present LCAO scheme. Total density of states is shown by the solid line indicated by the notation 'Total (not including spin-orbit)' in Fig. 2. We found, however, that the modification does not so drastically change the PES.

Recently, HASEGAWA has extended his APW calculation to take the spin-orbit interaction into account [16]. In the present LCAO band, we include the spin-orbit interaction of Sb as $\zeta_p \mathbf{sl}_p$ with ζ_p = 0.044 Ry. The spin-orbit interaction of the f-electron is neglected in the band calculation, because it is small. The total density of states, the partial density of states on p-orbits and that on d-orbits of this band are shown in Fig. 2. The total density of states has a shape considerably different from that of the bands calculated by not including the spin-orbit interaction, because the splitting energy $3\zeta_p/2$ is not small compared with the width of the valence bands.

When the APW bands are compared with the experimental results of the de Haas-van Alphen effect in LaSb, it suggests the necessity of shifting the 5d

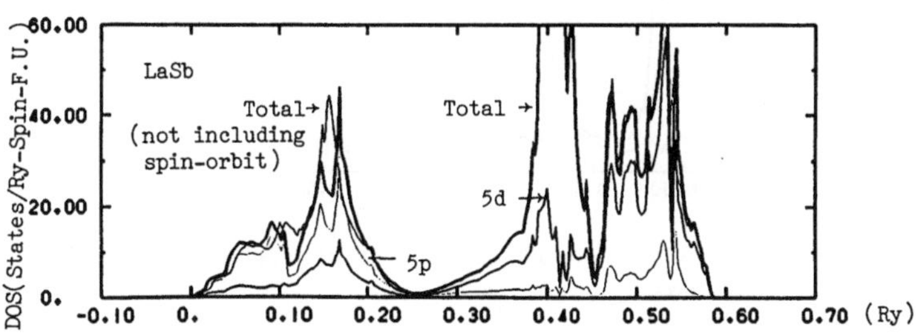

Fig. 2 Density of states for LaSb. The total density of states of bands calculated by neglecting the spin-orbit interaction is shown (Fermi energy = 0.253 Ry). The total density of states, the partial density of states on p-orbits and that on d-orbits of bands calculated by including the spin-orbit interaction are also shown (Fermi energy = 0.255 Ry).

and 4f bands to higher energy side [17]. If we shift the 5d band 0.065 Ry and 4f level to 0.6 Ry in LCAO, calculated Fermi surfaces show good agreement with the experiments. However, the shape of the density of states does not drastically change from the solid line indicated by the notation 'Total' in Fig. 2 in the valence band region. In this paper, we do not shift the 5d and 4f bands.

2b. Ce-metals

We have calculated energy bands for γ- and α- La [31] by APW method. Obtained results well correspond to those of PICKETT et al [18]. The band structure calculated by the LCAO-OPW scheme [19,20] is shown in Fig. 3 for γ-La. We use 5p, 5d and 4f orbits and four OPW states which are characterized by the reciprocal vectors [000], [100], [1/2,1/2,1/2] and [1/2,-1/2,-1/2] for the Bloch states in the trigonal pyramid of the Brillouin zone, $0 \leq k_z \leq k_y \leq k_x$. In this case, it is necessary to take account of both the two center bonding integrals between f-orbits and the

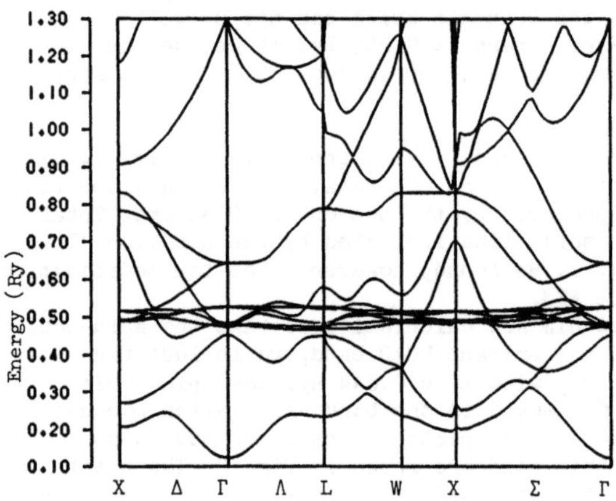

Fig. 3 LCAO-OPW band for γ-La. Fermi energy = 0.322 Ry.

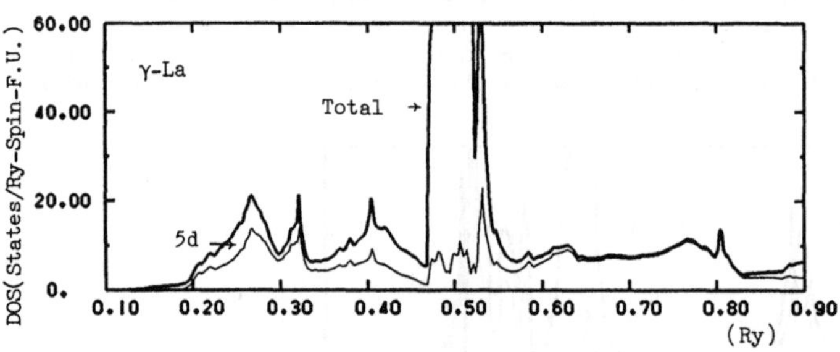

Fig. 4 Total density of states and the partial density of states on d-orbits for γ-La. Fermi energy = 0.322 Ry.

non-orthogonality effects between the atomic orbits and the OPW states. The total density of states and the partial density of states on d-orbits are shown in Fig. 4. The Fermi level is located at a sharp peak as in the calculation of PICKETT et al.

3. Hybridization Matrix and PES in Ce-monopnictides

In the following, we study the photoemission spectra based on the one impurity model. An attempt to calculate PES based on the lattice model is given in the separate paper [12].

The hybridization matrix elements usually appear through the following quantity,

$$M_\eta(\varepsilon) = \sum_{\lambda\mathbf{k}} |v_\eta(\lambda\mathbf{k})|^2 \delta(\varepsilon-\varepsilon_{\lambda\mathbf{k}}) \ , \tag{4}$$

which we call hereafter the hybridization matrix. In a literal sense, the unhybridized bands of Ce-compounds should be determined by removing the 4f states from the bands of La-compounds. However, in the impurity model it may be a rather good approximation to calculate the hybridization matrix, by using the Bloch states for La-compounds which has unoccupied f-states above the Fermi level [21].

The hybridization matrices in LaSb calculated for j = 5/2 Γ_7 and j = 5/2 Γ_8 states, and averaged value over all states in j = 7/2 are shown in Fig.5. We can see that M_η for Γ_7 is very small compared with that for Γ_8, especially, near the Fermi level.

Hybridization matrix averaged over all η is shown in Fig.6, which has a broad peak in the valence band region similar to that of the density of states. It has, however, relatively larger magnitude near the top of the valence band because the p-f hybridization is strong there [22].

Hybridization matrices averaged over all η for other La-monopnictides are also shown in Fig. 6. We used the modified parameters as described in the caption of Table 1 to calculate these hybridyzation matrices. Uncertainty of the bonding parameters increases as pnictogen goes to lighter ion, particularly for LaN. However, it may be true that the magnitude of the hybridization increases gradually when pnictogen goes from Sb to N.

We calculate the photoemission spectra by choosing the magnetic state [6],

$$|\Phi_g\rangle = f^+_{\eta g}|F\rangle C(g) + \sum_{\mathbf{k}2} c^+_{\mathbf{k}2}|F\rangle C(\mathbf{k}2) + \sum_{\eta\mathbf{k}1\mathbf{k}2} f^+_\eta c_{\mathbf{k}1} c^+_{\mathbf{k}2}|F\rangle C(\eta\mathbf{k}1\mathbf{k}2) \ , \tag{4}$$

as an initial state, where $|F\rangle$ is the Fermi vacuum, C(g), C($\mathbf{k}_2$) and C($\eta\mathbf{k}_1\mathbf{k}_2$) are the coefficients, and $\lambda\mathbf{k}$ is simply denoted as $\mathbf{k}$. This selection of the initial state may give good approximate spectra at high temperatures. In (5),

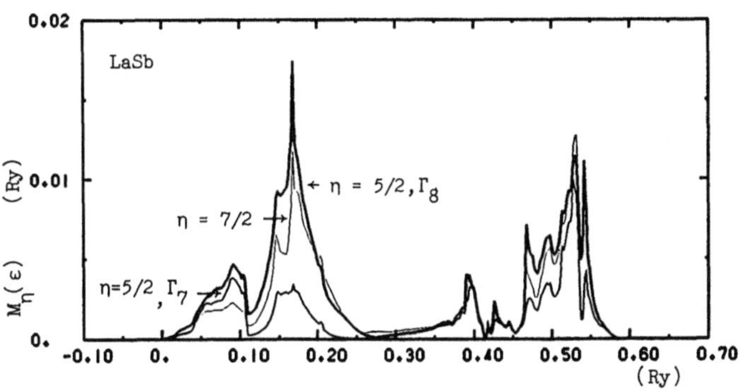

Fig. 5 Hybridization matrices for LaSb. That values for η = (j = 5/2, Γ_7) and η = (j = 5/2, Γ_8) and the value averaged over j = 7/2 are shown. The modified parameters as described in the caption of Table 1 are used in the calculation. The spin-orbit interaction of the 5p-electrons are also included. Fermi energy = 0.255 Ry.

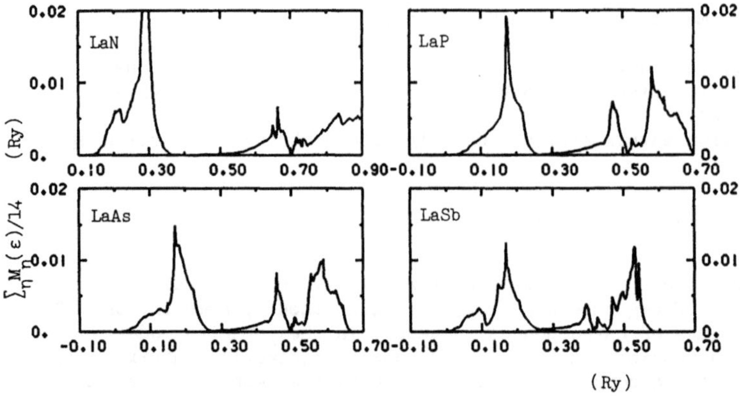

Fig. 6 Hybridization matrices averaged over all η for La-monopnictides. Fermi energy = 0.366, 0.256, 0.265 and 0.255 Ry for LaN, LaP, LaAs and LaSb, respectively.

we put as $\eta_g = (j = 5/2, \Gamma_8)$. In the final states of PES, we consider the screening process through the d-f Coulomb interaction as well as the relaxation of the band electron to the vacant f-hole state through the c-f hybridization. Neglecting the many electron-hole pair excited states as done previously [6], we calculate the final states in the manifold of states: $|F\rangle$, $f^+_\eta c_{\mathbf{k}1}|F\rangle$, $c^+_{\mathbf{k}2}c_{\mathbf{k}1}|F\rangle$ and $f^+_\eta c_{\mathbf{k}'1}c^+_{\mathbf{k}2}c_{\mathbf{k}1}|F\rangle$. The spectral function is given as follows [11]

$$F_{PES}(\varepsilon) = \frac{1}{\pi}\,\mathrm{Im}\,\frac{|C(g)|^2}{\varepsilon-\varepsilon_f-\Delta_g-\Sigma_c(\varepsilon-\varepsilon_f-\Delta_g-io)-\Sigma_0(\varepsilon-\Delta_g-io)-io}\,, \qquad (6)$$

where, Δ_g is the energy-shift of the initial state due to the c-f hybridization, which has the magnitude of 10^2 K. The quantities Σ_c and Σ_0 are the self-energies of the f-hole state originated from the d-f Coulomb and the c-f hybridization interactions, respectively [23],

$$\Sigma_{00}(z) \equiv \sum_\eta \int_{-\infty}^{\mu} d\varepsilon_1\,\frac{M_\eta(\varepsilon_1)}{z-\varepsilon_1+\delta_\eta}\,, \qquad (7)$$

$$\Sigma_\eta(z) \equiv \int_{\mu}^{\infty} d\varepsilon_2\,\frac{M_\eta(\varepsilon_2)}{z+\varepsilon_2-\varepsilon_f-\Sigma_{00}(z+\varepsilon_2)}\,, \qquad (8)$$

$$\Sigma_0(z) \equiv \sum_\eta \int_{-\infty}^{\mu} d\varepsilon_1\,\frac{M_\eta(\varepsilon_1)}{z-\varepsilon_1+\delta_\eta-\Sigma_\eta(z-\varepsilon_1)}\,, \qquad (9)$$

$$\Sigma_c(z) \equiv |u_{df}|^2 \sum_\zeta \int_{-\infty}^{\mu} d\varepsilon_1 \int_{\mu}^{\infty} d\varepsilon_2\,\frac{\rho_\zeta(\varepsilon_1)\rho_\zeta(\varepsilon_2)}{z-\varepsilon_1+\varepsilon_2-\Sigma_{00}(z-\varepsilon_1+\varepsilon_2+\varepsilon_f)}\,. \qquad (10)$$

Here we define as $\varepsilon = \varepsilon_K-\omega_0-\mu$, ω_0 and ε_K are the energies of photon and photo-electron, respectively, and μ is the Fermi level, $\delta_\eta = \varepsilon_\eta - \varepsilon_{\eta g}$, $\rho_\zeta(\varepsilon)$ is the partial density of states on the ζ-orbit of 5d state. In the derivation of

(6), terms which include the coefficient $C(\eta\mathbf{k}_1\mathbf{k}_2)$ are neglected because we consider the case that Ce^{3+} state is almost stable.

At first we show PES calculated by neglecting the dynamical screening effect due to the d-f Coulomb interaction. In this case we neglect Σ_c in (6), but replace the 4f level ε_f by the effective one which is well screened by the d-f Coulomb interaction. As shown in Fig.7, PES have three peaks, at 0.26 Ry near the bottom of the valence band, at about 0.02 Ry, and at the Fermi level. The peak at 0.26 Ry is due to the bonding effect of the c-f hybridization. The peak at 0.02 Ry is mainly due to the anti-bonding effect. In general, the relaxation to the j = 5/2 states and that to the j = 7/2 states causes peak structures at the Fermi level and at about 0.02 Ry, respectively [8]. The energy 0.02 Ry corresponds to the excitation energy from j = 5/2 to j = 7/2 states. The former is the Kondo peak, and the latter overlaps with the anti-bonding peak in CeSb. These two peaks are relatively small in CeSb because the Fermi edge effect is not large in semi-metals.

In the calculation, the effective 4f energy-level is chosen to be 0.17 Ry to reproduce the intensity ratio 2:1 between the deep and the shallow peaks. The location of the bonding peak, 0.26 Ry seems to be considerably deeper than 0.23 Ry in experiments. The anti-bonding peak also appears too near the Fermi level. These results may indicate the fact that the present hybridization matrix has too large magnitude. Calculations for other Ce-monopnictides also

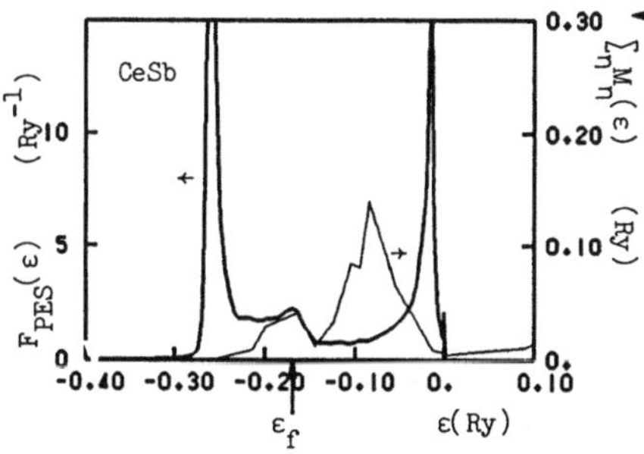

Fig. 7 PES for CeSb calculated by neglecting the dynamical screening effects of d-f Coulomb interactions. Approximated values of the total hybridization matrices are also shown.

Fig. 8 PES for Ce-monopnictides calculated by including the d-f Coulomb interactions. Approximated values of the total hybridization matrices and the effective f level are also shown.

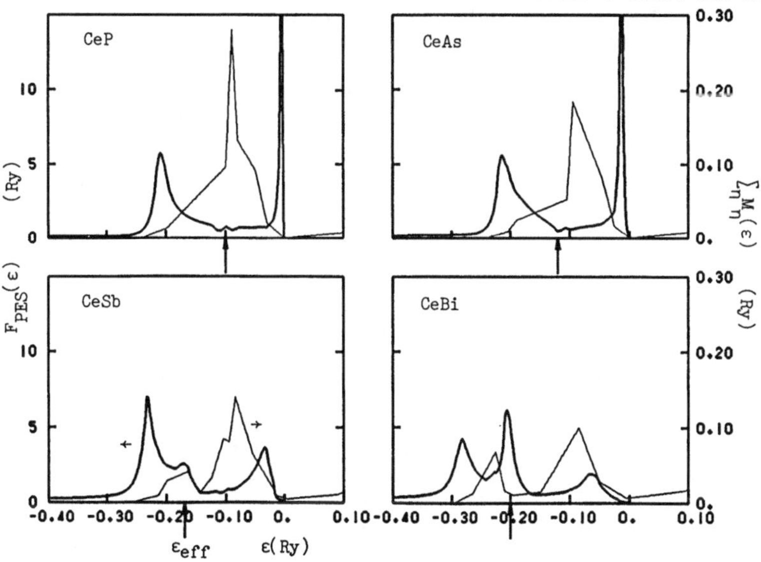

show this fact. For example, the anti-bonding peak appears above the Fermi level in CeP, which indicates the instability of the trivalent state, and contradicts the experiments.

In Fig.8, we show PES for Ce-monopnictides calculated by including Σ_c. The locations of the peaks and the intensity ratios of them are grossly well reproduced. The anti-bonding peak in CeN still appears above the Fermi level.

Let us see the effects of the d-f Coulomb interaction in detail. As shown in Fig. 2, the partial density of states on d-states has integrated intensity of about 0.1/state in the valence band region, and a broad peak at about 0.4 Ry above the Fermi level. Therefore main part of the electron-hole pair excitation spectrum has an effective energy-gap about 0.4 Ry. This is the one reason why we neglect the many-pair-excited states in the calculation of Σ_c.

The effective f level ε_{eff} which is well screened by the d-f Coulomb interaction is determined by the relation, $\varepsilon_{eff}-\varepsilon_f-\Delta_g-\mathrm{Re}\Sigma_c(\varepsilon_{eff}-\varepsilon_f-\Delta_g) = 0$. If we expand Σ_c around ε_{eff}, we have an expression,

$$F_{PES}(\varepsilon) \simeq \frac{1}{\pi}\,\mathrm{Im}\,\frac{|C(g)|^2}{Z[\varepsilon-\varepsilon_{eff}-\frac{1}{Z}\Sigma_0(\varepsilon-\Delta_g-io)-io]}\,, \qquad (11)$$

where $Z = 1-\partial\Sigma_c(\varepsilon_{eff}-\varepsilon_f-\Delta_g)/\partial\varepsilon_{eff}$ [11]. This indicates that the magnitude of the hybridization and the spectral intensity are reduced when we consider the dynamical effects of the d-f Coulomb screening. The reduction factor Z^{-1} is estimated to be about 0.6 for Ce-monopnictides by using the value u_{df} = 10 eV [24]. If we calculate PES by simply reducing the hybridization matrix to 0.6 of the original value, the locations of the peaks coincide with those in Fig. 8, but the widths of the bonding peaks seems to be narrow compared with those in Fig. 8. The effect of the decay of the f-hole state on the d-f Coulomb interaction is calculated in Σ_c through the Σ_{00} term in (10). Even when we consider this effect, the calculated peaks seem to be still narrow compared with those in experiments. This point will be studied in future.

In the present calculation, the self-energy shift due to the d-f Coulomb interaction is calculated to be about 10 eV, and the screening charge at the f-hole site is about 2. These indicate that the effect of the d-f Coulomb interaction is over-estimated. The latter fact shows the necessity to take into account the d-d Coulomb interaction, which will suppress the effect of the d-f Coulomb interaction.

The over-estimation of the self-energy shift may be corrected by adjusting the bare 4f level in the present calculation. It may be a rather good approximation to over-reduce the self-energy Σ_0, because the present c-f hybridization matrix is calculated from the band calculation for La-compounds which gives larger hybridization matrix about 1.2~1.4 of the Ce-compounds [25].

4. Ce-metals

There is some arbitrariness in the definition of the hybridization matrix in La-metals, because it depends on the definition of the f-state, which has uncertainty relating to the non-orthogonality to the OPW states. We use the original f-state as the f-state, and orthogonalize the OPW states to the f-state.

In Fig. 9, the hybridization matrices for γ-La are shown. The j = 5/2 Γ_7 state has a larger hybridization matrix than that for the j = 5/2 Γ_8 state in this case. We also calculated the hybridization matrices in α-La, and found that their magnitude is about 1.5 of that in γ-La.

The PES for γ- and α-Ce are shown in Fig. 10. The reduction factor Z^{-1} is estimated to be 0.6 also in this case. The peak positions change a little in

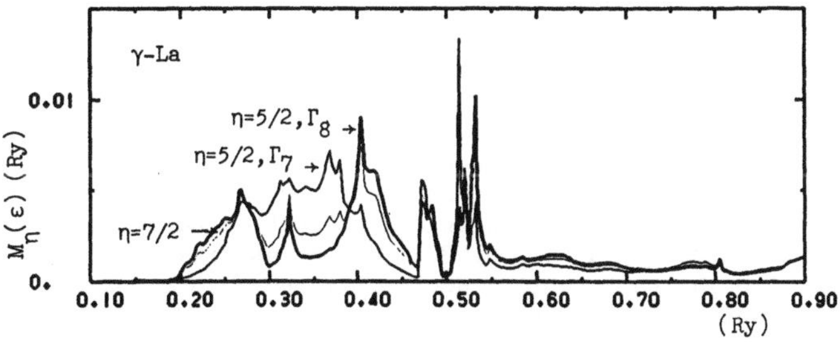

Fig. 9 Hybridization matrices for γ-La. That values for $\eta = (j = 5/2, \Gamma_7)$ and $\eta = (j = 5/2, \Gamma_8)$ and the averaged value over $j = 7/2$ are shown. Fermi energy = 0.322 Ry.

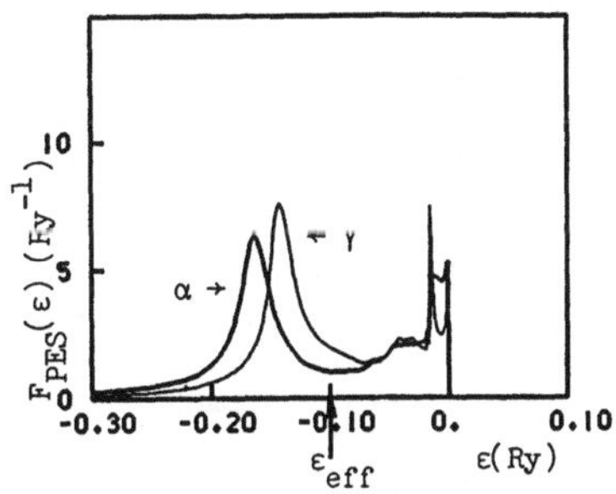

Fig. 10 PES for γ-and α-Ce.

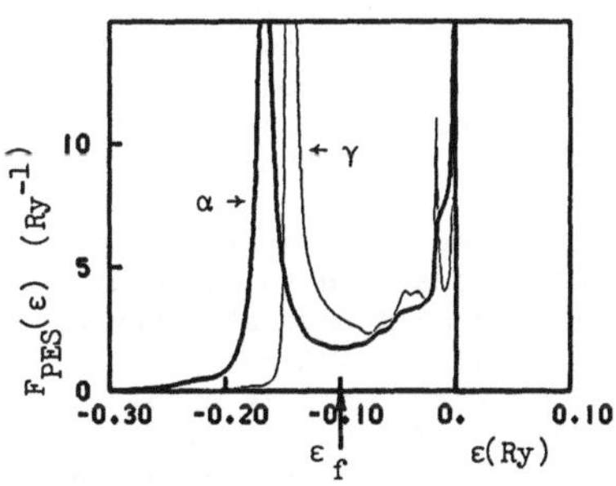

Fig. 11 PES for γ-and α-Ce calculated based on the Gunnarsson-Schonhammer model.

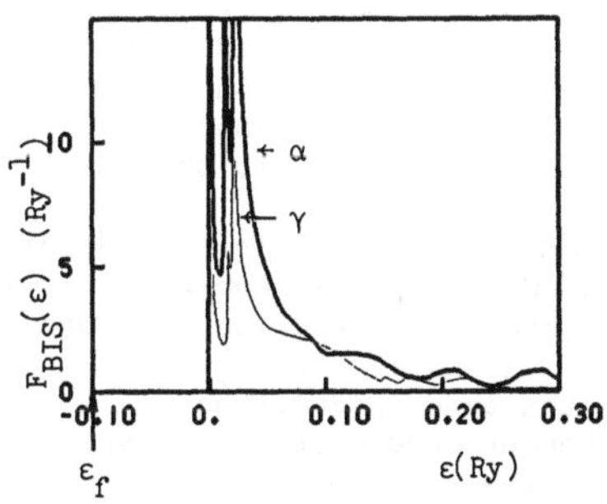

Fig. 12 BIS for γ- and α-Ce.

γ- and α- phases, in good agreement with experiment [2]. The upper peaks are composed of the anti-bonding peak and the peaks due to the Fermi edge effect. These are not separate in α-Ce. Again, the peak widths are too narrow in these calculations.

Up to now, we used the magnetic state as the initial state of the photoexcitation. This may be allowed at high temperatures. In Fig. 11, we show PES calculated based on the GUNNARSSON-SCHONHAMMER model [5] which uses a non-magnetic initial state and may be good at low temperature. Here, the hybridization matrix is reduced to 0.6 of the original magnitude and the d-f Coulomb interaction is neglected to simplify the calculation. Both calculations give similar results.

We show BIS for γ- and α-Ce in Fig. 12. The BIS spectrum is given by the following expression [11],

$$F_{BIS}(\varepsilon) = \frac{1}{\pi} \mathrm{Im} \sum_{\eta k2} \frac{|C(k2)|^2}{\varepsilon-\varepsilon_2-\delta_\eta+\Delta_g+\Sigma_\eta(-\varepsilon-\Delta_g+\varepsilon_2+io)-io}, \tag{12}$$

where the d-f Coulomb interaction is not included. In this calculation, the f^2 configuration is neglected. So the peak structures only just above the Fermi level in experiment [26] should be compared with the present calculation. The peak with relatively large intensity at $\varepsilon \sim 0.02$ Ry may be ascribed to the peak seen in experiment, though its intensity does not agree quantitatively. Extended calculations including f^2 configuration are necessary [27].

5. Summary and discussion

The c-f hybridization matrices in Ce-monopnictides and Ce-metals were calculated based on the detailed band structure. They have a broad peak at about 0.1 Ry below the Fermi level. The double peak structure of PES in these compounds was well reproduced by the calculation considering the dynamical screening effect through the d-f Coulomb interaction. The d-f Coulomb interaction reduces the magnitude of the hybridization and broadens the width of the peak.

The deeper peak is due to the bonding effect of the c-f hybridization. The shallower peak in Ce-monopnictides is mainly ascribed to the anti-bonding effect, while the shallower peak in Ce-metals is composed of the anti-bonding peak and those due to the Fermi edge effects. Even in Ce-metals,the fact that the hybridization matrix has a broad peak near the Fermi level is important to produce a noticeable peak just below the Fermi level. Concerning this point, $CePd_3$ may be interesting. It does not show the peak structure at the Fermi level,though the temperature characterizing the low-energy anomalies is higher than that in γ-Ce [28]. The broad peak corresponding to the valence-band states in $CePd_3$ appears deeper than that in Ce-metals,as seen from PES under off-resonance condition [29]. Therefore, the anti-bonding peak appears separately from the peaks due to the Fermi edge effects. The latter peaks are not so strong.

There are many ambiguities in the determination of the band parameters of the LCAO and LCAO-OPW scheme. Moreover, it is almost impossible to calculate the parameters quantitatively for compounds with complicated band structures, such as $CePd_3$. Systematic mehtods to calculate the hybridization matrix directly from the band calculation are necessary, and works on this problem based on the recent development [30] are in progress. The authors would like to express their sincere thanks to Professor A. Hasegawa for sending them his numerical data of band calculation. This work was greatly simplified by using the library subroutine TSPACE which was generously supplied by Professor A. Yanase, to whom the authors are grateful.

References

1. F.U. Hillebrecht, W. Gudat, N. Martensson, D.D. Sarma and M. Campagna: J. Magn. Magn. Mat. 47-48, 221 (1985), and references therein.
2. D.M. Wieliczka, C.G. Olson and D.W. Lynch:Phys. Rev. B29, 3028 (1984).
3. R.D. Parks, N. Martensson and B. Reihl: Valense Instabilities, eds. P. Wachter and H. Boppart (North-Holland, 1982) p. 239.
4. W. Gudat, M. Iwan, R.Pinchaux and F. Hulliger: Valence Instabilities, eds. P. Wachter and H. Boppart (North-Holland, 1982) p. 249.
5. O.Gunnarsson and K. Schonhammer:Phys. Rev. B28, 4315 (1983).
6. O. Sakai, H. Takahashi, M. Takeshige and T. Kasuya:Solid State Commun. 52, 997 (1984).
7. see for example, H. Kojima, Y. Kuramoto and M. Tachiki:Z. Phys. B Condensed Matter 54, 293 (1984).

8. O. Sakai, M. Takeshige and T. Kasuya:J. Phys. Soc. Jpn. 53, 3657 (1984).
9. A. Fujimori:Phys. Rev. B27, 3992 (1983).
10. recently, P.S. Riseborough hase studied the same effect, J. Magn. Magn. Mat. 47-48, 271 (1985).
11. M. Takeshige, O. Sakai and T. Kasuya:Proc. of the 5th Int. Con. on Crystalline Field and Anomalous Mixing Effects in f-Electron Systems, April, 15-18, 1985 Sendai, to be published in J. Magn. Magn. Mat.
12. O. Sakai, M. Takeshige, H. Harima, K. Otaki and T. Kasuya, in the same volume as ref. [11].
13. A. Hasegawa, J. Phys. C:Solid State Phys. 13, 6147 (1980).
14. O. Sakai, A. Yanase and T. Kasuya:J. Phys. Soc. Jpn. 42, 596 (1977).
15. H. Takahashi and T. Kasuya:J. Phys. C:Solid State Phys. 18, 2697, 2709, 2721, 2731, 2745 and 2755 (1985).
16. A. Hasegawa:J. Phys. Soc. Jpn. 54, 677 (1985).
17. H. Kitazawa, T. Suzuki, M. Sera, I. Oguro, A. Yanase, A. Hasegawa and T. Kasuya:J. Magn. Magn. Mat. 31-34, 421 (1983).
18. W.E. Pickett, A.J. Freeman and D.D. Koelling:Phys. Rev. B22, 2695 (1980).
19. L. Hodges, H. Ehrenreich and N.D. Lang:Phys. Rev. 152, 505 (1966).
20. F. Mueller:Phys. Rev. 153, 659 (1967).
21. The hybridization matrix is actually calculated by using the quantity, $v_\eta(\Lambda\mathbf{k}) = \sum_\lambda v_\eta(\lambda\mathbf{k})\langle\lambda\mathbf{k}|\Lambda\mathbf{k}\rangle + \sum_{\eta'}\sum_{\mathbf{h}(\neq 0)}\varepsilon_{\eta\eta'}(\mathbf{h})e^{i\mathbf{k}\cdot\mathbf{h}}\langle\eta'\mathbf{k}|\Lambda\mathbf{k}\rangle$, where $|\Lambda\mathbf{k}\rangle$ means the Bloch state of La-compounds.
22. K. Takegahara, H. Takahashi, A. Yanase and T. Kasuya:Solid State Commun. 39, 857 (1981).
23. Note that, the definitions of the quantities (7)~(10) are different from those in ref. [11] to fit them to the usual definition of the Fermion self-energies. We have relations, $\omega = -\varepsilon$, $\Sigma^m_\eta(z) = -\Sigma_\eta(-z)$, $\Sigma^m_{00}(z) = -\Sigma_{00}(-z)$, $\Sigma^m_0(z) = -\Sigma_0(-z)$ and $\Sigma^c(z) = -\Sigma_c(-z)$.
24. W.C. Martin, R. Zalubas and L. Hagen, Atomic Energy levels, The Rare-Earth Elements, National Bureau of Standards (Washington,1978).
25. H. Harima, K. Takegahara and T. Kasuya,in the same volume as ref. [11].
26. E. Wuilloud, H.R. Moser, W.D. Schneider and Y. Baer:Phys. Rev.B28, 7354 (1983).
27. O. Gunnarsson and K. Schonhammer:Phys. Rev.B31, 4815 (1985).
28. J.W. Allen, S.-J. Oh, I. Lindau, J.W. Lawrence, L.I. Johansson and S.B. Hagstrom:Phys. Rev. Lett. 46, 1100 (1981).
29. R.D. Parks, S. Raaen, M.L. denBoer, Y.-S. Chang and G.P. Williams:Phys. Rev. Lett. 52, 2176 (1984).
30. O.K. Andersen and O. Jepson:Phys. Rev. Lett.53, 2571 (1984).
31. In this paper we use the terminology α-La and γ-La as a analog of α-Ce and γ-Ce.

Part V

Theories for Periodic Systems and Heavy Fermions

Models for Intermediate Valence Systems and Kondo Lattices — Applications to Cerium and Ytterbium Compounds

A.C. Hewson, D.M. Newns, J.W. Rasul[+], *N. Read*[*], *and H.-U. Desgranges*
Department of Mathematics, Imperial College, London SW7 2BZ, U.K.

P. Strange
Department of Physics, Bristol University, Bristol, BS8 1TL, U.K.

1. Introduction.

For sometime now there has been a group of us at Imperial College (increasing in number but now dispersing) concerned with the theoretical problems of intermediate valency and Kondo lattices. Some of our recent results based on the 1/N expansion, the Bethe ansatz and mean field approach with applications to cerium and ytterbium will be presented here. However, by way of introduction I shall review some of the earlier work concentrating on its motivation and the way the early ideas have been developed and clarified with time.

Much of the motivation for the present work stems from the rather simple local fermi liquid theory put forward in 1980 [1]. This theory was the simplest phenomenological description we could conjecture which included the two basic features of rare earth intermediate valency, namely, the large total angular momentum degeneracy (2j+1) of the spin-orbit split f-states, and a narrow f-band or resonance at the Fermi-level, associated with the valence fluctuations. These seemed to be necessary ingredients to explain the high-temperature magnetic moments, and the low-temperature electronic properties such as the very large coefficients of specific heat γ. One of the merits of this approach was that it involved only two basic parameters, the position of the resonance ε_f and its width Δ. These are reduced to one if the f-occupation is known, by neglecting inter-site effects and making the theory a local or impurity one and then invoking the Friedel sum rule. Experimental evidence from alloying and neutron scattering suggested that the local approximation might be valid to a good approximation for many systems. The simplicity of the theory was appealing to the Amsterdam group (de Châtel, Mattens, De Boer) looking for suitable theoretical results to make a comparison with their measurements on a series of Yb compounds [2]. This theory gives a low-temperature maximum in the magnetic susceptibility $\chi(T)$ as is observed in Ce and Yb compounds. The low filling of the one-body resonance, which has a degeneracy 2j+1, by electrons in Ce (j=5/2) and holes in Yb (j=7/2) means that the resonance must span the Fermi-level asymmetrically, with the peak of the resonance well above the Fermi-level. The positive curvature of the density of states at the Fermi-level then leads to a positive T^2 contribution to χ at low temperatures, and a peak as χ eventually falls off at higher temperatures in a Curie-like law. This gave a reasonably consistent picture for the thermodynamic properties of YbCuAl, studied in detail by Mattens [2]. The low-temperature magnetic isotherm M(H) in an applied field H of YbCuAl has a marked positive curvature which could not be explained, however, without invoking a field dependence for the

[+] Present address: Department of Theoretical Physics, 1, Keble Road, Oxford.

[*] Present address: Department of Physics, Brown University, U.S.A.

chemical potential. Surprisingly, YbCuAl was found to be almost integral valent, implying that it is more of a Kondo lattice than a true mixed valence system. The quasi-particle resonance of the local fermi liquid theory in the almost integral valence regime could then, perhaps, be identified with the Kondo resonance [3] (it would also mean that this resonance should be asymmetric about the Fermi level for $j > \frac{1}{2}$). This carried the paradoxical implication that the many-body Kondo resonance could be treated, at least approximately, as a simple density of states effect at low temperatures. The aim, therefore, was to see whether or not the theory had any basis starting from a microscopic model.

Taking the local, almost integral valent, picture of YbCuAl as basically correct, we considered the Coqblin-Schrieffer or SU(2+j) Kondo model should be the appropriate starting point for developing a fully microscopic theory for this system. The success of the Bethe ansatz approach in diagonalizing the Kondo model to give exact results for the thermodynamic properties [4] lead to the interest in its extension to the degenerate SU(2j+1) model [5]. We were particularly interested to see whether the peak in $\chi(T)$ YbCuAl could be explained within the j=7/2 model. The ground state $\chi(H)$ or magnetic isotherm M(H) for this model is an easier quantity to calculate, as it requires the solution of a finite set of linear integral equations (an infinite non-linear set is involved in the calculation of the temperature dependence). Arguing that the effect of increasing magnetic field is similar to increasing temperature, both tending to release the moment from the magnetic screening cloud of the conduction electrons, we were led to speculate that a peak in $\chi(H)$ would indicate a peak in $\chi(T)$ as well as explain the positive curvature in the M(H) curve seen in YbCuAl. Neither of these is present in the solutions for the $j=\frac{1}{2}$ Kondo model, $\chi(T)$ in this case decreases at all temperatures and M(H) have a negative curvature for all fields so a qualitative change in the behaviour of the model with j would be required. They also do not occur in the $j=\frac{1}{2}$ local fermi liquid theory, as in this case the resonance is half full. Our calculations for M(H) [6] and those of Rajan [7] for $\chi(T)$ showed the emergence of both of these features for j=3/2, and in a much more pronounced form for higher j values. A detailed comparison of the results of the j=7/2 model for YbCuAL for M(H), $\chi(T)$ and the specific heat C_V is shown in figure 1. The overall agreement is remarkably good, particularly as there is only one free parameter, the Kondo temperature T_k, which is fitted to give the T=0 susceptibility. The curvature of M(H), which involves no adjustable parameter agrees well, so do the positions of the peaks in $\chi(T)$ and C_V though their magnitude is somewhat greater than those observed experimentally. Theoretical evidence considered later suggests that when crystal fields are included the agreement may be even better. The reduction of the high-temperature magnetic moment for $T > T_k$ (Figure 1(iv) compared with the free ion value is explained by the residual logarithmic terms from the conduction electron screening, which are negligible only in the extreme high-temperature range $\ell n(T/T_k) \gg 1$. A reduction in moment at high temperatures is to be expected in all Kondo lattice systems where the fully degenerate 2j+1 model is valid, even if the low-temperature properties are controlled by a model of low degeneracy because of significant crystal-field splitting. We conclude that the local j=7/2 model can explain the main features in the thermodynamic behaviour of YbCuAl at all temperatures.

A direct connection between the local fermi liquid theory and the Coqblin-Schrieffer model emerged on applying a large N (=2j+1) approach within the functional integral formalism [8]. The Coqblin-Schrieffer model is described by Hamiltonian,

$$H = \sum_{k,m} \varepsilon_k c^+_{km} c_{km} + J \sum_{k,k'} c^+_{fm} c_{fm'} c^+_{k'm'} c_{km} , \qquad (1.1)$$

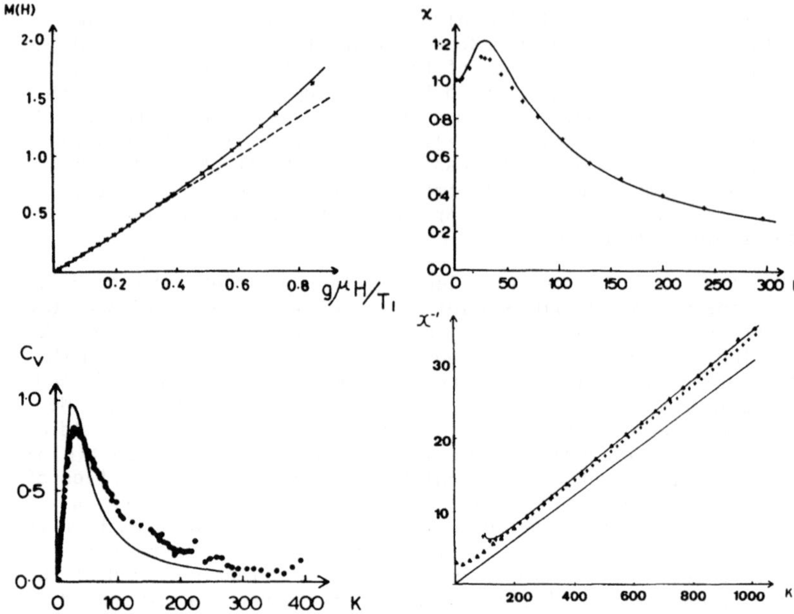

Fig. 1 Results for the j=7/8 Coqblin-Schrieffer model with $T_K = 66^0K$ compared with experimental results for YbCuAl (i) The magnetic isotherm M(H) in units of gμB with g = 8/7 compared with experimental results (crosses) taken at 4.2^0K. (ii) The impurity susceptibility scaled to the zero-temperature value compared with experimental results (crosses). (iii) The specific heat compared with experimental results (dots) for the specific heat difference between YbCuAl and LuCuAl. (iv) The asymptotic form of $\chi(T)^{-1}$ (crosses), applicable for $T \gg T_K$, compared with the experimental results (triangles) for YbCuAl. The full curve corresponds to an Yb ion with $f^{13.025}$. The straight line corresponds to the Curie law for the free ion (j=7/2, g=8/7, f^{13}) $\chi * 10^6$ is in units of m^3 per mole Yb atom.

where c^+_{km}, c_{km} are creation and annihilation operators for a conduction electron for a partial wave corresponding to an energy ε_k and a total angular momentum j,m. The corresponding operators for the localized f-state are c^+_{fm}, c_{fm} and J is the coupling constant. The f occupation is subject to the constraint $\sum_m c^+_{fm} c_{fm} + n_f = 1$.

In applying the functional integral approach, the interaction term in (i) is written as $J|\sum_m c^+_{fm} c_{km}|^2$, and so applying the Hubbard-Stratonovitch trick one is left to consider a non-interacting model with a coupling term term $\xi \sum_m c^+_{fm} c_{km} + \text{h.c.}$ to a time-dependent field ξ. Read and Newns [8] pursued this approach taking into account the f-occupation constraint by an integration over a further field λ. The simplest functional integral approximation is to treat the fields as static ones and estimate the value of the functional integral in a saddle-point approximation. Such an approximation corresponds to a mean field or generalized Hartree-Fock approach in which anomalous averages $< c^+_{fm} \, c_{km} >$ are calculated self-consistently. In this approximation the model corresponds to a non-interacting Anderson model with an effective f-level position $\tilde{\varepsilon}_f$ (dependent on λ) an effective hybridization (dependent on ξ) leading to a resonance in the one-particle density of states with an effective width $\tilde{\Delta}$. This then reproduces the local fermi liquid theory, apart from the fact that the resonance width $\tilde{\Delta}$ is not an independent parameter, but depends on the f-level position, temperature and field. The only qualitative change from the local fermi liquid theory that results from the self consistency is that the low-field curvature in the M(H) is positive in this case for $j > 1$ [9] as in the exact Bethe ansatz results. To prove that the mean-field result corresponds to the leading asymptotic term for large N, taken in such a way so that T_k remains finite, Gaussian fluctuations were taken into account. The Wilson or Sommerfeld 'χ/γ', $R = 4\pi^2 \chi / g^2 \mu^2 (N^2-1) \gamma$, known to be N/N-1 from Bethe ansatz [5] and fermi liquid calculations [10] corresponds to that of the non-interacting or local model in the limit $N \to \infty$. The leading order correction to the mean-field result, on taking into account the Gaussian fluctuations, was 1/N in agreement with the exact result to this order.

This left a number of questions to be resolved. There was some scepticism as to whether the large N expansion could be applied in the integral valence regime i.e. to the Coqblin-Schrieffer model. Calculations of Ramakrishnan [11] using the Keiter-Kimball diagrammatic perturbation expansion for the degenerate, U infinite, Anderson model indicated a breakdown on the approach to integral valence due to the anomalous Kondo scattering terms that produce the $\ell n \, T$ terms at high temperatures. There was also doubt expressed in some quarters as to whether the 1/N expansion had any validity in general, and also, even if valid, whether it would be a sufficiently good approximation to apply the results to real physical systems if only the first one or two terms in the expansion could be calculated in practice. This motivated two of us (Rasul and Hewson), having calculated systematically all the 1/N corrections for the degenerate Anderson model using the Keiter-Kimball technique, to compare them with the exact Bethe ansatz results recently derived [12] and so test the validity of the approach for all parameter regimes.

To make any comparison between Bethe ansatz and other results for degenerate (U-infinite) models, a technical problem of relating cut-offs must be solved. The Bethe ansatz is applied to the linear dispersion model, which has no cut-off, but after diagonalization, cut-off D' must be imposed on the charge excitations of the resulting charge excitation spectrum to obtain physically meaningful results. In the conventional model the cut-off on these excitations is due to the finite width 2D of the conduction band, and is equal to D for both electrons and holes for a half-filled band. As these cut-off procedures differ, one cannot equate D to D'. However, assuming a relation of the form $D = f_N(D')$ the form of f_N was deduced for the N-fold degenerate Anderson model by comparing terms in the expansion of the Bethe ansatz results and those of the conventional model in powers of Δ/ε_f. The resulting relation is

$$\left(\frac{\pi D}{N}\right)^{1-1/N} \leftrightarrow N^{1/N} \left(\frac{D'}{e}\right)^{1-1/N} \tag{1.2}$$

with this result a complete comparison can be made in all parameter regimes, and the Bethe ansatz results expressed in terms of the parameters of the conventional model [13]. The main points of interest to emerge are (i) the leading term in the 1/N expansion for the ground-state properties such as n_f, the f occupation, or χ, the susceptibility, are identical to the exact Bethe ansatz results for $N \to \infty$ in all parameter regimes, both Coqblin-Schrieffer and mixed valence, (ii) the exact ground-state results can be expressed as a 1/N expansion, the order 1/N results being in agreement with those calculated by diagram summation. The first two terms in the 1/N expansion give a relation between χ and n_f which is asymptotically correct as $n_f \to 1$ or $n_f \to 0$ and constitutes an excellent approximation over the whole regime $0 < n_f < 1$ for values for N = 6,8, and are very good even for N as low as 2 [13]. A comparison of some of the leading and next leading order results in 1/N with the corresponding exact ones is shown in figures 2(i) and 2(ii) for N = 6. The order 1/N correction is significant mainly in the approach of integral valence, where it gives a leading derivation of n_f from unity of the order of $\frac{\Delta}{N\varepsilon_f}$ rather than the exponential deviation $e^{-\pi\varepsilon_f/N\Delta}$ of the leading order result. This changes the asymptotic form for finite N of χ versus $n_f \to 1$ from the leading order result as shown in figure 2(ii) for N = 6.

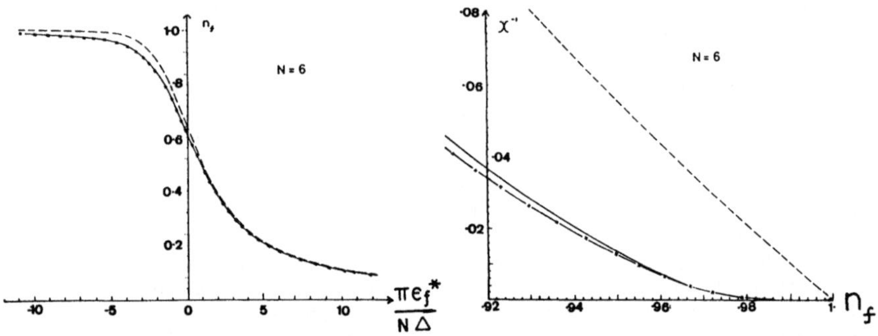

Fig. 2 Exact Bethe ansatz (full curve) compared with leading order (crosses) results of the 1/N expansion for the Anderson model with N=6 (i) The f-occupation n_f versus $\pi\varepsilon_f^*/N\Delta$ where ε_f^* is the effective f-level position. (ii) $1/\chi$ versus n_f in the approach to integral valence $n_f \to 1$.

One further result that emerged from this comparison, and also from earlier calculations on the Coqblin-Schrieffer model, was a general expression for the Wilson number W(N) which relates the ground-state susceptibility χ and the Kondo temperature T_k, as defined from the high-temperature series for $\chi(T)$, for the integral valence impurity models,

$$\chi = W(N)\,\frac{(g\mu)^2\, j\,(j+1)}{3T_k}\,, \qquad \text{where } W(N) = \frac{e^{(1+c-3/2N)}}{2\pi\,\Gamma(1+1/N)}\,, \tag{1.3}$$

where c is Euler's constant. This is a universal number for the model relating the behaviour in quite different regimes, the low-temperature or fermi liquid regime, where the moment is quenched, and the high-temperature regime where the moment is asymptotically free. For the Kondo model case $j=\frac{1}{2}$ (1.3) simplifies to $W(2) = e^{c+1/4}/\pi^{3/2}$. The precise form of T_k depends on whether one is in the regime where the Schrieffer-Wolff transformation to the Coqblin-Schrieffer model is valid when the f-level lies below the conduction band, or the regime $D > -\varepsilon_f \gg N\Delta$ corresponding to the near integral valence within the Anderson model. Due to use of the linear dispersion model, these regimes have to be considered separately within the Bethe ansatz approach, but are encompassed within the single 1/N calculation.

A generalization of the functional integral and mean-field approach to the Anderson model proved possible by use of the Coleman boson representation [14] to restrict the f-occupancy to the subspaces with $n_f = 0$ and $n_f = 1$. In this representation the infinite U Anderson model corresponds to the Hamiltonian,

$$H = \sum_{k,m} \varepsilon_k c^+_{km} c_{km} + \sum_{k,m} (v_{km} c^+_{fm} c_{km} b + v^*_{km} c^+_{km} c_{fm} b^+) + \varepsilon_f \sum c^+_{fm} c_{fm} \tag{1.4}$$

with the restriction to the subspace $\sum_m c^+_{fm} c_{fm} + b^+ b = Q = 1$, where b^+, b are the creation and annihilation operators for the Coleman boson, and v_{km} the hybridization matrix element of the f-level and conduction states related to the noninteracting (U=0) resonance width Δ by $\Delta = \pi\rho_0|\bar{V}|^2$ where ρ_0 is the conduction band density of states. The constraint $Q = I$ can again be imposed by an integration over an additional field λ as in the calculation for the Coqblin-Schrieffer model.

The static of mean-field approximation corresponds to assuming a finite expectation value for the boson operator in $\langle b \rangle$, which is then calculated variationally. Denoting $|\langle b \rangle|^2 = z$, then the subspace restriction gives $z = 1 - \langle n_f \rangle$, where $n_f = \sum_m c^+_{fm} c_{fm}$. Minimization of the free energy with respect to z and λ leads to self-consistent equations for the renormalized resonance width $\tilde{\Delta} = |z|\Delta$, and renormalized f-level position $\tilde{\varepsilon}_f = \varepsilon_f + i\lambda$,

$$\frac{\tilde{\Delta}}{\Delta} = 1 - \frac{N}{\pi} \tan^{-1}\left(\frac{\tilde{\Delta}}{\tilde{\varepsilon}_f}\right) \tag{1.5}$$

and

$$\ell n\left\{\frac{\tilde{\varepsilon}_f^{\,2} + \tilde{\Delta}^2}{D}\right\} = \frac{\pi(\tilde{\varepsilon}_f - \varepsilon_f)}{N\Delta} \tag{1.6}$$

at T = 0. This very simple approximation reproduces the leading order in 1/N results obtained by diagrammatic summation [11, 13 and references therein]. It goes over to the earlier mean-field theory for the Coqblin-Schrieffer model on the approach to integral valence $\langle n_f \rangle \to 1$. Full details are given in the paper of Read and Newns [15]. Once the mean-field approximation has been made, with the addition of the constraint field, Hamiltonian [14] corresponds to the non-interacting Anderson model (U=0) with a renormalized resonance width $\tilde{\Delta}$ and position $\tilde{\varepsilon}_f$. It is essentially equivalent to the simple local fermi-liquid theory with the additional selfconsistency condition on the resonance width $\tilde{\Delta}$ [15]. In the approach to integral valence the quasi-particle resonance $\tilde{\varepsilon}_f$ lies above the Fermi level, though the actual f-state lies well below. This is required to satisfy the Friedel sum rule, essentially equation (1.5), when the resonance is only fractionally occupied. This resonance is identical with the Kondo many-body resonance, asymmetric with respect to the Fermi-level when N is large, and agrees with the earlier calculations for the Coqblin-Schrieffer model. Thus a simple non-interacting resonance, albeit with selfconsistently determined parameters, can precisely mimic the many-body resonance in the limit of large N. The fact that as $\langle n_f \rangle \to 1$ the actual f-levels lie well below the Fermi level and there can be little f-charge character in the excitations in the resonance is apparent when the true one-electron f-Green's function is calculated. In the mean-field approximation this Green's function $\langle\langle b c_f^+ : b^+ c_f \rangle\rangle$ is $|\langle b \rangle|^2 \langle\langle c_f^+ : c_f \rangle\rangle$, and hence is equal to the quasi-particle Green's function multiplied by the weight function $z = 1 - \langle n_f \rangle$, which is very small on the approach to integral valence.

The calculation of the leading order fluctuations about the mean-field result presents some technical problems due to infra-red divergences. These problems have been resolved by Read [16]. The mean-field results for both the Coqblin-Schrieffer and the Anderson model involve a breaking of symmetry and anomalous expectation values. For a local model such broken symmetry states can only exist in the limit $N \to \infty$. For finite N the symmetry cannot be broken, so inclusion of the fluctuations about the mean-field result must restore the symmetry. In taking the expectation value $\langle b \rangle$ for the Anderson model (4) the phase symmetry of the bose operators is broken, and the infra-red divergences occur due to the soft excitation corresponding to a change of this phase. Read has shown that if a low momentum cut-off Λ is taken, the fluctuations lead to a logarithmic contribution to the expectation value of $|\langle b \rangle|^2$, i.e.

$$|\langle b \rangle|^2 \sim (1 - n_f)\left(1 - \frac{n_f^2}{N} \ln \Lambda\right) \tag{1.7}$$

On the assumption that higher-order terms will give an exponential series (c.f. x-ray edge problem)

$$|\langle b \rangle|^2 \sim c \Lambda^{n_f^2/N} \tag{1.8}$$

where c_2 is some constant (equal to $1 - n_f$ to leading order in N) then as $N \to \infty$ $|\langle b \rangle|^2 \to 0$ and the symmetry is restored by the fluctuations, in accord with the general theorems. In the calculation of physical quantities the infra-red divergent terms cancel, and the results for n_f and χ on taking into account the Gaussian fluctuations agree with the calculations to order 1/N of Rasul and Hewson [13].

It might be worthwhile to make a few comments on the relative merits of the Keiter-Kimball diagrammatic and functional integral approaches to the 1/N calculations. Both approaches have proved extremely useful and we have derived much benefit from not relying exclusively on one or the other. In the Keiter-Kimball method the restriction to the subspace $n_f = 0$, 1 is built into the formalism and retained at all stages of the calculation. The terms in the expansion are readily classified in their N dependence (note in the Anderson model $N\Delta$ is taken to be of zero order in 1/N, this ensures finite T_k as $N \to \infty$). Calculations extending this approach to the two-impurity problem and the spectral density are described in the next section. In the functional integral approach the classification of terms in their dependence on N is not easy, though it has recently been solved by Read [11]. The restriction to the subspaces $n_f = 0$,1 is not strictly adhered to in the mean-field calculation. It is sufficient in practice to treat the field as a chemical potential or Lagrange multiplier, any errors due to this are higher order in 1/N. The simple mean-field picture that then emerges in the large N limit from this approach is appealing, and is used as the starting point for tackling the challenging lattice problem in a later section.

An interesting limitation of the Keiter-Kimball approach is that so far it has not proved possible to derive a 1/N expansion by direct application to the Coqblin-Schrieffer model. The Coqblin-Schrieffer results for the ground state have been derived by using the Anderson model in the regime with the f-level below the conduction band and then using the Schrieffer-Wolff transformation to relate the parameters to the Coqblin-Schrieffer model (i.e. $J = \Delta/\varepsilon_f$). The difficulty in obtaining the expansion more directly is due to the fact that one works entirely within the subspace $n_f = 1$ and there is no initial singlet state that can be renormalized to become the singlet ground state of the interacting model. It is only by breaking this restriction that the 1/N expansion is derived for the Coqblin-Schrieffer model by the functional integral approach described earlier.

2. The 1/N Expansion for the two Impurity Problem

Part of the motivation for developing 1/N expansion and mean-field technique as described in the previous section lies in the ease in which they may be extended to treat problems involving many impurities. Of these, one may distinguish two important lines of enquiry: for periodic systems, such as the Kondo and Anderson lattices at very low temperatures, coherence effects will be important. These are most easily treated by extending the mean-field approach, as will be seen later. The other aspect concerns the competition between magnetic interactions between the impurities and the kind of screening of local moments described in the previous section. This will be treated by examining the two impurity Anderson model from the standpoint of the diagrammatic 1/N procedure. Details of the lowest-order diagrammatic contributions can be found in an earlier paper [18] and we shall discuss results before proceding to examine higher-order corrections. For two degenerate Anderson impurities separated by a distance $\underline{R}$ the leading intersite contribution to the partition function, corresponding to figure 3(i) may be written

$$-Z_{of}^{-2} \sum_{m=1}^{N} T \sum_{\omega_n} \{ |v|^2 \sum_k G_k(i\omega_n) \exp(\underline{ik}\cdot\underline{R}) \}^2 I_m(\omega_n)^2 \tag{2.1}$$

where Z_{of} is the bare f-electron partition function, $G_k(i\omega_n)$ is the bare conduction electron Green's function and the vertex term $I_m(\omega_n)$ may be written

$$I_m(\omega_n) = \frac{1}{2\pi i} \int_C \frac{dz\, e^{-\beta z}}{(z - \varepsilon_{fm} + i\omega_n)(z - P(z))} \tag{2.2}$$

where $P(z) = |v|^2 \sum_k \frac{\varepsilon_{km}}{(z - \varepsilon_{fm} + \varepsilon_k)}$, the remaining notation being as in

[13]). The function $I_m(\omega_n)$ is dominated by pole contributions, each of which may be written as a product of an exponential factor and an effective t-matrix. In the Kondo regime $-\varepsilon_f \gg N\rho_0|v|^2$ the relative magnitudes of the pole contributions differ by a term $e^{-T_k/T}$ where $T_k = D\exp(-\pi|\varepsilon_f|/N\Delta))$ is the pole at $z = \varepsilon_{fm} - i\omega_n$ survives, and we obtain, after factorising out the leading order single impurity partition functions, a contribution to the free energy equal to:

$$\sum_{m=1}^{n} T \sum_{\omega_n} J_{eff}^2(m, \omega_n, T) \left(\sum_{\underline{k}} G_{\underline{k}}(i\omega_n) e^{\underline{ik}\cdot\underline{R}}\right)^2 = I(R) \tag{2.3}$$

where

$$J_{eff}(m,\omega_n,T) = \frac{\Delta}{\pi(\epsilon_f - i\omega_n - P(\epsilon_f - i\omega_n))} \sim \frac{N\Delta}{\pi|\epsilon_f|} \quad (\text{for } T \gg T_k).$$

This is the usual form of the RKKY interaction between local moments. The integral I(R) is of the order of $D(\rho_0 J)^2 \frac{\cos 2k_F R}{(k_F R)^3}$ and for a magnetic ordering in the ground state should exceed $2T_k$, the binding energy of the two Kondo singlets. This criterion corresponds to that given in references [19,20] for the lattice problem.

For the low-temperature range $T \ll T_k$ the lowest pole of (2.2) corresponding to the lowest root $z_c = P(z_c)$ dominates and we may write

$$I(\omega_n) = e^{-\beta z_c} t_m^0(i\omega_n) = e^{-\beta z_c} \frac{\Delta}{\pi} \frac{\left[1 + \frac{N\Delta}{\pi T_k}\right]^{-1}}{(i\omega_n - T_k)} \qquad (2.4)$$

where $t_m^0(i\omega_n)$ is the t-matrix for an effective singlet at each site. This is the single impurity f-Green's function, which at this level of approximation corresponds to a body excitation energy T_k and a z-factor $(1 + \frac{N\Delta}{\pi T_k})^{-1}$ which is equal to $1 - \langle n_f \rangle$ in leading order. The reduction in ground state energy ΔE_{12} due to the interaction of the two singlet states is given by

$$\Delta E_{12} = \begin{cases} \dfrac{-\pi \epsilon_F n_f^2 \cos(2k_F R)}{4N(k_F R)^3} & R \gg R_A \\[2ex] \dfrac{-\pi T_A n_f^2 \sin(2k_F R)}{4N(k_F R)^2} & R \ll R_A \end{cases} \qquad (2.5)$$

where the characteristic length $R_A = \frac{E_F}{k_F T_A}$, where $T_A = T_K$ in the Kondo regime, but is also valid in the mixed valence regime where $T_A \sim \Delta$ [13].

At the next order in our expansion procedure we can separate the resulting diagrams into (i) those involving 4 intersite Green's functions (higher-order in v) and (ii) those involving corrections to the single-site parts of equation (2.1). In the dominant pole approximation ($T \ll T_k$) the form (ii) gives terms corresponding to a renormalized hybridization to an f-quasi-particle as in the mean-field approach. For a lattice these terms lead to a renormalized band structure, which is discussed in detail in the next section. Here we shall concentrate on the latter (ii), confining our attention to the lowest pole contribution to the partition function, to obtain the zero-temperature limit.

The relevant diagrams are shown in figures 3(ii) - (iv). The diagram figure 3(ii) corresponds essentially to renormalizations of the resonance position T_k and the z factor in (2.4). The contributions from diagrams in figure 3(iii) (a) and (b) we denote by $\Phi^{(a)}$ and $\Phi^{(b)}$ respectively.

They are given by

$$\Phi^{(a)}(i\omega_n) = -\frac{T_k}{N(i\omega_n - T_k)^2}\int_0^d dx\, \tilde{J}[(x+1)T_k + i\omega_n]$$

and

$$\Phi^{(b)}(i\omega_n) = \frac{-1}{\mu N T_k}\int \frac{dx}{(x+1)^2}\,\tilde{J}[(x+1)T_k + i\omega_n]$$

where $\tilde{J}(y) = [\ln(\frac{y}{T_k})]^{-1}$.

The sum $\sum_{\underset{\sim}{k}} \underset{\sim}{G}(i\omega_n)\, e^{i\underset{\sim}{k}\cdot\underset{\sim}{R}}$ in (2.1) provides an exponential frequency cut-off inversely proportional to $\underset{\sim}{R}$. The contributions $\Phi^{(a)}$ and $\Phi^{(b)}$ are most significant for $\omega \gg T_k$ where they contribute to the interaction between the two impurity sites at short distances apart. In this frequency range

$$\Phi^{(b)}(\omega_n) \sim \frac{\pi}{N\Delta}\,\tilde{J}[T_k + i\omega_n] \qquad (2.6)$$

and that of $\Phi^{(a)}$ is the same as the contributions from figure 3(ii). The contribution from figure 3(iv) is higher order in $\tilde{J}$.

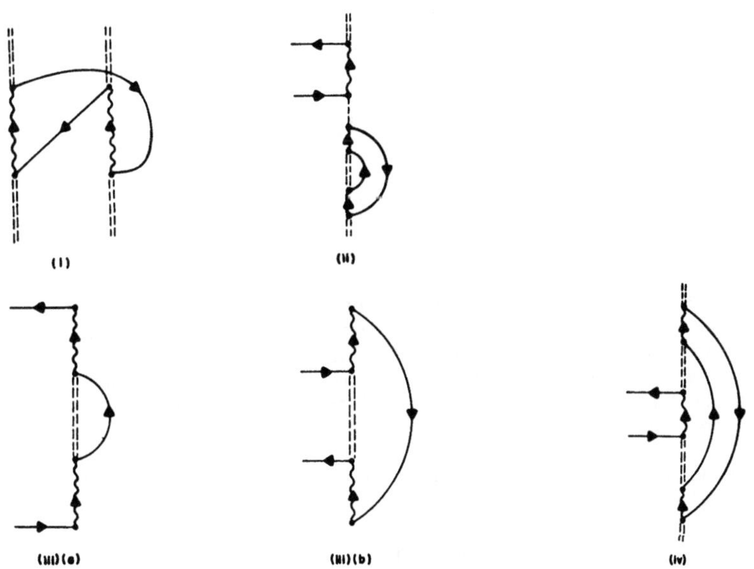

Fig. 3 Terms arising in the perturbation theory for the two-site partition function. The double dashed line corresponds to a singlet propagator dressed to leading order in 1/N. (i) is the lowest-order inter-site diagram. (iii) (a) (b) and (iv) correspond to higher-order dressing of a single-site propagator in (i).

When substituted into the single-site part of (1.1), these terms lead to different qualitative behaviour to that obtained in [18]. In particular, inserting a $\Phi^{(b)}$ for both impurity sites leads to a term proportional to I(R) at short distances, but with an extra factor $1/N^2$. It can be interpreted as due to an interaction between the moments at the two sites when they are virtually excited out of the screened singlet state. At short distances and smaller values of N this may disrupt the picture of coherently scattering singlet states. The effects of these terms is to be investigated more fully.

Figures 3(ii) to (iv) correspond to 1/N corrections to the f-electron Green's function or conduction electron t-matrix if we make the analytic continuation $i\omega_n \to \omega + i\delta$. The imaginary part of the contributions from figures 3(iii) (a) and (b) are

$$\mathrm{Im}\,\Phi^{(a)}(\omega + i\delta)\ \frac{\Delta\Theta(\omega)}{(\omega - T_M)^2(1+\mu)}\cdot\left\{\frac{-1}{(1+\mu)} - \Theta(\omega - T_A)\, I^+\left(\mu, \frac{\omega}{T_M}\right)\right\}$$

$$\mathrm{Im}\,\Phi^{(b)}(\omega + i\delta) = \frac{\Delta\Theta(-\omega)}{(1+\mu)}\left\{\frac{-1}{(1+\mu)(\omega - T_A)^2} - \Theta(|\omega| + T_A)\, I^-\left(\mu, \left|\frac{\omega}{T_A}\right|\right)\right)$$

where $\Theta(x)$ is the step function, $\mu = \dfrac{N\Delta}{\pi T_A}$, and

$$I^{\pm}(\mu, y) = \int_0^{\ln y} \frac{e^{\pm t}\, dt}{((e^t - 1 - y + \mu \ln|e^t - y|)^2 + \mu^2\pi^2)}$$

The contribution of these terms to the f-electron Green's function corresponds to the results of Gunnarsson and Schönhammer [22] to the valence and BIS spectra to leading order in 1/N.

Defining an imaginary part for a self-energy Σ_f for the f-Green's function to first order in 1/N via

$$\mathrm{Im}\,\Sigma_f(\omega + i\delta) = (\omega - T_A)^2(1+\mu)\{\mathrm{Im}(\Phi^{(a)} + \Phi^{(b)})\} \tag{2.7}$$

we find a constant part corresponding to the resonance at T_A together with threshold contributions beginning at $\omega = \pm T_A$. When evaluated at the Fermi-level (2.7) is $-\Delta/(1+\mu)$. Since the leading order self-energy and z factor are given by T_A and $(1+\mu)^{-1}$ $(= 1 - \langle n_f \rangle)$ respectively [13] we have that $\mathrm{Im}\,G_f^{-1}(\omega + i\delta) = \Delta$. This is valid for both the mixed valent and Kondo regimes and agrees with the exact fermi liquid sum rule (see Müller-Hartmann [23]). This suggests that the non-analyticity of the NCA is a spurious feature arising from the self-consistent solution of the integral equations.

A further threshold contribution $\Phi^{(c)}$ starting at $\omega = T_A$ arises from diagram 3(iv). For large $\omega/T_A = y$ this can be written as

$$\mathrm{Im}\,\Phi^{(c)}(\omega + i\delta) = \frac{-\mu^2\pi}{(1+\mu)NT_A}\ [y - 1 + \mu \ln y]^{-2}$$

As μ is varied we can see a cross-over from y^{-2} behaviour in the mixed valence regime to $(\ln y)^{-2}$ in the Kondo regime. Its contribution to the non-spin flip t-matrix in the local moment regime is given by

$$\mathrm{Im}\, t_c(\omega + i\delta) = |V|^2\, \mathrm{Im}\, \Phi^c(\omega + i\delta) = \frac{-\pi}{N^2 \rho_0} \ln^{-2}\left(\frac{\omega}{T_k}\right)$$

which for high energies has the same energy dependence as obtained from earlier approaches to the Kondo problem [24].

3 Mean-Field Calculations for the Lattice: Results for $CeSn_3$.

There are two main schools of thought in dealing with f-electrons in rare earth mixed valence lattices, Kondo lattices and metallic actinides compounds. One is the model Hamiltonian approach using one of the impurity type models such as the infinite U Anderson model (1.4), discussed earlier, generalized to the lattice. The second approach is to calculate the self-consistent band structure using the local density approximation (LDA). This approach has been applied successfully to the calculation of the ground-state properties of transition metals, leading to reasonable Fermi surface cross-sections, effective masses and band structure. In principle the ground-state properties can be calculated exactly via a density functional approach, but the local density approximation to this, based on a uniform electron liquid, may not be adequate for dealing with the tightly bound f-electrons, and hence, in practice, the LDA may not be good enough to describe all ground-state properties. There is no rigorous reason for the density functional to give the correct excitation spectrum. The two approaches are not easy to reconcile, the LDA gives a wide strongly hybridized f-band in contrast with the narrow peaks in the density of states which emerge from the model Hamiltonian approach and required to interpret the experimental data. Here we modify the LMTO (linear muffin orbital) method for carrying out the LDA calculations as a way of carrying out mean-field calculations for the lattice in line with the approach used for the impurity problem (Strange and Newns [25]).

The infinite U Anderson model in the Coleman boson representation generalized to describe a lattice is

$$H = \sum_{\underline{k}} \varepsilon_{\underline{k}} c^+_{\underline{k}} c_{\underline{k}} + \varepsilon_f \sum_i c^+_{fi} c_{fi} + \sum_{m,i,\underline{k}} (V_{\underline{k}m} c^+_{fim} c_{\underline{k}} b_i e^{i\underline{k}\cdot\underline{R}_i} + \text{h.c.}) \tag{3.1}$$

The Block states $|\underline{k}\rangle$ are for a lattice with no 4f states and $\underline{k}$ includes the band index, $|i,m\rangle$ are the 4f states for a unit cell i with position vector $\underline{R}_i$, and $V_{\underline{k}m}$ is the coupling matrix element of the 4f states to the rest of the system. The restriction to the subspaces $n_{fi} = 0,1$ for all i corresponds to imposing the constraint $Q_i = n_{fi} + b^+_i b = I$. Pursuing the mean-field approach as described earlier for the impurity problem, the Bose operators b_i and b^+_i are replaced by their expectation values, i.e. $\langle b^+_i \rangle = \langle b_i \rangle = z^{\frac{1}{2}}$ where $z = 1 - \langle n_f \rangle$, $\langle n_f \rangle$ being the average occupation per site. On incorporating the constraint $Q = I$ via a Lagrange multiplier $\tilde{\varepsilon}_f - \varepsilon_f$ the mean-field Hamiltonian, which describes a renormalised single-particle model, takes the form

$$H = \sum_{\underline{k}} \varepsilon_{\underline{k}} c^+_{\underline{k}} c_{\underline{k}} + \tilde{\varepsilon} \sum_i c^+_{fi} c_{fi} + z^{\frac{1}{2}} \sum_{mi\underline{k}} (V_{\underline{k}m} c^+_{fim} c_{\underline{k}} e^{i\underline{k}\cdot\underline{R}_i} + \text{h.c.})$$

$$+ \sum_i (\varepsilon^*_f - \varepsilon_f)(z - 1) \tag{3.2}$$

The values of $\tilde{\varepsilon}_f$ and z are then determined by requiring the free energy corresponding to (3.2) to be stationary. The mean-field Hamiltonian corresponds to summation of the diagrams discussed in section 2 which are dressed only with the leading order in 1/N single-site insertions, and then evaluated for $T << T_A$ in the dominant pole approximation. This Hamiltonian may then be compared with the LMTO Hamiltonian [26] in its simplest form, which at Bloch vector $\underline{k}$ (neglecting spin-orbit terms) is

$$H^{\underline{k}}_{L'L} = c_\ell \, \delta_{L'L} + \frac{S}{2}\Phi_{\ell'}(-) \, [S^{\underline{k}}(1-\gamma S^{\underline{k}})^{-1}]_{L'L} \, \Phi_\ell(-) \ , \qquad (3.3)$$

where $L = (\ell, m)$, S = atomic sphere radius, $S^{\underline{k}}_{L'L}$ are the structure constants, c_ℓ is the effective centre of the ℓth band while $S\Phi_\ell$ is a band width determining the potential parameter for the ℓth band. The structure matrix $S^{\underline{k}}$ depends only on the crystal structure of the system.

It will be assumed that the non-f electrons are adequately described by the LDA, assuming a definite f-charge distribution so the potential may be defined, and that the uncertainty of the LDA lies chiefly in the dynamics of the f-electrons. This will be modified by changing the parameters c_3, $\Phi_3(-)$, which is the only way the f-electron dynamics enters, to renormalized values $\tilde{c}_3$ and $\tilde{\Phi}_3(-)$, in analogy with the derivation of (3.2). The shift $\tilde{c}_3 - c_3$ is analogous to $\tilde{\varepsilon}_f - \varepsilon_f$ and $z^{\frac{1}{2}}$ is analogous to the ratio $\tilde{\Phi}_3(-)/\Phi_3(-)$, so we then expect

$$[\tilde{\Phi}_3(-)/\Phi_3(-)]^2 = 1 - \langle n_f \rangle \ . \qquad (3.4)$$

This can only be expected to be approximate as it is difficult to define $\langle n_f \rangle$ uniquely within the LMTO formalism, because in a material with several atoms per unit cell we may expand an $\ell = 2$ spherical harmonic on one site about a second site and this will give some $\ell = 3$ character in general.

This procedure has been carried out for the mixed valence compound $CeSn_3$, including the spin-orbit coupling in the whole self-consistency process, both with the LDA formalism and in the form analogous to (3.2) where the f-electron dynamics have been modified. The LDA results are broadly in agreement with the earlier calculations of Koelling [27] and Yanase [28]. There is a tall f-peak just above the Fermi level with a band width of about 2ev (see figure 4(i)). The Fermi energy falls in the tail of the 4f peak giving a density of states at $\rho(E_F)$ at the Fermi level, in reasonable agreement with previous calculations but in very poor agreement with $\rho(E_F)$ deduced from specific heat measurements. The calculated Fermi surface results are similar to those of Koelling [27[and Yanase [28], and like theirs compare well with the experimental results. The calculated band masses, however, are not in agreement with experiment, being nearly an order of magnitude too small in most cases.

In carrying out the modified calculations, the shift $c_3 - \tilde{c}_3$ and the factor $[\tilde{\Phi}_3/\Phi_3]^2$ are determined to satisfy equation (3.4) and to fit the density of states at the Fermi-level $\rho(E_F)$ determined from the specific heat coefficient γ via the Sommerfeld formula $\gamma = \pi^2/3 \, k_B^2 \, \rho(E_F)$ for the renormalized single-particle Hamiltonian (3.2). Within the probable margin of error $\tilde{c}_3 = c_3$ is adequate, and the factor $[\tilde{\Phi}_3/\Phi_3]^2$ required to bring the calculated density of states into agreement with the experimental value is 0.1. The resulting renormalized band (figure 4(ii)) differs significantly from the unrenormalized one (figure 4(i). In the renormalized band the f-band width is much narrower than the spin-orbit splitting (0.41 ev) and splits into two narrow bands, a $j = 5/2$ band near E_F, and a wider $j = 7/2$ band about 0.25 ev further up.

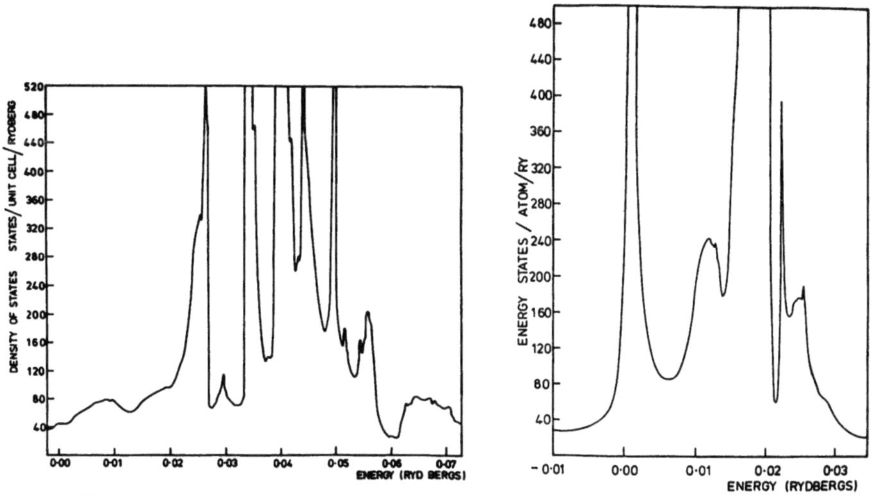

Fig 4 The density of states for $CeSn_3$ close to E_F (i) calculated within the Local Density Approximation (LMTO method). (ii) calculated by modification of the LMTO method in the spirit of mean-field theory.

The Fermi surfaces do not undergo much change either in shape or area. However, the rescaling of the f-band produces dramatic effects on the band masses, illustrated in Table I. The renormalized masses are seen to be in much better agreement with experiment [29] than the LDA masses, being in general much heavier. However the mass for the small orbit around the Γ point is decreased relative to the LDA mass, in agreement with experiment, due to changes in hybridization removing f-character from a band which crosses E_F. This detail is a significant confirmation of the correctness of the approach.

The electronics pressure = ∂(Total energy)/ ∂(Volume) has been estimated in both calculations. It must be zero in equilibrium. For the unrenormalized case a pressure of 144 k bars is found to be compared with 39 k bars for the renormalized case. The renormalized value is clearly more satisfactory, and is similar though slightly larger, than values obtain for transition metals. These results provide evidence that the low-temperature experimental data for $CeSn_3$ are consistent with a heavy fermi liquid describable by a band structure program, in which many-body effects are taken into account by modifying the $\Phi_3(-)$ parameter in the LMTO theory. This suggests that some explicit ℓ-dependence is required in the exchange and correlation energy in the density functional theory to give a description of the narrow f-bands to give agreement with experiment.

4 Crystal Field Effects

In this final section we consider the effects of crystal fields using the exact solution of the impurity model. These have to be included for any realistic comparison of theory with experimental results on the wide variety of rare earth i.v. and Kondo impurities. The effects on ground-state properties have already been considered by Schlottman [30] using the Anderson model. Here we concentrate on the thermodynamic properties of the Coqblin-Schrieffer model and, in particular, the specific heat. If the crystal-field states $|r\rangle$ are

TABLE I: BAND MASSES AND FERMI SURFACE AREAS IN $CeSn_3$

Central Point	our calculated area (au)	our calculated band mass $\frac{m^*}{m}$	our rescaled band mass $\frac{m^*}{m}$	area from experiment
Γ	0.003	4 - 12	0.5 - 0.6	0.0035
Γ	0.264	0.3 - 1.7	3.2 - 12	
M	0.045	0.2 - 1.2	1.0 - 4.4	0.045
X	0.198	0.3 - 1.2	6 - 12	0.166 or 0.216
R	0.227	0.3 - 1.1	2 - 6	0.22

Central Point	$\frac{m^*}{m}$ mass from experiment [29]	area from D. D. Koelling's calculation [27]	$\frac{m^*}{m}$ mass from D. D. Koelling's calculation [27]
Γ	.45		
Γ		0.27	3 - 5
M	4.6	0.04	0.6 - 1.5
X	4 - 9		
R	4.2 - 4.4	0.23 - 0.24	-0.075 - 1.0

characterized by the quantum number r and energy level E_r then the Hamiltonian (1.1) is modified by replacing the angular momentum quantum number m by r. The effective coupling $J_{rr'}$, which can be derived by applying the Schrieffer-Wolff transformation to the Anderson model [10], will be non-diagonal and dependent on r. However, this may be neglected for $|\varepsilon_f| \gg |E_r - E_{r'}|$, which is satisfied in the Coqblin-Schrieffer regime, so we take $J_{rr'} = J\delta_{rr'}$. If the crystal field term extended to the conduction electrons integrability by the Bethe ansatz is unaffected, the only point at which the crystal field enters is in the boundary conditions,

$$\text{limit } (g_{n+1}^{(r)} - g_n^{(r)}) = h^{(r)}/T \tag{4.1}$$

where $h^{(r)} = E_{r+1} - E_r$, similar to an applied magnetic field, where

$g_n^{(r)} = \ln(1 + \rho_{nh}^{(r)}/\rho_n^{(r)})$ and $\rho_{nh}^{(r)}$, ρ_n and the particle and hole densities for the nth string excitations in the rth channel. The application of the crystal field to the conduction electrons only affects the impurity via the cut-off D in $T_k(D)$ and is not of any significance. The thermodynamic equations in the presence of a crystal field have also been considered by Kawakami and Okiji [31], in their case for the Anderson model. They have derived certain analytic relations corresponding to generalizations of the Wilson ratio, but have not so far given any numerical results. The $g_n^{(r)}$ for values of the argument at $\pm\infty$ are required for the numerical iteration of the thermodynamic equations. These are determined by a set of algebraic recurrence equations subject to condition (4.1). An analytic solution of these equations has not proved possible except for strong and weak fields. A numerical solution has however generated for the special case of an SU(4) multiplet split into two doublets such that $h^{(1)} = h^{(3)} = 0$ and $h^{(2)} = A$, and then used for the iteration of the thermodynamic equations. Preliminary results for the impurity specific heat C are shown in figure 5 (Degranges [32]). In a finite field A at very low temperatures the numerical procedure becomes inaccurate due to the finite grid size in the numerical integrations, so this regime has not been shown. It is being covered by an alternative algorithm. However, the results so far are sufficiently extensive for one to draw a number of physical conclusions. For relatively small crystal fields $A/T_k \lesssim 1$ the specific heat peak as a function of $T/T_k(A=0)$ shifts to higher temperatures, is reduced in height, and spreads to give more of a high-temperature tail. This is just the trend required to improve on the earlier comparison of the SU(8) results for YbCuAl shown in figure 1(iii). If this result is general, then crystal-field splittings of this order might explain the discrepancy in YbCuAl. For larger crystal-fields, $A/T_k = 4$, the peak is not unlike a Schottky peak due to a crystal-field splitting in the absence of the Kondo effect, which is shown in figure 5 for comparison. It is, however, significantly lower in height and at a slightly lower temperature. For $A/T_k = 4$ the crystal field is large enough for a lower temperature Kondo peak associated with the ground-state doublet to be seen to emerge. Using poor man's scaling the Kondo temperature T_k corresponding to this is given by $T_k = T_k^2(A=0)/(A+T_k)$, or as $T_k/A >> 1$, $T_k \sim T_k^2(A=0)/A$. Results so far indicate an SU(2) peak of magnitude and position consistent with this estimate of T_k.

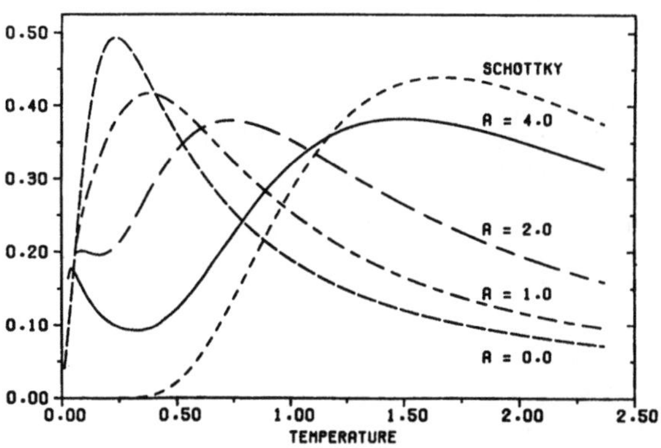

Fig. 5 Results for specific heat C obtained for various crystal-field splittings. The Schottky curve compounds to a crystal-field A in the absence of the exchange interaction.

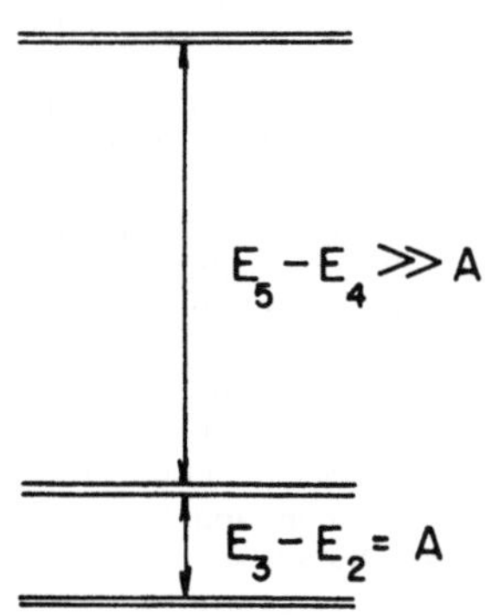

Fig. 6 A crystal field splitting of SU(6) states.

The results may be applicable to cerium impurities in $La_{1-x}Ce_xB_6$. The dominant level splitting due to the crystal field in the cerium ion in this compound is into a Γ_7 and Γ_8 with the Γ_8 quarter lying lowest [33]. If the Γ_8 is further split into two doublets, as it is in CeB_6 where $\Gamma_7 - \Gamma_8 \sim 530K$, and there is a further Γ_8 splitting ~ 30K [34] as indicated in figure 6, then the higher crystal-field level (Γ_7) may be neglected and the results applied to the lower Γ_8 level. The further splitting could explain why the observed peak is lower than that given by the SU(4) model [33].

We gratefully acknowledge the support of a grant from the S.E.R.C.

1. D.M. Newns and A.C. Hewson: J. Phys. F: Met. Phys. 10, 2429 (1980)
2. W.C.M. Mattens, H. Holscher, G.E.M. Tuin, A.C. Moleman and F.R. de Boer: J. Magn. and Magnetic Materials 15-18, 982 (1980): W.C.M. Mattens: Ph.D. Thesis, University of Amsterdam (1980)
3. K. Yamada: Prog. Theor. Phys. 53, 1286 (1975)
4. N. Andrei, Phys. Rev. Lett. 45, 379 (1980): P.B. Wiegmann: Pisma v Zh. Eksp. Teor. Fiz. 31, 392 (1980): J. Phys. C: Solid State Phys. 14, 1463 (1981)
5. A.M. Tsvelick and P.B. Wiegmann, J. Phys. C15, 1707 (1982): J. W. Rasul: Proceedings of the Zurich Conference, p. 49 N. Andrei, K. Furuya and J.H. Lowenstein: Rev. Mod. Phys. 55, 331 (1983)
6. A.C. Hewson, J.W. Rasul and D.M. Newns: Solid State Commun. 47, 59 (1983): A.C. Hewson and J.W. Rasul: J. Phys. C16, 3275 (1983)
7. V.T. Rajan: Phys. Rev. Lett. 51, 308 (1983)
8. N. Read and D.M. Newns: J. Phys. C16, 3273 (1983)
9. D.M. Newns, N. Read and A.C. Hewson: in "Moment Formation in Solids" Proceedings of the NATO Summer School, Vancouver Island. ed. W.J.L. Buyers. NATO ASI Series B vol 117 (1984)
10. P. Nozieres and A. Blandin: J. Physique 41 193 (1980)
11. T.V. Ramakrishnan in "Valence Fluctuations in Solids", eds. L.M. Falicov, W. Hanke and M.S. Maple. (North Holland, Amsterdam, 1981, p. 27)
12. E. Ogievestski, A.M. Tsvelick and P.B. Wiegmann: J. Phys. C16, L797 (1983): P. Schlottman: Phys. Rev. Lett. 50, 1697 (1983) and Z. Phys. 51, 3 (1983)
13. J.W. Rasul and A.C. Hewson: J. Phys. C16, L933 (1983): J. Phys. C17, 2555 and 3337 (1984)
14. P. Coleman: Phys. Rev. B29, 3035 (1984)
15. N. Read and D.M. Newns: J. Phys. C16, L1055 (1983)
16. N. Read: to appear in J. Phys. C (1985)
17. N. Read: Ph.D. Thesis, University of London (1985)
18. J. W. Rasul and A.C. Hewson: Solid State Comm. 52, 217 (1984)
19. N. Read, D.M. Newns and S. Doniach: Phys. Rev. B30, 384 (1984)
20. P. Coleman, Phys. Rev. B28, 5255 (1983)
21. C. Jayaprakash, H.R. Krishna-Murthy and J.W. Wilkin: Phys. Rev. Lett. 47, 737 (1981)
22. O. Gunnarsson and K. Schönhammer, Phys. Rev. B28, 4315 (1983)
23. E. Müller-Hartmann
24. D.R. Hamman: Phys. Rev. 158, 570 (1967)
25. P. Strange and D.M. Newns: Preprint (1985)
26. O.K. Andersen: Phys. Rev. B12, 3060 (1975)
27. D.D. Koelling: Solid State Commun. 43, 247 (1982)
28. A. Yanase: Jour. Magn. Magn. Mat. 31-34, 453 (1983)
29. W.R. Johanson, G.W. Crabtree, A.S. Edelstein, and O.D. McMasters: Jour. Magn. Magn. Mat. 31-34, 377 (1983): W.R. Johanson, G.W. Crabtree, A.S. Edelstein and O.D. McMasters:

Phys. Rev. Letts. 46, 504 (1984)
K. Ikeda and K.A. Gscheider; Phys. Rev. B25, 4623 (1982).
30. P. Schlottmann: Phys. Rev. B30, 1454 (1984)
31. N. Kawakami and A. Okiji: Preprint (1984)
32. H-U. Desgranges: Preprint (1985)
33. H.J. Ernst, H. Grure, T. Krug and K. Winzer: Proc. of LT17, Karlsruhe 1984, eds. U. Eckern and A. Schmidt.
34. E. Zirngieb , H. Hillebrand , S. Blumenr der, G. Güntherodt, M. Loewenhanpt, J. M. Carpenter, K. Winzer and Z. Fisk: Phys. Rev. B30, 4052 (1984).

Self-Consistent Perturbation Theory for Valence-Fluctuating Lattice Systems

Y. Kuramoto

Department of Applied Physics, Tohoku University, Sendai 980, Japan

This paper presents an extension of the self-consistent perturbation theory for the Anderson model to periodic valence-fluctuation systems. The extended scheme takes the reciprocal of the number of effective neighbors as a proper expansion parameter for intersite interactions. Self-consistent account is taken of the Kondo effect, the intersite exchange interaction and the intersite coherence. Simple illustration is given how the heavy-fermion behavior emerges in the present framework.

1. Introduction

Construction of a reliable microscopic theory for valence-fluctuating lattices has been a difficult but attractive subject in recent years. Many authors agree [1-6] that there are two distinct effects both of which should be included in the theory. One is the Kondo effect or the valence-fluctuation effect which is present even in the dilute limit, and suppresses the moment formation at low temperatures. The other effect is the intersite magnetic interaction which gives rise to various types of magnetic ordering in many rare-earth systems without valence fluctuations. These two effects are competing in valence-fluctuating systems, and experiments show that the moment suppression effect dominates in many cases. We notice that an approach from the itinerant f-electron model can hardly deal with the intersite exchange interaction such as the RKKY interaction. The purpose of this paper is to present a contrary approach from the strong correlation limit (atomic limit) which takes account of both the Kondo effect and the intersite interaction.

Another distinct feature of periodic valence-fluctuation systems is the formation of highly correlated bands at low temperatures [7-13]. Certain compounds show extremely large specific heat but low resistivity. These are the clearest indication of the heavy-fermion behavior. The present theory identifies a main element leading to the heavy-fermion behavior.

In a strongly correlated system, there is in general no small parameter in terms of which one can make a perturbation expansion. In recent years it has become clear that the reciprocal of the 4f-state degeneracy n serves as a valid expansion parameter in dilute valence-fluctuation systems [14-23]. The validity continues down to the smallest possible case $n = 2$, if one does not stop at the lowest order of the expansion [17]. It has been suggested that the RKKY interaction is of lower order in $1/n$ and that because of this the interaction is not important [5,11]. However, the RKKY interaction is long-ranged with a large number z_n of effective neighbors. Thus z_n/n can be larger than unity in practical cases. This paper takes $1/z_n$ as an expansion parameter for intersite interactions. The expansion parameters $1/n$ and $1/z_n$ do not involve temperature. Therefore a possibility arises that the theory is valid both above and below the crossover temperature which separates the local-moment-type and band-type behaviors. In the dilute system, it has turned out that $1/n$ is a good expansion parameter at all temperatures [18-23].

The present theory, however, does not use a direct expansion in terms of these parameters. Instead, the parameters are used to classify the skeleton diagrams. Resolvents play a role of the Green functions for f electrons in this diagram method. We then renormalize the resolvents in order to take self-consistent account of the processes involved. The single-site version of this self-consistent perturbation theory has been highly successful in giving unified understanding of both statics and dynamics at all temperatures [19-25]. In the next section we provide a general scheme to extend the resolvent method to many-site systems. Section 3 deals with the $1/z_n$ classification of intersite interactions. Then we formulate in Section 4 the simplest self-consistent approximation which includes the Kondo effect, the intersite exchange and the intersite coherence. Section 5 illustrates how the heavy-fermion behavior emerges in the present scheme. Some discussions are given in Section 6.

2. Resolvent Method Extended to Many-Site Systems

The self-consistent perturbation theory [19-25] utilizes resolvent operators which incorporate the strong 4f-shell correlation in zeroth order. To extend this method to periodic valence fluctuation systems, we introduce a projection operator which furnishes a linked-cluster expansion. This device is necessary because direct application of the resolvent method leads to perturbation expansion of the grand partition function [26], instead of the thermodynamic potential in the standard Green function method. The grand partition function is not proportional to the size of the system. Therefore the perturbation series involves complicated volume dependence, which renders it impossible to deal with macroscopic number of 4f shells.

We consider an extended version of the periodic Anderson model [27]. The Hamiltonian H consists of three parts: $H = H_c+H_f+H_{hyb}$. Here H_c describes the conduction-electron motion and is given by

$$H_c = \sum_{\underline{k},\nu} \varepsilon_{\underline{k}\nu} c^+_{\underline{k}\nu} c_{\underline{k}\nu} , \tag{2.1}$$

where $\underline{k}$ denotes the momentum and ν the set of quantum numbers such as band index and spin. The energy $\varepsilon_{k\nu}$ is measured from the Fermi level. We can also represent H_c using the tight-binding basis:

$$H_c = \sum_{i,j} \sum_{\kappa,\kappa'} \langle\kappa i|t|\kappa' j\rangle c^+_{\kappa i} c_{\kappa' j} . \tag{2.2}$$

Here i and j are lattice-site indices, and κ and κ' are atomic quantum numbers. The unit cell of the lattice may contain arbitrary number of non-rare-earth ions. For the 4f part H_f, we choose for definiteness a Ce-type model where only $4f^0$ and $4f^1$ configurations are taken into account. Configuration with more 4f electrons are neglected because of the strong Coulomb repulsion. Then we have

$$H_f = \sum_i H_f^{(i)}, \quad H_f^{(i)} = \sum_\beta \varepsilon_\beta X^{(i)}_{\beta\beta}, \tag{2.3}$$

where ε_β is the energy of a $4f^1$ configuration β with spin-orbit and crystalline-field splittings and $X_{\beta\beta}^{(i)}$ is the Hubbard projection operator [28] to this configuration at a lattice site $\underline{R}_i$. The hybridization part H_{hyb} is given by

$$H_{hyb} = \sum_j H_{hyb}^{(j)} , \quad (2.4)$$

$$H_{hyb}^{(j)} = \sum_i \sum_{\kappa,\beta} [\langle i\kappa|V|j\beta\rangle c_{\kappa i}^{+} X_{0\beta}^{(j)} + \text{h.c.}], \quad (2.5)$$

where $X_{0\beta}^{(j)}$ indicates a conversion $4f^1 \to 4f^0$ at a lattice site $\underline{R}_j$. Notice that $X_{0\beta}^{(j)}$ is not a fermion annihilation operator because of the constraint

$$\sum_\beta X_{\beta\beta}^{(j)} + X_{00}^{(j)} = 1. \quad (2.6)$$

Here $X_{00}^{(j)}$ is the projection operator to the $4f^0$ configuration.

The motivation of using this general model instead of a simpler version [27] is that the band structure and anisotropy around a rare-earth ion is very important in actual intersite interaction processes. It is therefore useful to provide a scheme which is capable of practical application.

Let Z be the grand partition function of the total system. We then consider a fictitious system $H-H_{hyb}^{(0)}-H_f^{(0)}$ in which the 4f states are absent at a particular site $\underline{R}_0 = \underline{0}$. The grand partition function of this fictitious system is written as Z_{N-1}. Then we define Z_f by

$$Z = Z_{N-1} Z_f . \quad (2.7)$$

This Z_f may be regarded as the grand partition function of the 4f states at $\underline{R}_0$ with inclusion of intersite interactions. The definition (2.7) turns out to be a natural generalization of the previous single-site theory [23]. Namely, using the propagator matrix $\bar{R}(\tau)$ [23] at the site $\underline{R}_0$, we obtain

$$Z_f = \text{Tr}_f \, \bar{R}(\beta), \quad (2.8)$$

$$\bar{R}(\beta) = \exp[-\beta H_f^{(0)}] P T_\tau \exp[-\int_0^\beta H_{hyb}^{(0)}(\tau) d\tau], \quad (2.9)$$

where Tr_f means the trace over the 4f states at $\underline{R}_0$, T_τ is the time-ordering symbol, and $\beta = 1/T$ with T being the temperature. The difference from the single-site theory [23] appears first in the projection operator P and second in the interaction representation of $H_{hyb}^{(0)}(\tau)$. Namely, P implies not only the statistical average over the conduction-electron states but also the average over all the other 4f states than those at $\underline{R}_0$. On the other hand, $H_{hyb}^{(0)}(\tau)$ is defined as

$$H_{hyb}^{(0)}(\tau) = \exp(\tau \tilde{H}_0) H_{hyb}^{(0)} \exp(-\tau \tilde{H}_0), \quad (2.10)$$

$$\tilde{H}_0 = H - H_{hyb}^{(0)} . \quad (2.11)$$

In short, the present generalization consists of regarding $\tilde{H}_0$ as the unperturbed Hamiltonian. It is straightforward to check that the product Z_{N-1} Z_f indeed gives the grand partition function Z.

Let us consider the perturbation expansion of Z_f. In contrast to the single-site theory the conduction-electron propagator now involves hybridization processes at other sites $\underline{R}_i \neq 0$. We first regard this conduction-electron propagator $\tilde{G}_c(z)$ as a given one, and carry out the skeleton-diagram expansion of the self-energy matrix $\Sigma(z)$ of the resolvent matrix R(z). The latter is the Laplace transform of $\overline{R}(\tau)$ [23]. Let us take for example the $4f^0$ component $\Sigma_0(z)$. The lowest-order skeleton diagram is shown in Fig.1 (a). This graph has the same structure as the single-site counterpart [23]. A new feature due to the intersite interaction appears in the next-order skeleton diagram shown in Fig.1(b). From all skeleton diagrams kept up to n-th order, we obtain a set of integral equations to determine the resolvents self-consistently. Thus we can get the n-th order self-consistent perturbation theory. This and other formal properties of the perturbation expansion can be taken over from the single-site formulation [23] without modification.

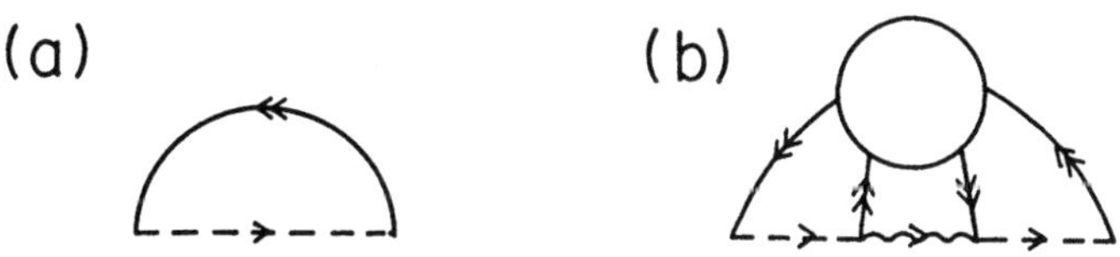

Fig.1 Skeleton diagrams for $\Sigma_0(z)$. The line with a double arrow represents $\tilde{G}_c(z)$. The circle in (b) is the vertex part in $\tilde{H}_0$. The $4f^1$ and $4f^0$ resolvents are indicated by a dashed line and a wavy line, respectively.

Another favorable feature of the present extention is that it provides a linked-cluster expansion without explicit use of Wick's theorem. Namely, perturbation processes which do not interfere with those at $\underline{R}_0$ are automatically left out by the projection operator P.

The formal development up to now has not relied on the periodicity of the rare-earth sites. Therefore if a proper way to find the conduction-electron propagator without $H^{(0)}_{hyb}$ exists, the theory can also be used for random systems [29] or for a small cluster of impurities [30]. In the latter system, the self-energy of Fig.1(b) and even higher-order ones seem to be important. In a dense system, on the other hand, the self-energy of Fig.1(a) is of most importance. Before taking a specific approximation, we proceed to a general analysis of intersite interactions.

3. Classification of Intersite Interactions

Since Wick's theorem is applicable to conduction electrons, the thermodynamic potential Ω of the whole system can be expanded formally in terms of the bare conduction-electron propagator and vertex parts associated with each rare-earth site. We take a representation [28] where none of the vertex parts have the same site index. Figure 2 shows some examples of intersite contribution to $\partial\Omega/\partial\lambda$ where λ is a parameter to scale the hybridization: $H_{hyb} \to \lambda H_{hyb}$.

In order to classify intersite processes we define an "intersite bond" as a pair of antiparallel conduction-electron propagators connecting two different sites. The propagators here should be bare ones. In successive replacement of a propagator pair by an intersite bond, we encounter with a diagram where a group of intersite bonds makes a loop structure. Figure 2(b)

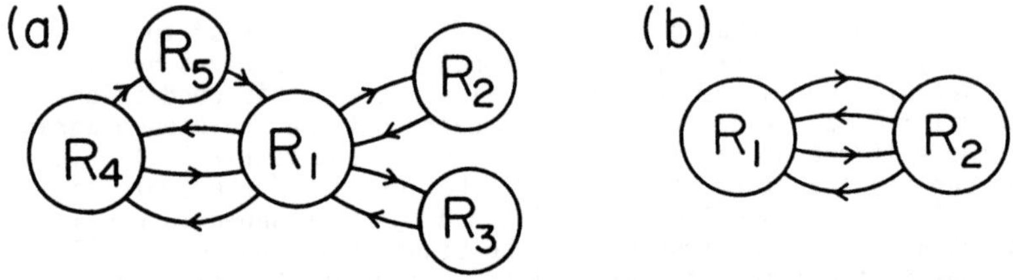

Fig.2 Examples of intersite contribution to the thermodynamic potential. The line with an arrow represents the bare conduction-electron propagator.

shows such an example. By comparison with the right portion of Fig.2(a) one sees that a loop reduces the site summation once for each order of the hybridization interaction. Thus if a diagram has m loops of intersite bonds, the diagram is at most of relative order z_n^{-m} in the present classification. There is a case where an intersite bond is created after removal of a self-energy part of conduction electrons, as in Fig.2(a). If a loop of intersite bonds emerges after such removal, an additional factor $1/z_n$ is associated to the diagram. Thus the diagrams in Fig.2 are both $O(z_n^{-1})$.

In the limit $1/z_n \to 0$, significant diagrams for $\partial\Omega/\partial\lambda$ are those which do not contain any loop of intersite bonds, even after elimination of self-energy parts. These are called tree diagrams by the analogy of a similar classification in spin systems [31]. We expect that summation of all the tree diagrams provides a reliable account of intersite interactions in periodic valence-fluctuating systems with large z_n.

There are at least two important effects due to intersite interactions. One is the indirect exchange which drives the system to a magnetic ordering. The other is the development of coherence which leads to formation of f bands. Both effects are taken into account in the tree diagrams. For example, the characteristic coherence energy $E_{coh} > 0$ per site is given by

$$\lambda\frac{\partial E_{coh}}{\partial\lambda} = \left|\int_{-\infty}^{\infty}\frac{d\varepsilon}{4\pi}\sum_{\kappa}\langle 0\kappa|g_c(i\varepsilon)T(i\varepsilon)g_c(i\varepsilon)T(i\varepsilon)|0\kappa\rangle\right|, \tag{3.1}$$

where $g_c(i\varepsilon)$ is the bare conduction-electron propagator in the matrix form, and $T(i\varepsilon)$ is the t-matrix at a rare-earth site with hybridization λH_{hyb}. One must exclude the same site index for the two t-matrices in the site summation. As an estimate one can use for $T(i\varepsilon)$ the single-site result at zero-temperature [32]. Further analysis of (3.1) will be reported elsewhere. Analogous calculation for a pair of Anderson impurities has been carried out [26, 33]. The formula for the two-site energy in [26, 33] lacks the λ derivative in contrast to (3.1).

The intersite exchange energy E_{exch} per site is obtained from a formula very similar to (3.1). The only difference is that each of the t-matrices now corresponds to a hypothetically polarized $4f^1$ configuration denoted by $|\beta_j\rangle$. With the 4f-electron number n_f per site, we obtain for the site $\underline{R}_j$

$$T(i\varepsilon) = V|j\beta_j\rangle n_f(i\varepsilon-\varepsilon_f^*+i\Gamma^*\mathrm{sgn}\varepsilon)^{-1}\langle j\beta_j|V, \tag{3.2}$$

where ε_f^* is the renormalized $4f^1$ level. If n_f is close to unity, ε_f^* is much below the Fermi level. Then the width Γ^* of the resonance is n times larger than the case of nondegenerate $4f^1$ states. The intersite exchange has been discussed by other methods as well [4, 34, 35]. It is known that the dependence on distance is rather different from the RKKY behavior when ε_f^* lies in the conduction band.

There are three extreme cases concerning relative magnitudes of E_{coh}, E_{exch} and the Kondo energy E_K: (i) $E_{exch} \gg E_{coh}, E_K$. Some magnetic ordering should occur in this case. (ii) $E_K \gg E_{exch}, E_{coh}$. The Kondo effect takes place as the temperature decreases. Then the band formation follows at lower temperature. The intersite exchange should not be effective even in the case of $E_{exch} > E_{coh}$. (iii) $E_{coh} \gg E_{exch}, E_K$. With decreasing temperature the system should go from the local-moment regime to the band regime without showing the Kondo effect.

4. Extended Non-Crossing Approximation (XNCA)

With the general analysis developed in the previous two sections, we are in a position to formulate a specific approximation for periodic valence-fluctuation systems. We neglect all diagrams with any loop of intersite bonds. Then the skeleton diagrams of the type shown in Fig.1(b) should also be neglected. This selection of diagrams has been guided by the smallness of $1/z_n$. As in the single-site theory, we then keep only the lowest-order skeleton diagram for $\Sigma_0(z)$ and the one for $\Sigma_\beta(z)$. The latter selection has been guided by the smallness of $1/n$. With the resolvents at hand one can construct the t-matrix which suffices to determine the renormalized conduction-electron propagator in the tree approximation. Thus we should be able to obtain a closed set of equations which are doubly self-consistent: with respect to intrasite and intersite interactions. We call this approximation scheme the extended non-crossing approximation (XNCA) in the following.

Integral equations to determine the resolvents in the XNCA are very similar to those in the single-site NCA. The only modification lies in the hybridization intensity. We represent by $\langle i\kappa|\tilde{G}_c(z)|j\kappa'\rangle$ the conduction-electron propagator of the fictitious system $\tilde{H}_0$. Then the hybridization intensity

$$\tilde{W}_{\beta\beta'}(\varepsilon) = -\pi^{-1}\mathrm{Im}\,\langle 0\beta|V\tilde{G}_c(\varepsilon+i0_+)V|0\beta'\rangle, \tag{4.1}$$

replaces the single-site counterpart defined in [23]. The point-group symmetry around the rare-earth site restricts the independent components of (4.1) to a small number. We refer to [23] for the symmetry analysis and the integral equations with inclusion of crystalline-field effects.

The renormalized conduction-electron propagator $G_c(i\varepsilon_n)$ of the whole system H is obtained from the Ω diagram by cutting a $g_c(i\varepsilon_n)$ line in all possible ways. With the tree diagrams for Ω, the resultant $G_c(z)$ has the self-energy which is $\underline{k}$-dependent only through the hybridization matrix element:

$$\langle \underline{k}\nu|G_c(z)|\underline{k}\nu\rangle = [z-\varepsilon_{\underline{k}\nu}-\sum_\beta |\langle \underline{k}\nu|V|0\beta\rangle|^2 Y_\beta(z)]^{-1}, \tag{4.2}$$

where the site-diagonal irreducible part $Y_\beta(z)$ has been introduced. The irreducible part $Y_\beta(z)$ is common to all sites because of the periodicity and the site index 0 has been arbitrarily chosen. Then we obtain

$$G_c(z) = \tilde{G}_c(z)+\sum_\beta \tilde{G}_c(z)V|0\beta\rangle Y_\beta(z)\langle 0\beta|VG_c(z), \tag{4.3}$$

which is equivalent to

$$\langle 0\beta|V\tilde{G}_c(z)V|0\beta\rangle^{-1} = \langle 0\beta|VG_c(z)V|0\beta\rangle^{-1}+Y_\beta(z). \tag{4.4}$$

The XNCA is completed if $Y_\beta(z)$ is given in terms of the resolvents $R_0(z)$ of $4f^0$ and $R_\beta(z)$ of $4f^1$ configurations. To derive the relation we notice that the on-site 4f Green function is given by

$$F_\beta(i\varepsilon_n) \equiv \langle 0\beta|G_f(i\varepsilon_n)|0\beta\rangle = Z_f^{-1}\int_C \frac{dz}{2\pi i}\, e^{-\beta z}R_0(z)R_\beta(z+i\varepsilon_n), \tag{4.5}$$

where $i\varepsilon_n$ is the Fermion Matsubara frequency and the contour C encircles counterclockwise all singularities of the integrand. Considering that the on-site Green function involves scattering of conduction electrons at many different sites, we obtain the desired relation:

$$F_\beta(z) = \langle 0\beta|Y(z)[1-Vg_c(z)VY(z)]^{-1}|0\beta\rangle, \tag{4.6}$$

where the matrix representation is used.

In concluding this section we give the equation for $G_f(z)$ in the momentum representation:

$$\langle \underline{k}\beta|G_f(z)^{-1}|\underline{k}\beta'\rangle = \delta_{\beta\beta'}Y_\beta(z)^{-1}-\sum_\nu \langle \underline{k}\beta|V|\underline{k}\nu\rangle g_c(\underline{k}\nu,z)\langle \underline{k}\nu|V|\underline{k}\beta'\rangle, \tag{4.7}$$

where $g_c(\underline{k}\nu,z) = (z-\varepsilon_{\underline{k}\nu})^{-1}$. Note that the 4f Green function is no longer diagonal in β indices because of the intersite interaction.

5. Heavy-Fermion Behavior

The self-consistent set of equations in the XNCA can be solved only by numerical methods. In the extreme case $E_K \gg E_{exch}$, E_{coh} mentioned in Section 3, we may neglect the influence of intersite interactions back on the resolvents. Then the t-matrix becomes the same as that in the single-site system. The scheme by GREWE [36] corresponds to this approximation which seems to be a reasonable starting point to discuss the heavy-fermion behavior. In the Grewe approximation one can use the analytical result of the single-site NCA at zero temperature [32]. The analytical treatment makes clear how the heavy-fermion behavior arises in periodic valence-fluctuating systems. To simplify the analysis we take in the following the Anderson lattice with a single conduction band. The orbital degeneracy or split-off $4f^1$ multiplets are important in making $E_K \gg E_{exch}$ [5,6]. Assuming that the crystalline-field ground state is a doublet, however, we do not consider the higher $4f^1$ states explicitly.

The retarded 4f Green function of the Anderson impurity system is given approximately by [32]

$$F_\sigma(\varepsilon) = a_f[\varepsilon-\tilde{\varepsilon}_f\exp(-i\alpha)]^{-1}, \tag{5.1}$$

$$a_f = \tilde{\varepsilon}_f(\pi W_0)^{-1}\sin\alpha, \qquad \alpha = \pi n_f/2, \tag{5.2}$$

with $|\varepsilon|<\tilde{\varepsilon}_f$ and W_0 being the hybridization intensity [32]. The Fermi-liquid relation [37] is satisfied in (5.1) and (5.2). The energy $\tilde{\varepsilon}_f$ is a generalization of the Kondo energy for arbitrary value of n_f [32]. The relation between $\tilde{\varepsilon}_f$ and n_f is derived from the Bethe-Ansatz theory [17,38] or from the 1/n expansion method [32] to a good accuracy. In the Grewe approximation [36], one uses (5.1) as the on-site Green function of the periodic system. The relation between $F_\sigma(z)$ and the irreducible part $Y_\sigma(z)$ is different from (4.6), since no other sites than $\underline{R}_0$ enter into $F_\sigma(z)$ in the Grewe approximation. Thus we obtain instead

$$Y_\sigma(z)^{-1} = F_\sigma(z)^{-1}-V_0^2N^{-1}\sum_{\underline{k}} g_c(\underline{k},z) = a_f^{-1}(z-\tilde{\varepsilon}_f\cos\alpha). \tag{5.3}$$

Here we have used a $\underline{k}$-independent hybridization constant V_0 and

$$V_0^2N^{-1}\sum_{\underline{k}} g_c(\underline{k},z) \simeq -\pi i W_0. \tag{5.4}$$

With (5.3) the Green functions are easily obtained as

$$G_f(\underline{k},z) = a_f[z-\tilde{\varepsilon}_f\cos\alpha-a_fV_0^2(z-\varepsilon_{\underline{k}})^{-1}]^{-1}, \tag{5.5}$$

$$G_c(\underline{k},z) = [z-\varepsilon_{\underline{k}}-a_fV_0^2(z-\tilde{\varepsilon}_f\cos\alpha)^{-1}]^{-1}. \tag{5.6}$$

The Green functions are very similar to those of a trivial two-band system, except for the presence of the renormalization factor a_f. Expressions similar to (5.5) and (5.6) have been obtained by NEWNS et al. [39] who have used a functional integral method combined with the slave-boson representation [21]. However, the renormalization factor by these authors is $a_f = 1-n_f$ which corresponds to zeroth-order result in the 1/n expansion, and which is much larger than our result of a_f (5.2) in the Kondo regime $n_f \lesssim 1$. BRANDOW [13] has also arrived at (5.5) and (5.6) with $a_f = 1-n_f$ by a variational method. He noticed that a_f becomes smaller in his second approximation, which did not lead to an analytical expression.

We now derive the effective mass m* of the quasi-particle on the basis of (5.5) and (5.6). The T-linear specific heat is determined by m* [40]. The quasi-particles have infinite life-time in the present approximation. The energy of a quasi-particle is indicated by $E_k^\pm$ where + and - represent upper and lower hybrid bands, respectively. They correspond to the solutions of the equation

$$(z-\tilde{\varepsilon}_f\cos\alpha)(z-\varepsilon_{\underline{k}})-a_fV_0^2 = 0. \tag{5.7}$$

Let us assume a metallic ground-state where the Fermi level lies near the top of the lower hybrid band. According to the Luttinger theorem [40], this situation is realized when the number of electrons per unit cell is slightly smaller than two. We follow the method of BRANDOW [13] in relating m* to the parameters in (5.7). The quasi-particle effective mass is defined by

$$\partial E_k^-/\partial k\Big|_{k=k_F} = k_F/m^*, \tag{5.8}$$

where E_k^- is assumed to be isotropic and k_F is the Fermi wave-number. From (5.7) we obtain

$$m^*/m_c = 1+a_f^{-1}[(E_k^- - \varepsilon_k)/V_0]^2_{k=k_F}, \tag{5.9}$$

where m_c is the effective mass of the conduction band. It can be seen that $|E_k^- - \varepsilon_k|$ at $k = k_F$ is of the order of the half-width D of the conduction band, provided that $\sqrt{a_f}V_0$ is much smaller than D. Then as a rough estimate we obtain

$$m^*/m_c \simeq a_f^{-1}(D/V_0)^2. \tag{5.10}$$

It should be emphasized that the renormalization factor a_f is very important in giving rise to a heavy-fermion mass of $m^*/m_c \sim O(10^3)$. The enhancement $(D/V_0)^2$ alone is not sufficient according to band-structure calculations [41]. We suggest that the zeroth order result $a_f = 1-n_f$ is not small enough either.

6. Discussion and Outlook

We have described how the intersite interaction is incorporated into the self-consistent perturbation theory. The present generalization called the XNCA takes into account the Kondo effect, the indirect exchange interaction, and the intersite coherence. As in the case of the single-site NCA, the present theory is capable of treating the crossover from the local-moment regime to the Kondo compensated regime with decreasing temperatures. In the case of $E_K >> E_{coh}$, there must appear another crossover to the f-band regime, which can also be treated by the XNCA. At zero temperature the XNCA may show a pathology in the low-frequency limit of dynamics as in the single-site case [20, 42]. Here the coarse-graining procedure [32] to remove the pathology may be employed.

A magnetic ordering is signaled by a divergence of the magnetic susceptibility. The calculation of it in the XNCA is possible but complicated. The complexity arises from ladder-type vertex corrections which are absent in the single-site NCA for Ce and Yb systems. More detailed discussion on the magnetic susceptibility in the XNCA will be reported elsewhere.

The XNCA can be applied to electronic-structure calculation of actual heavy-fermion systems. In doing this, one needs specific values of the parameters in the Hamiltonian H. The tight-binding fit [43] to a result of standard band calculation is useful here. With the parameters determined, one can also estimate E_K, E_{coh} and E_{exch} according to the formulae in Section 3. It seems possible to classify various heavy-fermion systems according to relative magnitudes of the three characteristic energies. This classification should be useful in understanding, for example, different behaviors of the resistivity. If the coherence energy is not small as compared with E_K, the intersite coherence plays an important role in determining the quasi-particle renormalization factor a_f. Recently, FULDE and coworkers have developed a scheme [44] to calculate the band structure of heavy-fermion systems. They calculate the quasi-particle bands with infinite lifetime. On the other hand, the XNCA has a scope of including damping and finite-temperature effects. We notice that GREWE [36] has already initiated a finite-temperature calculation using his own approximation scheme.

In addition to extensive application of the XNCA, it is also desirable to get more insight into the equations obtained in this paper. Investigations toward both directions are currently in progress.

References

1. S. Doniach: Physica 91B, 231 (1977).
2. R. Jullien, P. Pfeuty, A.K. Bhattacharjee and B. Coqblin: J. Appl. Phys. 50, 7555 (1979).
3. C. Lacroix and M. Cyrot: Phys. Rev. B20, 1969 (1979).
4. Y. Kuramoto: Z. Phys. B40, 293 (1981).
5. P. Coleman: Phys. Rev. B28, 5255 (1983).
6. K. Yamada, K. Yosida and K. Hanzawa: Prog. Theor. Phys. 71, 450 (1984).
7. R.M. Martin: Phys. Rev. Lett. 48, 362 (1982).
8. A. Yoshimori and H. Kasai: J. Magn. Magn. Mat. 31-34, 475 (1983). F.J. Ohkawa: J. Phys. Soc. Japan 53, 1828 (1984). H. Jichu, T. Matsuura and Y. Kuroda: Prog. Theor. Phys. 72, 366 (1984).
9. C.M. Varma and Y. Yafet: Phys. Rev. B13, 2950 (1976). B.H. Brandow: Int. J. Quantum Chem. Symp. 13, 423 (1979). D.L. Huber: Phys. Rev. B28, 860 (1983) and B29, 456 (1984). Y. Kuramoto and C. Horie: J. Magn. Magn. Mat. (in press).
10. A.J. Fedro and S.K. Sinha: Moment Formation in Solids, (W.J.L. Buyers ed., Plenum Press, New York 1984), p.135.
11. N. Read, D.M. Newns and S. Doniach: Phys. Rev. B30, 3841 (1984).
12. P. Coleman; T.M. Rice and K. Ueda: in these proceedings.
13. B. Brandow: preprint submitted to Phys. Rev. B (1985).
14. T.V. Ramakrishnan and K. Sur: Phys. Rev. B26, 1798 (1982).
15. O. Gunnarsson and K. Schönhammer: Phys. Rev. B28, 4315 (1983).
16. F.C. Zhang and T.K. Lee: Phys. Rev. B28, 33 (1983) and B30, 1556 (1984).
17. J.W. Rasul and A.C. Hewson: J. Phys. C17, 2555 and 3337 (1984).
18. G. Czycholl: Phys. Rev. B31, 2867 (1985).
19. H. Kojima, Y. Kuramoto and M. Tachiki: Z. Phys. B54, 293 (1984).
20. Y. Kuramoto and H. Kojima: J. Magn. Magn. Mat. (in press).
21. P. Coleman: Phys. Rev. B29, 3035 (1984).
22. N.E. Bickers, D.L. Cox and J.W. Wilkins: Phys. Rev. Lett. 54, 230 (1985).
23. Y. Kuramoto: Z. Phys. B53, 37 (1983).
24. N. Grewe: Z. Phys. B53, 271 (1983).
25. H. Keiter and G. Morandi: Phys. Rep. 109, 227 (1984). This is a review article with extensive list of earlier references.
26. N. Grewe and H. Keiter: Phys. Rev. B24, 4420 (1981).
27. H-J. Leder and B. Mühlschlegel: Z. Phys. B29, 341 (1978).
28. J. Hubbard: Proc. Roy. Soc. A296, 82 and 100 (1966).
29. K. Hoshino and Y. Kurata: J. Phys. F9, 131 (1979).
30. C. Jayaprakash, H.R. Krishnamurthy and J.W. Wilkins: Phys. Rev. Lett. 47, 737 (1981).
31. R.B. Stinchcombe, G. Horwitz, F. Englert and R. Brout: Phys. Rev. 130, 155 (1963).
32. Y. Kuramoto and E. Müller-Hartmann: Proc. The 5th Int. Conf. on Crystalline Field and Anomalous Mixing Effects in f-Electron Systems, Sendai 1985 (to be published in J. Magn. Magn. Mat.).
33. U. Brandt, H. Keiter and F-S. Liu: Z. Phys. B58, 267 (1985).
34. A.M. Tsvelick: Sov. Phys. JETP 49, 1142 (1979).
35. C. Proetto and A. Lopez: Phys. Rev. B25, 7037 (1982).
36. N. Grewe: Solid State Commun. 50, 19 (1984).
37. D.C. Langreth: Phys. Rev. 150, 516 (1966).
38. See the papers by Okiji and Kawakami and by Schlottmann in these proceedings.

39. D.M. Newns, N. Read and A.C. Hewson: in ref.10, p.257.
40. J.M. Luttinger: Phys. Rev. 119, 1153 (1960).
41. See the papers by Yanase, by Oguchi and by Harima in ref.32.
42. E. Müller-Hartmann: Z. Phys. B57, 281 (1984).
43. O. Sakai, M. Takeshige, H. Harima, K. Otaki and T. Kasuya: in ref.32.
44. P. Fulde: in these proceedings.

Modelling Mixed Valence Systems Using the Generalized Anderson Model

P. Coleman

Institute for Theoretical Physics, University of California, Santa Barbara, CA 93106, USA

With the aid of a generalized version of the infinite U Anderson model the strong coupling properties of mixed valence systems may be modelled by means of an expansion in 1/degeneracy about a broken symmetry state. The technique leads to new conceptual picture where the strong coupling divergence of the Kondo coupling constant is a direct consequence of Goldstone mode fluctuations about the broken symmetry state. The crossover to strong coupling is a vestage of the broken symmetry phase transition and we are able to relate the interactions and the charge and spin correlations of the heavy Fermi liquid to the fluctuations about this broken symmetry state.

1. Introduction.

The great diversity and richness of the physical properties of heavy Fermion and mixed valence systems at low temperatues raise a great many theoretical questions. At the heart of the problem posed by these systems is the isue of how to consistently model the low temperature, strong coupling properties. There are two prevailing conceptual frameworks within which we are accustomed to treating these systems: the renormalization group and the Fermi liquid picture [1,2,3].

In renormalization group language, the mixed valence and Kondo systems involve spin and charge correlations spanning many decades of frequency: each decade contributing a logarithmic correction to the renormalized parameters such as the Kondo coupling constant $J(\Lambda)$, at lower frequencies. At frequency scales the size of the Kondo temperature T_K there is a crossover into strong coupling where correlation effects of the interactions can no longer be treated as a perturbation to electronic motions. In the one impurity problem there is just one relevant parameter associated with the strong coupling fixed point but in the many site case there is a relevant, diverging interaction parameter for each wavevector. If we wish to continue scaling of the Hamiltonian below the Kondo temperature, we are obliged to use numerical renormalization group methods [2] and in practical terms, this limits us to the single impurity problem and to calculations of purely static correlations.

Nozieres [3] recognized that the low temperature fixed point of the renormalization group approach for the single impurity is a local Fermi liquid, and for this case he deduced the low temperature Fermi liquid properties using general arguments. Metallic mixed valence and heavy Fermion systems can also be viewed as the Fermi liquid "fixed-point" of a model such as the Anderson lattice model. However, if we wish to deduce Fermi liquid properties which depend on the intersite coherence of the heavy Fermi liquid (such as the triplet interaction parameter), then we require more than Fermi liquid phenomenology, for there are simply too many degrees of freedom to to constrain the Landau parameters using general arguments.

In this paper we wish to show how we can link these two pictures by incorporating the singular low energy behaviour of strong coupling into our model as a broken symmetry feature. The broken symmetry state is the exact solution at infinite spin degeneracy $N = 2j+1$. We can model the properties of a finite degeneracy system by means of a $1/N$ expansion about this broken symmetry limit. Our approach has grown from Anderson's suggestion of using large degeneracy as an expansion parameter in mixed valence [4] and the pioneering functional integral treatments of the Kondo problem by Chakravarty, Newns and Read [5,6] where the broken symmetry state appears as a saddle point of the path integral.

We can motivate our approach using Fermi liquid arguments. Consider the Coqblin Schrieffer model for one f-state with $N = 2j + 1$ fold spin degeneracy

$$\hat{H} = \sum E(k) c^{+}_{km} c_{km} - \frac{J}{N} \sigma^{+} \sigma \tag{1.1}$$

where $\sigma = \sum_{km} f^{+}_{m} c_{km}$. Suppose there are Q f-electrons in the local moment and consider the action of $\hat{\sigma}$ on the groundstate. The state $\sigma^{+}(\tau)|Q> = |\psi(\tau)>$ contains one less f-electron. From Nozieres' Fermi liquid picture, we know that the initial groundstate scatters quasiparticles elastically, without changing their spins and that the elastic scattering is determined by a f-channel phase shift $\delta_f = \pi Q/N$, using Friedel's sum rule. The action of $\sigma^{+}|Q>$ is to project one into the new space of states where the final state groundstate is governed by an f-channel phase shift $\delta_f = \pi(Q-1)/N$. The situation is identical to X-ray absorption, where a change in the number of localized electrons induces a change in the local scattering phase shift. From the X-ray catastrophy we know that the response of the system to the suddenly modified scattering potential is governed by a powerlaw relaxation of the wavefunction $\langle \psi(\tau) \mid \psi(0) \rangle \sim \tau^{-\alpha}$ where α is the Nozieres de Dominicis' X-ray exponent $\alpha = \sum(\Delta\delta_f/\pi)^2$, $= 1/N$ in this case[7]. Consequently at long times

$$\langle Q \mid \sigma(\tau)\sigma^{+}(0) \mid Q \rangle \sim 1/\tau^{1/N} \quad (\tau \to \infty) \tag{1.2}$$

This singular behaviour at long times is the essence of strong coupling in the Kondo problem, as was realized very early on in Anderson's and Yuval's scaling treatment[1]. Analagous behaviour occurs in 1 dimensional quantum systems with a tendency towards a broken symmetry ground state [8,9]. For instance, in the 1d Heisenberg antiferromagnet, the groundstate is antiferromagnetic in the limit of infinite spin degeneracy S. For finite S the long wavelength spin waves restore the $SU(2)$ symmetry associated with angular momentum conservation, leading to powerlaw decay in the order parameter $\vec{S}$ at large distances.

$$\langle \vec{S}_n \cdot \vec{S}_m \rangle \sim (-1)^{n-m} |n-m|^{-1/2S} \qquad (|n-m| \to \infty) \tag{1.3}$$

Of course these fluctuations only vanish in the large S limit, but even in a finite S system with no broken symmetry, the Goldstone fluctuations in the spin are sufficiently weak that many qualitative features (collective excitations, short range antiferromagnetic order) of the broken symmetry state are preserved. This is the basis of Anderson's semiclassical treatment of this model [8].

These considerations enable us to regard the strong coupling regime of the Kondo model as an almost broken symmetry, in which the system with a finite spin degeneracy

N has a tendency towards a broken symmetry state with $\langle\sigma\rangle \neq 0$, forming a state of "almost broken symmetry." In finite degeneracy systems, Goldstone mode fluctuations restore the symmetry and are *responsible* for the strong coupling divergence of the Kondo coupling constant.

True mean field behavior only occurs when $N \to \infty$ with $q = Q/N$ finite, and by preserving the full functional dependences in q in a $1/N$ expansion about the broken state a useful $1/N$ expansion is developed. In a physical language: we are examining the properties of mixed valence or Kondo ions in the limit where we macroscopically occupy them, and we are using this limit to extract information about the behaviour of finite degeneracy ions. This is a conceptually different way of taking the large N limit to that where where Q is constrained to one, and true meanfield behaviour does not develop.

This charge in the way the large N-limit is taken has enabled Andrei and Coleman to extract meanfield behavior from the large N limit of exact Bethe Ansatz equations [10]. They have solved the Coqblin Schrieffer model and the analogous generalization of the infinite U Anderson model for general Q, and demonstrated that when $N \to \infty$ with Q/N finite, the meanfield equations derived by direct means, appear as the large N limit of the Bethe Ansatz solutions.

By inserting $<\sigma>/\sqrt{N} = \tilde{\sigma}$ into [1] we obtain the mean field Hamiltonian

$$\hat{H}^* = \sum E(k) c^+_{km} c_{km} + \tilde{V} \sum \left\{ c^+_{km} f_m + \text{H.C} \right\} + E\left[\hat{Q} - Q_0\right] \tag{1.4}$$

where $\tilde{V} = -J\tilde{\sigma}$ and we have associated a Lagrange multiplier of E with the conserved quantity Q to aid in constraining it to the value Q_0. This Hamiltonian describes a resonance level at position E with width $\tilde{\Delta} = \pi\tilde{V}^2\rho = [\pi J^2\rho]\tilde{\sigma}^2$ where ρ is the density of states in the band, which we assume to be slowly varing on the scale of $\tilde{\Delta}$. The mean field theory breaks the symmetry associated with Q conservation, and as in BCS theory it only gives answers accurate to $O(1/N)$ due to the fluctuations $\langle\delta q^2\rangle = O(1/N)$ present in the broken symmetry state. It is straightforward to calculate $\tilde{\sigma}$ and $q = 1/N\langle\hat{n}_f\rangle$ from the mean field Hamiltonian. We find $\tilde{\sigma} = \tilde{V}\rho\text{Re }I(E + i\tilde{\Delta})$ and $q = Im\, I(E + i\Delta)$ where

$$I(\zeta) - \int \frac{d\epsilon}{\pi} f(\epsilon) \frac{\Phi(\epsilon)}{\epsilon - \xi} \tag{1.5}$$

and $\phi(\epsilon) = \rho(\epsilon)/\rho(\epsilon_F)$ is the ratio of electronic density of states at ϵ to the value at the Fermi energy. Employing a Lorentzian cutoff function $\Phi(\epsilon) = D^2/[\epsilon^2 + D^2]$ we find

$$I(\xi) = \frac{1}{\pi}[\tilde{\psi}(\xi) - \ell n(D\beta/2\pi i)] \tag{1.6}$$

where $\xi = E + i\tilde{\Delta}$, $\tilde{\psi}(z) = \psi(\frac{1}{2} + \frac{z\beta}{2\pi i})$ and $\psi(z)$ is the digamma function. This enables us to impose self consistency in one compact equation

$$\ell n \frac{T_0}{T} + iq\pi = \tilde{\psi}(\xi) + \frac{i\pi}{2} \tag{1.7}$$

where $T_0 = D\, e^{-\frac{1}{J\rho}}$ is recognized as the leading order expression in the $1/N$ expansion Kondo temperature . At $T = 0$ $\psi(\beta z) \to ln(z\beta)$ and (1.7) becomes

$$\xi = T_0\, e^{iq\pi} \tag{1.8}$$

elegantly demonstrating how the broken symmetry state is a resonance level whose position is tuned through the Fermi level by the value of the phase shift $\delta = \pi q$. This resonance level is to be identified with the Abrikosov Suhl resonance, and it is satisfying to discover that this strong coupling feature appears naturally in our mean field treatment. Chakravarty, Newns and Read obtained this expression for the case when $q = 1/N$ from the saddle point of their functional integral formation[5,6].

Equation (1.7) clearly shows how the strong coupling properties are all expressed in terms of one temperature scale T_0. Furthermore, setting the imaginary part of (1.7) to zero determines the temperature T_c at which there is a sharp phase transition from the broken symmetry state where $<\sigma> \neq 0$ to the weak coupling regime where $<\sigma> = 0$ and the moment is unquenched. We shall see later how fluctuations smooth away this sharp transition in a finite degeneracy system.

2. Generalized Anderson Model.

We have recently generalized the infinite U Anderson model in a way which exploits the underlying symmetry associated with restricting charge fluctuations to a few valence states [11], enabling us to use the above approach for mixed valence. One introduces "slave boson fields" b which describe the availability of an f-state for hybridization. For the single impurity the generalized Anderson model is written

$$\hat{H}(\lambda) = \sum_{k,m} E(k) c^+_{km} c_{km} + \sum E_f f^+_m f_m + \hat{H}_{\text{mix}} + \lambda \hat{Q} \tag{2.1}$$

where

$$\hat{H}_{\text{mix}} = \sum V(k) \left\{ c^+_{km} f_m b^+ + b f^+_m c_{km} \right\} \tag{2.2}$$

induces mixing between the localized f-states and band electrons, whilst

$$\hat{Q} = \sum f^+_m f_m + b^+ b \tag{2.3}$$

is a conserved charge which plays the analagous role to Q in the Coqblin Schrieffer model. The sums over m run from $-j$ to j. The operators c^+_{km} create band electrons in the partial wave state of energy $E(k)$ with the same total angular momentum numbers $|jm>$ as an f-state. b^+ creates a local boson, and b^+ satisfies canonical commutation laws $[b, b^+] = 1$.

In the case $Q = 1$, the generalized Anderson model reverts to the infinite U Anderson model, with the identification $b^+|0> = |f^0>$ and $f^+_m|0> = |f^1 : j_m>$. Perhaps the easiest way of seeing this is by recognizing that the operators$X_{oo} = b^+ b$, $X_{m'm} = f^+_{m'} f_m$ $X_{om} = b^+ f_m$, and $X_{mo} = f^+_m b$ obey the same commutation algebras as Hubbards operators introduced for the infinite U Anderson model. For the case $E_f << 0$ (taking the Fermi energy = 0), we can carry out a canonical transformation to integrate out the Bose field, in which case we arrive back at the Coqblin Schrieffer model with $J = -V^2/E_f$ and $Q = n_f$.

A natural interpretation of b [12], in analogy with liquid Helium, is as a collective backflow associated with the motion of the electrons. Since Q is conserved, the inflow of an f-electron is always accompanied by a backflow of the Bose field. When the bose field has a small amplitude, the backflow is restricted and the f-electrons become highly localized. The strong correlations of the f-electrons are acutally built into the model by the Q conservation symmetry. This is easily seen by noting that the matrix element

$$\langle f^{n+1};Q \mid H_{\rm mix} \mid k, f^n;Q\rangle = V(k)\sqrt{\frac{Q-n}{N}} \tag{2.4}$$

showing that the absorption of a band electron into the f-even depends on the occupation and becomes zero when $n_f = Q$. To constrain $q = Q_0$ in thermodynamic ensemble, a Lagrange multiplier has been associated with Q.

The generalized Anderson model can be written down for many impurities, or the lattice, by introducing a conserved charge Q_i and slave Boson field for each site at position R_i. In a real electron band the electrons have spin half, whereas the f-electrons in a typical mixed valence or achnide ion are spin-orbit coupled into a state of definite j . To develop the formalism with maximum clarity, we shall forego a realistic angular momentum structure until later discussion.We suppose that both the f-electrons and the band electrons have a spin degeneracy $N = 2j + 1$. In this case, the mixing at each site can be written

$$\hat{H}_{\rm mix}(\vec{R}_j) = \sum_{\vec{k},m} \frac{V(k)}{\sqrt{N}}\left\{c^+_{\vec{k}m} f_m(\vec{R}_j) b^+_j\ e^{-i\vec{k}\cdot\vec{R}_j} + f^+_m(R_j) c_{km} b_j e^{i\vec{k}\cdot\vec{R}_j}\right\} \tag{2.5}$$

enabling us to write down the many site version of the generalized Anderson model

$$\hat{H} = \sum_{\vec{k}m} E(k) c^+_{\vec{k}m} c_{\vec{k}m} + \sum_{\vec{R}_i m} E_f \hat{n}_{fm}(\vec{R}_i) + \sum_{\vec{R}_i} H_{\rm mix}(\vec{R}_i) + \sum_{\vec{R}_i} \lambda_i \hat{Q}_i \tag{2.6}$$

where $\hat{Q}_i = \sum n_{fm}(\vec{R}_i) + b^+_i b_i$ and we associate a chemical potential λ_i with each conserved charge Q_i.

We are interested in ultimately including the fluctuations about meanfield theory, for which a path integral formalism is helpful, so we shall begin by showing how meanfield theory emerges as a saddle point of a path integral formalism.

Following earlier work [6,11], the partition function is written as a constrained functional integral

$$Z_{MV} = \int_0^{2\pi i/\beta} \prod_{R_j} \frac{d\lambda_j \beta}{2\pi i}\ Z(\lambda_j)\ exp\ \beta\left\{\sum \lambda_j Q_0\right\} \tag{2.7}$$

where the Fourier transforms in the chemical potentials constrain $\hat{Q} = Q_0$, and

$$Z(\lambda_j) = {\rm Tr}\ e^{-\beta\hat{H}(\lambda_j)} = \int \mathcal{D}[\bar{\psi},\psi]\mathcal{D}[\bar{b},b]\ e^{-\beta F[\bar{\psi},\psi,\bar{b},b]} \tag{2.8}$$

with

$$\beta F[\bar{\psi},\psi;\bar{b},b] = \int_0^\beta d\tau\left\{\left[\sum_i \bar{\psi}_i(\tau)\frac{\partial}{\partial\tau}\psi_i(\tau) + \sum_i \bar{b}_i\frac{\partial}{\partial\tau}b_i\right] + H[\bar{\psi},\psi,\bar{b},b,\lambda_i]\right\} \tag{2.9}$$

Here $\bar{\psi}_i$ and ψ_i denote the Fermi fields (formally represented by anticommuting variables) and $\bar{b}_i$ and b_i are independent functions representing the Bose fields at site i. The measures $\mathcal{D}[\bar{\psi},\psi]$ and $\mathcal{D}[\bar{b},b]$are canonical Fermi and Bose measures. τ is the imaginary time variable and implicitly contained in the functional integral are periodic boundary conditions for the Bose fields and antiperiodic boundary conditions for the Fermi fields $\psi(\tau+\beta)=-\psi(\tau)$.

To determine the saddle point equations we "freeze" a given Bose field configuration and calculate the effective Free energy functional for that particular configuration, writing

$$\exp\{\beta\sum_j \lambda_j Q_0\} Z(\lambda_j) = \int \mathcal{D}[\bar{b},b]\ exp\ -\beta\Omega_{eff}[\bar{b},b] \tag{2.10}$$

Ω_{eff} is the effective free energy functional for the bose fields with the Fermions integrated out of the dynamics. This integral is quadratic in the Fermi field so can evaluate it to gain

$$\beta\Omega_{\text{eff}}[\bar{b},b] = \int_0^\beta d\tau \cdot \sum_i \left\{\bar{b}_i(\tau)\left(\frac{\partial}{\partial\tau}+\lambda_i\right)b_i(\tau) - \lambda_i Q_0\right\} + \beta F_\psi[\bar{b},b] \tag{2.11}$$

$F_\psi[\bar{b},b]$ is the Fermion free energy for the frozen Bose field configuration $b(\tau)$. This is expressed compactly as a trace over the Fermion Fock space and imaginary time

$$\beta F_\psi[\bar{b},b] = \underset{(\tau,\bar{\psi},\psi)}{-\text{Tr}}\ \ell n\left[\hat{1}\frac{\partial}{\partial\tau} + \hat{H}^*[\bar{b},b]\right] \tag{2.12}$$

where $\hat{H}^*[\bar{b},b]$ and is the Fermion part of the Hamiltonian for a given classical field configuration $[b(\tau),\bar{b}(\tau)]$.

The saddle points of the full path integral are determined by the stationary point of the effective free energy

$$\frac{\delta\Omega_{\text{eff}}}{\delta\bar{b}_j(\tau)} = \left[\frac{\partial}{\partial\tau}+\lambda_j\right]b_j(\tau) + \frac{1}{\sqrt{N}}\sum_{\vec{k}m} V(k)e^{-i\vec{k}\vec{R}_j}\langle c^+_{\vec{k}m}(\tau)f_m(\vec{R}_j,\tau)\rangle = 0$$

$$\frac{\delta\Omega_{\text{eff}}}{\delta\lambda_j} = \mid b_j(\tau)\mid^2 + \langle\hat{n}_f(R_j,\tau)\rangle - Q_0 = 0 \tag{2.13}$$

In thermal equilibrium we seek a static solution to the Bose field $\tilde{b}_j(\tau)=\tilde{b}_j=\langle\hat{b}_j\rangle/\sqrt{N}$ leading to the simple meanfield equations

$$0 = \lambda_j\tilde{b}_j + \frac{1}{N}\sum_{\vec{k}m} V(k)\langle c^+_{\vec{k}m} f_m(\vec{R}_j)\rangle e^{-i\vec{k}\vec{R}_j}$$

$$|\tilde{b}_j\mid^2 + \tilde{n}_f(R_j) = q_0 \tag{2.14}$$

where $\tilde{n}_f = n_f/N$. We can set all of the phases equal, because of Q conservation symmetry. In the case of the two impurity model, there is inversion symmetry which ensures that the $\tilde{b}_j = \tilde{r}_0$ and $\lambda_j = \lambda$ are all equal in magnitude and in the case of the lattice, translation symmetry has the same effect, enabling us to write down a simple meanfield Hamiltonian governing the motion of the Fermions.

$$H^* = \hat{H}_{\text{band}} + \sum_{R_i}\tilde{E}_f\hat{n}_f(R_i) + \sum_{\vec{R}_i} H^*_{\text{mix}}(\vec{R}_i) \tag{2.15}$$

where $\hat{H}_{\rm band}$ describes the band electrons and

$$\hat{H}^*_{mix}(\vec{R}_j) = \sum_{\vec{k}m} \tilde{V}(k)\left\{c^+_{\vec{k}m} f_m(\vec{R}_j)\; e^{-i\vec{k}R_j} + H.C.\right\} \tag{2.16}$$

and $\tilde{E}_f = E_f + \lambda$, $\tilde{V}(k) = V(k)\tilde{r}_0$ are the renormalized f-level and hybridization. This Hamiltonian is self consistently determined by the meanfield equations (2.14). The most straightforward way to determine the expectation values in (2.14) is by calculating the equal time values of the meanfield propagators $G_f(\tau) = \langle f_m(\tau) f^+_m(0)\rangle$ and $G_{fb}(\tau) = \sum_{\vec{k}} \tilde{V}(k)\langle T f_m(\tau) c^+_{\vec{k}m}(0)\rangle$. To illustrate the application of these meanfield equations we shall consider the case of one and two impurities. For one impurity, in the large bandwidth limit we can write in frequency space

$$\begin{aligned} G_f(i\omega_n) &= [i\omega_n - \tilde{E}_f + i\tilde{\Delta} sgn(\omega_n)]^{-1} \\ G_{fb}(i\omega_n) &= -\tilde{\Delta} sgn(\omega_n) G_f(i\omega_n) \end{aligned} \tag{2.17}$$

where $\tilde{\Delta} = \Delta r_o^2$ is the renormalized resonance level width and $\Delta = \pi V(k_F)^2\rho$ is the bare resonance level width enabling us to determine

$$\begin{aligned} \langle \tilde{n}_f\rangle = \frac{1}{\beta}\sum_n G_f(i\omega_n) &= \frac{1}{\pi} Im[\tilde{\psi}(\xi) - ln\frac{D}{2\pi i}] \\ \sum \tilde{V}(k)\langle c^+_{\vec{k}m} f_m\rangle &= \frac{\tilde{\Delta}}{\pi} Re[\tilde{\psi}(\xi) - ln\frac{D}{2\pi i}] \end{aligned} \tag{2.18}$$

where $\xi = \tilde{E}_f + i\tilde{\Delta}$ defines the renormalized resonance level and we have asssumed a Lorentzian cutoff as before. This enables us to write the meanfield equations compactly as

$$\frac{\Delta}{\pi}\left[\tilde{\psi}(\xi) + \frac{i\pi}{2}\right] + \xi = E^*_f + i\Delta q + \ell n\frac{\Delta\beta}{2\pi^2} \tag{2.19}$$

where

$$E^*_f = E_f + \frac{\Delta}{\pi}\ell n\left(\frac{D\pi}{\Delta}\right) \tag{2.20}$$

is recognized as the leading expression for the Haldane scale invariant f-level in the $1/N$ expansion. These equations revert to (1.7) in the Kondo limit.

It turns out to be remarkably easy to extend these results to two impurities. Suppose we have two impurities at $\pm\vec{R}/2$. The meanfield Hamiltonian then becomes

$$\begin{aligned} \hat{H}^* &= \hat{H}_{\rm band} + \tilde{E}_f(\hat{n}_{f1} + \tilde{n}_{f2}) \\ &+ \sum_{\vec{k}m} \tilde{V}(k)\left[c^+_{\vec{k}m}\left\{f_{1m}e^{i\vec{k}.\vec{R}/2} + f_{2m}e^{-i\vec{k}.\vec{R}/2}\right\} + H.C.\right] \end{aligned} \tag{2.21}$$

where $\tilde{E}_f = E_f + \lambda$ and $\tilde{V}(k) = V(k)\tilde{r}_0$. This Hamiltonian can be split up in to two parts where the Fermi fields are symmetry or antisymmetric eigenstates of the parity operator by defining, for $P = \pm 1$

$$\left.\begin{aligned}
f_m^P &= \frac{1}{\sqrt{2}}\left\{f_{m1} + P f_{m2}\right\}\\
c_{km}^{(+)} &= \frac{1}{\sqrt{\lambda_+(k)}}\int \frac{d\Omega_k}{4\pi}\cos\left(\frac{\vec{k}\cdot\vec{R}}{2}\right)c_{\vec{k}m}\\
c_{km}^{(-)} &= \frac{i}{\sqrt{\lambda_-(k)}}\int \frac{d\Omega_k}{4\pi}\sin\left(\frac{\vec{k}\cdot\vec{R}}{2}\right)c_{\vec{k}m}
\end{aligned}\right\} \tag{2.22}$$

where $\lambda_P(k) = \frac{1}{2}(1 + P \cdot \frac{\sin kR}{kR})$. The antisymmetric and symmetric states anticommute $\{f_m^P, f_{m'}^{P'+}\} = \delta_{pp'}\delta_{mm'}$ and $\{c_{km}^P, c_{k'm'}^{P'+}\} = \delta_{pp'}\delta_{mm'}\delta_{kk'}$, enabling us to write $H^* = \sum_{P=\pm1} \hat{H}^P$, where

$$H^P = \sum_{km} E(k) c_{km}^{P+} c_{km}^P + \tilde{E}_f n_f^P + \sum_{km} \tilde{V}^P(k)\left[c_{km}^{P+} f_m^P + H.C.\right] \tag{2.23}$$

Here, $\tilde{V}^P(k) = V^P(k)\tilde{r}_0, V^P(k) = V(k_f)\sqrt{2\lambda_P(k)}$, and $\tilde{E}_f = E_f + \lambda$. The states of opposite parity are not mixed, enabling us to write

$$\begin{aligned}
&\frac{1}{N}\sum_{km} V(k)\langle c_{km}^+ f_m(\pm\vec{R}/2)\rangle e^{\pm ikR/2} = \frac{1}{2}\frac{1}{N}\sum_{k,P=\pm1,m} V^P(k)\langle c_{km}^{P+} f_{km}^P\rangle\\
&\tilde{n}_{f1} = \tilde{n}_{f2} = \frac{1}{2}\left(\tilde{n}_f^{(+)} + \tilde{n}_f^{(-)}\right)
\end{aligned} \tag{2.24}$$

Just as in the one impurity case, the f-electron propagator and the mixing propagator can be written

$$\begin{aligned}
G_f^P(i\omega_n) &= (i\omega_n - \tilde{E}_f - \Sigma_f^P(i\omega_n))^{-1}\\
G_{fk}^P(i\omega_n) &= \Sigma_f^P(i\omega_n) G_f^P(i\omega_n)
\end{aligned} \tag{2.25}$$

where $G_f^P(\tau) = \langle f_m^P(\tau) f_m^{+P}(0)\rangle$ and $G_{fk}^P(\tau) = \sum \tilde{V}^P(k)\langle T f_m^P(\tau) c_{km}^{P+}(0)\rangle$ and $\Sigma_f^P(i\omega_n) = \sum_k |\tilde{V}^P(k)|^2/(i\omega_n - E(k))$. Now providing that $R/v_F << \tilde{\Delta}^{-1}$, where v_F is the Fermi velocity, then the expectation values entering the mean field equations are dominated by electrons with wavevectors very close to the Fermi wavevector, and we can approximate $\tilde{V}^P(k)$ by its value at the Fermi energy $|\tilde{V}(k_F)|^2$. Within this "near field" region the impurities essentially interact through band electrons at the Fermi energy leading to only one relevant lengthscale: the Fermi wavelength. With this approximation

$$\Sigma_f^P(\omega) = -isgn(\omega_I)\Delta^P \tilde{r}_0^2 \Phi(\omega) \qquad R << v_F/\tilde{\Delta} \tag{2.26}$$

where $\Delta^P = \Delta(1 + P.(\sin(k_F R)/k_F R))$, $\omega_I = \text{lm}\omega$ and $\Phi(\omega)$ is the bandwidth cutoff function as before. We see that the symmetric and antisymmetric states form *two* renormalized levels, with the same renormalized position $\tilde{E}_f$, and different renormalized widths $\Delta^P \tilde{r}_0^2 = \tilde{\Delta}^P$. We can calculate

$$\begin{aligned}
\frac{1}{N}\sum_{m,k}\tilde{V}^P(k)\langle c_{km}^{P+} f_m^P\rangle &= \frac{\tilde{\Delta}^P}{\pi}\text{Re}\tilde{\psi}(\tilde{E}_f + i\tilde{\Delta}^P)\\
\frac{1}{N}\sum\langle f_m^{+P} f_m^P\rangle &= \text{lm}\tilde{\psi}(\tilde{E}_f + i\tilde{\Delta}^P)
\end{aligned} \tag{2.27}$$

and substituting in the meanfield equations we obtain

$$\begin{aligned}\tilde{E}_f + \frac{1}{2}\sum_{P=\pm1} \mathrm{Re}\left\{\frac{\Delta^P}{\pi}[\tilde{\psi}(\xi^P) - \ell n(\Delta\beta/2\pi^2 i)]\right\} &= E_f^* \\ \tilde{r}_0^2 + \frac{1}{2}\sum_{P=\pm1} \mathrm{Im}\left\{\frac{1}{\pi}[\tilde{\psi}(\xi^P) - \ell n(\Delta\beta/2\pi^2 i)]\right\} &= q_0\end{aligned} \tag{2.28}$$

where $\xi^P = \tilde{E}_f + i\tilde{\Delta}^P\tilde{r}_0^2$.

Clearly ,we have found a coherent meanfield solution characterized by a Fermi liquid involving two energy scales $\tilde{\Delta}^{\pm} = \tilde{\Delta}(1 \pm \sin(k_F R)/\sin(k_F R))$. Within this coherent Fermi liquid intersite propagation is related to the *difference* between the antisymmetric and symmetric propagators $G_f^{12}(\omega) = \frac{1}{2}[G_f^+(\omega) - G_f^-(\omega)]$ which reaches a maximum when $|\sin(k_F R)/(k_F R)|$ is maximized. This actually corresponds to placing the second impurity at a maximum of the spin cloud from the first.

3. 1/N Fluctuations about meanfield theory.

Our broken symmetry saddle point equations furnish the starting point for a controlled 1/N expansion. At finite degeneracy, the bose field is not rigid as in the meanfield limit, and it undergoes small fluctuations about the classical trajectory which we shall relate to collective properties of the Fermi liquid . In addition, motions of the quasiparticles generate additional fluctuations which create interactions between quasiparticles. We will calculate the Landau (or Nozieres) Fermi liquid parameters by calculating the (linear) response of the bose field to the quasiparticles.

Extending our calculations entails expanding the effective free energy functional in terms of small fluctuations $b(\tau) = b_o + \delta b(\tau)$ about the saddle point. In weak coupling, this is very straight forward, for the equilibrium value of the Bose field is zero. Strong coupling presents a problem because Goldstone mode fluctuations lead to unbounded phase fluctuations, so that $\langle\delta\theta^2\rangle$ is divergent.

The problem of dealing with the Goldstone modes has been elegantly solved by Read and Newns [6]. Read and Newns exploit the Q conservation symmetry to carry out a simple gauge transformation on the fields, in which the phase associated with the Bose field is factored out of both the Bose and f-electron fields, writing schematically

$$\begin{aligned} b(\tau) &= r(\tau)e^{i\theta(\tau)} \\ f_m(\tau) &= f'_m(\tau)e^{i\theta(\tau)} \end{aligned} \tag{3.1}$$

By factoring out the phase fluctuations the divergences are removed and one obtains an equivalent description of the dynamics which is appropriate to the Fermi liquid. Whereas the untransformed fields have divergent strong coupling interactions at low energies, the transformed Fermi fields have well behaved and finite interactions at low energies. Two aspects of this transformation must be considered: (i) the changes in measure which must be introduced into the path integral and (ii) the modified Lagrangian.

The Jacobian associated with the transformation of the Fermi fields involves a product of phase factors and is therefore 1. The new measure for the transformed Bose fields is now

$$\mathcal{D}[r,\theta] = \prod_{\tau} \frac{du(\tau)d\theta(\tau)}{2\pi} \tag{3.2}$$

where $u(\tau) = r^2(\tau)$.

The transformed Lagrangian contains a new interaction term $i\dot{\theta}(Q - Q_o)$ coming from the derivative terms in the Free energy functional.

$$\int_0^\beta d\tau \left\{\bar{f}_m(\tau)\partial f_m(\tau) + \bar{b}(\tau)\partial b(\tau)\right\} = \int_0^\beta d\tau \left\{\bar{f}'_m(\tau)\partial f'_m(\tau) + i\dot{\theta}[Q(\tau) - Q_0]\right\} \tag{3.3}$$

where ∂ denotes $\dfrac{\partial}{\partial\tau}$ and $\dot{\theta} = \partial\theta(\tau)/\partial\tau$. The periodic boundary conditions in the Bose field ensure that we can remove total derivative terms from the functional.

When we integrate over θ we must be careful to allow for paths where θ winds many times around the origin. We will factor out the integration over the constant phase and this enables us to express the angular part of the measure in terms of the angular velocity $\dot{\theta}$. Expanding the angular velocity as a Fourier series $\dot{\theta}(\tau) = \sqrt{k_B T}\sum_n \dot{\theta}_n e^{-i\nu_n\tau}$ then in the periodic boundary conditions restrict the static part of $\dot{\theta}$ to integral multiples of $2\pi/\beta$. Now in the constraint integral over the chemical potential, $Im\lambda = \Omega_0$ runs from 0 to $2\pi/\beta$. In the Lagrangian, Ω_0 and $\dot{\theta}$ enter in the same way so if we define $\dot{\theta}'(\tau) = \dot{\theta} + \Omega_0$, then the new function can have an arbitrary static value, allowing us to write

$$\frac{d\lambda\beta}{2\pi i}\mathcal{D}[\theta] = \mathcal{N}\prod_{\tau}\left\{\frac{d\dot{\theta}(\tau)\beta}{2\pi}\right\} = \mathcal{D}[\dot{\theta}] \tag{3.4}$$

where $\mathcal{N} = \prod_{n\neq 0}(\beta i\nu_n)^{-1}$ is the normalization factor. Thus, by including paths with any winding number, we are able to incorporate the constraint naturally as part of the dynamics.

The partition function in the transformed fields is then written

$$Z_{MV} = \int \mathcal{D}[r,\dot{\theta}]\mathcal{D}[\bar{\psi},\psi]e^{-\int_0^\beta d\tau\mathcal{L}(\tau)} \tag{3.5}$$

where $\mathcal{L}(\tau)$ is the transformed Lagrangian, given by

$$\mathcal{L}(\tau) = \sum_i \bar{\psi}_i\frac{\partial}{\partial\tau}\psi_i + H'[\bar{\psi},\psi,r,\dot{\theta}] + \lambda_0[r^2(\tau) - Q_0] \tag{3.6}$$

and $\hat{H}' = \hat{H}_{\mathrm{band}} + \sum_m \tilde{E}_f\hat{n}_{fm} + \hat{H}_{int}$ with

$$\hat{H}_{int} = r(\tau)\sum_{k,m}\frac{V(k)}{\sqrt{N}}\{c^\dagger_{km}f_m + f^\dagger_m c_{km}\} + i\dot{\theta}(\tau)[\hat{Q} - Q_0] \tag{3.7}$$

where $f_m(\tau)$ refers to the gauge transformed f electron field.

In the transformed coordinates the Fermion part of the Lagrangian no longer contains the explicit Q conservation symmetry $b \to be^{i\theta}, f \to fe^{i\theta}$, and now the integral over $\dot{\theta}$ imposes Q conservation on the dynamics. This clever transformation is due to Read and Newns. Unlike their derivation we have shown that it is unnecessary to promote the chemical potential to a dynamical variable if one incorporates all paths with multiple winding number.

For small fluctuations about our meanfield limit, the Bose measure can be approximated as locally Cartesian

$$\mathcal{D}[r,\dot{\theta}] = \mathcal{N}'\prod_{\tau} dr(\tau)d\dot{\theta}(\tau) \tag{3.8}$$

where $\mathcal{N}'$ is the appropriate normalization constant(we will never need to know this constant). The calculation of the small Gaussian fluctuations provides us with a sort of semi-classical theory of mixed valence which contains much of the important physics of strong coupling. We shall develop the approach using the one impurity model as our example and then discuss how it is generalized to include spatial fluctuations in the many site case.

When we expand the Lagrangian about the saddle point we find

$$\mathcal{L}(\tau) = \bar{\psi}_i \frac{\partial}{\partial\tau}\psi_i + H^*[\bar{\psi},\psi] + N\lambda_0[\tilde{r}_0^2 - q_0] + H_{\rm int} \tag{3.9}$$

where $H^*[\bar{\psi},\psi]$ is the mean field Hamiltonian and

$$\hat{H}_{\rm int} = \sum_{km} V(k)\left[f_m^+ c_{km} + c_{km}^+ f_m\right]\delta\tilde{r}(\tau) + i\dot{\theta}\hat{n}_f$$
$$+ N\left[\lambda\delta\tilde{r}(\tau)^2 + 2i\delta\tilde{r}(\tau)\tilde{r}_0\dot{\theta}(\tau)\right] \tag{3.10}$$

We shall introduce the nomenclature $\vec{r}(\tau) = (\delta\tilde{r}(\tau), i\tilde{r}_0\dot{\theta}(\tau))$ where $\sqrt{N}\delta\tilde{r} = (r(\tau) - r_0(\tau))$ and $\sqrt{N}\tilde{r}_0 = r_0$ are rescaled fields. Then to second order

$$\beta\Omega_{\rm eff} = \beta\Omega_0 + \int_0^\beta d\tau\vec{j}(\tau)\cdot\vec{r}(\tau) + \frac{N}{2}\int_0^\beta d\tau_1 d\tau_2 \vec{r}^T(\tau_2)\Gamma(\tau_2-\tau_1)\vec{r}(\tau_1) \tag{3.11}$$

where a source term has been added to enable us to generate the Bose propagator from the partition function. Carrying out the path integral over the Gaussian fluctuations we find

$$\ell n Z_{MV}[\vec{j}] = -\beta[\Omega_0 + \Omega_1] + \frac{1}{2N}\int_0^\beta d\tau_1 d\tau_2\, \vec{j}^T(\tau_2)\Gamma(\tau_2-\tau_1)^{-1}\vec{j}(\tau_1) \tag{3.12}$$

where $\Omega_1 = \frac{1}{2}k_B T {\rm Tr}\ell n N\Gamma$. From (3.12) we can read off the Bose propagator as $\langle T\delta\vec{r}(\tau)\ \delta\vec{r}(0)\rangle = \delta^2 \ell n Z_{MV}[\vec{j}]/\delta\vec{j}(\tau)\delta\vec{j}(0)\Big|_{\vec{j}=0} = 1/N\Gamma^{-1}(\tau)$ and from (3.9) we see that $\Gamma(\tau)$ is divided into two parts $\Gamma(\tau) = \Gamma^{(0)}(\tau) + \Pi(\tau)$ where

$$\Gamma^{(0)}(\tau) = 2\begin{bmatrix}\lambda & 1\\ 1 & 0\end{bmatrix}\delta(\tau) \tag{3.13}$$

is the bare Bose contribution to the Gaussian terms and Π is the self energy of the Bose field generated from the RPA polarization graphs formed by contracting two meanfield Fermion propagators with H_{int} (Fig. 1.)

From the RPA diagrams, using the large bandwidth approximation we find that

$$\Pi_{rr}(i\nu_n) = 2\left[\frac{1}{g(i\nu_n)}\mathrm{Re}\mathcal{L}(i\nu_n) + \frac{\Delta}{\pi}\left[\mathrm{Re}\tilde{\psi}(\xi) - \ell n\frac{D\beta}{2\pi}\right]\right]$$
$$\Pi_{r\dot{\theta}}(i\nu_n) = 2\frac{1}{\nu_n}lm\mathcal{L}(i\nu_n) - 2 \tag{3.14}$$
$$\Pi_{\dot{\theta}\dot{\theta}}(i\nu_n) = -2\left(\frac{g(i\nu_n)}{\nu_n^2}\right)\mathrm{Re}\mathcal{L}(i\nu_n) \qquad (n > 0)$$

where $\Pi(i\nu_n) = \Pi^*(-i\nu_n)$ for $n < 0$ and

$$g(i\nu_n) = \frac{i\nu_n}{i\nu_n + 2i\tilde{\Delta}}$$
$$\mathcal{L}(i\nu_n) = \frac{\Delta}{\pi}\left[\tilde{\psi}(\xi + i\nu_n) - \tilde{\psi}(\xi)\right] + i\nu_n \tag{3.15}$$

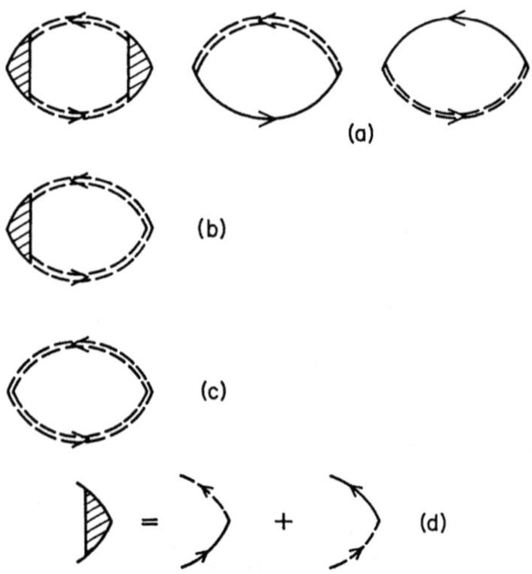

Fig. 1.: RPA diagrams contributing to single impurity Bose propagator. Dashed lines refer to f- propagator, continuous lines to the band electron propagator. (a) Π_{rr} ,(b) $\Pi_{r\dot{\theta}}$ and (c) $\Pi_{\dot{\theta},\dot{\theta}}$. This becomes

$$\mathcal{L}(i\nu_n) = \frac{\Delta}{\pi}\ell n\left(\frac{\xi + i\nu_n}{\xi}\right) + i\nu_n \tag{3.16}$$

in the $T \to 0$ limit. Adding $\Pi(i\nu_n)$ and $\Gamma^0(i\nu_n)$ at the saddle point and inverting the matrix yields

$$\Gamma^{-1}(i\nu_n) = R(i\nu_n) = \frac{1}{2\mathcal{L}(i\nu_n)\mathcal{L}^*(i\nu_n)}\begin{bmatrix} g(i\nu_n)\mathrm{Re}\mathcal{L}(i\nu_n) & \nu_n lm\mathcal{L}(i\nu_n) \\ \nu_n lm\mathcal{L}(i\nu_n) & \frac{-\nu_n^2}{g(i\nu_n)}\mathrm{Re}\mathcal{L}(i\nu_n)\end{bmatrix} \tag{3.17}$$

Equivalent results can be obtained by carefully using the Cartesian Bose coordinates

[14] . In particular one finds that

$$\tilde{r}_0^2\langle\theta(i\nu_n)\theta(-i\nu_n)\rangle = -\frac{1}{N\nu_n^2}R_{\dot\theta\dot\theta}(i\nu_n) \tag{3.18}$$

The frequency denominator present in the transverse fluctuations is the Goldstone divergence associated with the low frequency spin fluctuations from which we deduce that at long times the transverse fluctuations are given by

$$\langle\delta\theta(\tau)\delta\theta(0)\rangle \sim \frac{\alpha}{N}\ell n\tau \tag{3.19}$$

where $\alpha = [Re(\Delta/(\Delta+\pi\xi))]^2 \sim 1$ enabling us to determine the asymptotic behaviour of the Bose propagator

$$\langle b(\tau)b^+(\tau)\rangle \sim r_0^2 e^{\langle\delta\theta(\tau)^2\rangle/2} \sim \tau^{-\alpha/N} \tag{3.20}$$

which fits in well with our picture of the Goldstone mode destroying the broken symmetry.

In the Kondo regime, the first term in $\mathcal{L}(i\nu_n)$ dominates and provided one is interested in properties at frequencies $\nu << \Delta$, the linear frequency term can be neglected. We can analytically extend these results to the real axis, for instance the radial propagator is given by

$$\langle\delta\tilde{r}(\nu+i\delta)\delta\tilde{r}(\nu+i\delta)\rangle = \frac{1}{2Ng(\nu)}\left[\frac{1}{\mathcal{L}(\nu)}+\frac{1}{\mathcal{L}^*(\nu)}\right],$$

$$\mathcal{L}^*(\nu) = \frac{\Delta}{\pi}\left[\tilde\psi(\nu-\xi^*)-\tilde\psi(-\xi^*)\right]-\nu \tag{3.21}$$

The bulk of the radial oscillations occur at a much higher frequency $\sim E_f^*$ than the renormalized resonant level width $\tilde\Delta$. We shall see that we can associate these fluctuations with the high frequency charge fluctuations. At low temperature the quasiparticles move in the average field of the Boson, and fail to see these high frequency fluctuations.

From $\Gamma(i\nu_n)$ we can also determine the shift in the Free energy due to zero point and thermal fluctuations

$$\Omega_1 = \int_{-\infty}^{+\infty}\frac{d\nu}{\pi}(n(\nu)+\frac{1}{2})\mathrm{Im}[\ell n\mathcal{L}(\nu)] \tag{3.22}$$

4. Dynamic susceptibilies.

The collective properties of the Fermi liquid are related to the Bose fluctuations and we can relate the dynamic correlations in the spin, charge and $q(\tau) = Q(\tau)/N$ to the Gaussian fluctuations.

To determine the collective spin and charge fluctuations of the f-electrons we associate a source term with the f-occupations which we add to the Free energy functional $\beta F_s[\phi_m] = \sum_{m=-j}^{j}\int_0^\beta d\tau\phi_m(\tau)n_{fm}(\tau)$. This enables the determination of the dynamic susceptibility of the f-electrons from a functional derivative of the full free energy

$$\chi_{mm'}(\tau) = \langle T\delta\hat{n}_{fm}(\tau)\delta\hat{n}_{fm'}(0)\rangle = \frac{\delta^2\ell nZ_{MV}[\phi]}{\delta\phi_m(\tau)\delta\phi_{m'}(0)} \tag{4.1}$$

Diagrammatically this is simply the RPA susceptibility. Now since the source term enters into the effective energy in the same way as the angular velocity, to second order the effective free energy is given by

$$\beta\Omega_{\text{eff}} = \beta\Omega_0 + \frac{N}{2}\sum_m (\vec{r}+\vec{\phi}_m)^T \Pi (\vec{r}+\vec{\phi}_m) + \vec{j}\cdot\vec{r} \tag{4.2}$$

where we have used a shorthand notation for the time integrations $\vec{j}\cdot\vec{r} = \int_0^\beta d\tau \vec{j}(\tau)\vec{r}(\tau)$ and put $\vec{\phi}_m = (0, \tilde{r}_0\phi_m)$. The Gaussian integral over the Bose fields then introduces the following source terms to the free energy

$$\frac{1}{2N}[\vec{j}^T + \sum_m \vec{\phi}_m^T \Pi]\Gamma^{-1}[\vec{j} + \Pi\sum_m \vec{\phi}_m] - \sum_m \vec{\phi}_m^T \Pi \vec{\phi}_m \tag{4.3}$$

so that

$$\chi_{mm'} = \tilde{r}_0^2 \left[\frac{1}{N}(\Pi\Gamma^{-1}\Pi) - \delta_{mm'}\Pi\right]_{\dot{\theta}\dot{\theta}} \tag{4.4}$$

where $\dot{\theta}\dot{\theta}$ denotes the $(i\tilde{r}_0\dot{\theta}, i\tilde{r}_0\dot{\theta})$ element of the matrix operator. It is now straightforward to derive the spin and charge susceptibilities $\chi_s(\tau) = \langle TM_z(\tau)M_z(0)\rangle$ and $\chi_c(\tau) = \langle T\delta\hat{n}_f(\tau)\delta\hat{n}_f(0)\rangle$ where $M_z = \sum_{mm} n_{fm}$. Putting $\chi_s = \sum m^2\tilde{\chi}_s$ and $\chi_c = N\tilde{\chi}_c$ we find

$$\begin{aligned}\tilde{\chi}_c(i\nu_n) &= \tilde{r}_0^2\left[\Pi(i\nu_n)\Gamma^{-1}(i\nu_n)\Pi(i\nu_n) - \Pi(i\nu_n)\right]_{\dot{\theta}\dot{\theta}} \\ &= N\left\{4\tilde{r}_0^2\langle\delta\tilde{r}(i\nu_n)\delta\tilde{r}(i\nu_n)\rangle\right\} = 2\tilde{r}_0^2 g(i\nu_n)\text{Re}[\mathcal{L}(i\nu_n)^{-1}]\end{aligned} \tag{4.5}$$

whilst

$$\tilde{\chi}_s(i\nu_n) = -\tilde{r}_0^2\left[\Pi(i\nu_n)\right]_{\dot{\theta}\dot{\theta}} = \left(\frac{2\tilde{r}_0^2}{(\nu_n + 2\tilde{\Delta})\nu_n}\right)\text{Re}\mathcal{L}(i\nu_n) \tag{4.6}$$

Taking $i\nu_n \to 0$ we find that the static susceptibilities are

$$\begin{aligned}\tilde{\chi}_c(0) &= \frac{1}{\Delta} lm\left(i + \frac{\Delta}{\pi}\tilde{\psi}'(\xi)\right)^{-1} \\ \tilde{\chi}_s(0) &= -\text{Im}\left(\frac{1}{\pi}\tilde{\psi}'(\xi)\right)\end{aligned} \tag{4.7}$$

Analytically extending the spin susceptibility to the real axis we find at $T = 0$

$$\tilde{\chi}_s(\nu + i\delta) = \frac{-1}{(\nu + 2i\tilde{\Delta})\nu}\frac{\tilde{\Delta}}{\pi}\left[\ell n\left(\frac{(\xi+\nu)(\xi^* - \nu)}{\xi\ \xi^*}\right)\right] \tag{4.8}$$

Taking the $\nu \to 0$ limit we find

$$\lim_{\nu\to 0}\text{Im}\frac{\tilde{\chi}_s(\nu + i\delta)}{\nu} = \pi\left[\frac{\tilde{\Delta}}{\pi(\tilde{E}_f^2 + \tilde{\Delta}^2)}\right]^2 = \pi\left[\tilde{\chi}_s(0, T=0)\right]^2 \tag{4.9}$$

This is the Korringa relation derived by Shiba for the infinite U Anderson model [14]. Satisfaction of this equality in the leading fluctuations is an indication that our approximation is conserving the Ward identities associated with spin conservation. Earlier work has found difficulties in eliminating infrared divergences at zero frequency and has failed to satisfy this relationship .

From (4.3) we can also calculate the correlation between the radial fluctuations and the charge fluctuations

$$\begin{aligned}\langle T\delta\tilde{r}(\tau)\delta\tilde{n}_{fm}(0)\rangle &= \frac{1}{N}\frac{\delta^2 \ell n Z[\vec{j},\phi_m]}{\delta j_r(\tau)\delta\phi_m(0)} = -4N\tilde{r}_0\langle\delta\tilde{r}(\tau)\delta\tilde{r}(0)\rangle \\ \langle T\delta\tilde{n}_{fm}(\tau)\delta\tilde{r}(0)\rangle &= \langle T\delta\tilde{r}(\tau)\delta\tilde{n}_{fm}(0)\rangle \end{aligned} \tag{4.10}$$

where $\tilde{n}_{fm} = n_{fm}/N$, from which we see thay are anticorrelated. We can sum all these fluctuations to determine the fluctuations in $q(\tau)$

$$\begin{aligned}\langle T\delta q(\tau)\delta q(0)\rangle = & 4\tilde{r}_0^2\langle\delta\tilde{r}(\tau)\delta\tilde{r}(0)\rangle + 2\tilde{r}_0\langle\delta\tilde{r}(\tau)\delta\tilde{n}_f(0)\rangle + \\ & 2r_0\langle\delta\tilde{n}_f(\tau)\delta\tilde{r}(0)\rangle + \langle\delta\tilde{n}_f(\tau)\tilde{n}_f(0)\rangle \\ = & 0\end{aligned} \tag{4.11}$$

and we see that the anticorrelations between the Bose and f-electron fields lead to a vanishing correlation function. This is a crucial result, for it explicitly shows that we are consistently maintaining q conservation at all times to leading order in the Gaussian fluctuations. We note that the way our constraint is built in derives from the underlying symmetry of the dynamics, rather than a careful selection of time ordered Goldstone diagrams as in perturbative schemes [15,16].

5. Non local fluctuations.

By building in the constraint as an underlying symmetry, our approach extends naturally to spatially extended systems. By way of example we shall discuss the two impurity model, where there are two non-degenerate modes of oscillation in the Bose field corresponding to symmetric and antisymmetric spin and charge fluctuation modes. Two conserved charges Q_i means two Goldstone modes leading to a strong coupling divergence in both inter and intrasite interactions.

We refer the reader to other work [15] for details of the two impurity calculation. Here we shall merely note that because reflection parity is conserved in the interaction between the Fermions and Bose fields, we can diagonalize the Gaussian fluctuations in terms of the even and odd parity fluctuation modes $\vec{r}^{\pm} = \{\vec{r}_1 \pm \vec{r}_2\}/\sqrt{2}$. and using the nearfield approximation mentioned above, we are able to explicitly calculate the Gaussian expansion of the effective Free energy

$$\frac{\beta\Omega_{\rm eff}}{N} = \frac{\beta\Omega_0}{N} + \frac{1}{2}\sum_{P=\pm}\int_0^\beta d\tau' d\tau_2 \vec{r}^{(P)T}(\tau_2)\Gamma(\tau_2-\tau_1)\vec{r}^P(\tau_1) \tag{5.1}$$

Corresponding to the Γ^P are the even and odd parity Bose propagators, given by

$$\langle\delta\vec{r}^P(i\nu_n)\delta\vec{r}^P(-i\nu_n)\rangle = \frac{1}{N}R^P(i\nu_n) = \frac{1}{N}[\Gamma^P(i\nu_n)]^{-1} \tag{5.2}$$

We can repeat the derivations for the single mode in the one impurity model for both modes in the two impurity model from which we automatically know that fluctuations in q are zero to Gaussian order and that the even and odd spin and charge susceptibilites are given by

$$\begin{aligned} \tilde{\chi}_c^P(\tau) &= 4\tilde{r}_0^2 R_{rr}^P(\tau) \\ \tilde{\chi}_s^P(\tau) &= -\tilde{r}_0^2 \Pi_{\dot\theta\dot\theta}^P(\tau) \end{aligned} \tag{5.3}$$

The intersite spin susceptibilities are then given by $\tilde{\chi}_{12} = \{\tilde{\chi}^+ - \tilde{\chi}^-\}/2$, whilst the intrasite susceptibilities are the average of the even and odd parity susceptibilies. Figure 2. shows an example of the even and odd dynamical spin susceptibility calculated using these results, clearly indicating how the intersite coherence leads to a marked splitting of the degeneracy between even and odd parity spin correlations at energies comaparable to $\tilde{\Delta}$. In the case shown $q = \frac{1}{6}$ and the even parity susceptibility is smaller than the odd parity susceptibility leading to antiferromagnetic intersite spin correlations.

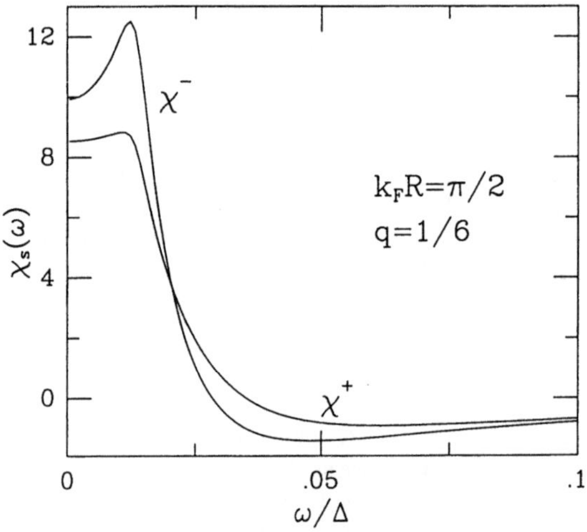

Figure 2. : showing the real parts of the symmetric (+) and antisymmetric spin susceptibilities in the two impurity model.

For the static intersite susceptibility we find($T = 0$)

$$\tilde{\chi}_{12} = \frac{1}{2}\left[\frac{1}{2\pi}\sum_{P=\pm}\frac{\tilde{\Delta}^P}{|\xi^P|^2} - \frac{1}{\pi\tilde{\Delta}(\sin k_F R/k_F R)}\ell n\left|\frac{\xi^+}{\xi^-}\right|\right] \tag{5.4}$$

There are two competing terms in this susceptibility of opposite sign: infact we shall be able to interpret each term in terms of the Fermi liquid properties. In the case where $q = \frac{1}{2}$ in the Kondo regime we actually find that the second term dominates, leading to a parallel pairing tendency between sites. This triplet pairing feature also emerges when we calculate the Fermi liquid properties.

6. Fermi Liquid Properties.

The low energy static Fermi liquid properties are directly related to the Bose propagators. Thus the f-electron part of the quasiparticle interactions in the one impurity case is given by

$$\hat{H}_{\rm int} = \frac{1}{2}U^* \sum_{m,m'} f^+_m f^+_{m'} f_{m'} f_m \tag{6.1}$$

where

$$U^* = \langle i\dot{\theta}(\nu = 0) i\dot{\theta}(\nu = 0)\rangle = \frac{1}{Nr_0^2} R_{\dot{\theta}\dot{\theta}}(\nu = 0) = \rightarrow \frac{1}{N}\pi\tilde{\Delta} \qquad (\tilde{\Delta} << \Delta) \tag{6.2}$$

is the renormalized interaction strength in the Kondo regime. What we are doing is using the response function of the Bose fields to additional quasiparticles to determine the interaction that they mediate. By summing up the interaction terms generated from each component of $\mathbf{R}(0)$ we can determine the net interactions between the quasiparticles. The results are most elegantly expressed in terms of the change in the scattering phase shifts when we add an additional quasiparticle. Due to lack of space we shall report on the result we obtain for the two impurity model. Suppose we add $n^{P'}_{m'}$ quasiparticles of parity P' and spin component m' near the Fermi energy and examine the perturbation to the phase shift in the channel of parity P, spin m. We find that

$$\delta^P(\omega, n^{P'}_{m'}) = \delta^P_m(\omega) + \sum_{P'm' \neq Pm} \phi^+_{PP'} n^{P'}_{m'} - \phi^-_{P,-P} n^{-P}_m \tag{6.3}$$

where

$$\phi^{\pm}_{PP'} = \frac{1}{N\rho}(\pi\rho_{fP}\rho_{fP'})\Phi^{\pm} \tag{6.4}$$

where $\rho_{fP} = \tilde{\Delta}^P/\pi|\xi^P|^2$ is the quasiparticle density of states in the channels with parity P and Φ^P is an interaction strength determined by

$$\Phi^P = \frac{1}{\bar{r}_0^2}\left[(2\tilde{E}_f)^2 R^P_{rr} - (4\tilde{E}_f)R^P_{r\dot{\theta}} + R^P_{\dot{\theta}\dot{\theta}}\right] \tag{6.5}$$

In this expression the propagators are evaluated at zero frequency. We can actually write

$$\hat{H}_{int} = \frac{1}{2}\Phi^+ \sum_{Pm \neq P'm'} \hat{n}^P_m \hat{n}^{P'}_{m'} - \sum_m \Phi^- \hat{n}^+_m \hat{n}^-_m \tag{6.6}$$

as the effective interaction between the quasiparticle states at the Fermi energy. Φ^+ and Φ^- are the direct and exchange interactions between the quasiparticles and they only become equal at large seperations of the impurities, where even and odd fluctuations become degenerate. In this limit we find that

$$\phi^+ = \phi^- = \frac{1}{N\rho} lm\left\{\left(1 + \frac{\pi\xi}{\Delta}\right)\xi\right\}^{-1} \quad (R >> 0) \tag{6.7}$$

and we find we have derived the simple Nozieres Fermi liquid picture. In fact, in the Kondo regime where $\tilde{\Delta} << \Delta$, we find $\phi \approx \frac{1}{N\rho} lm \xi^{-1}$, so that in this limit

$$N\phi\rho = -\pi\rho_f \tag{6.8}$$

which is the Nozieres Fermi liquid identity written to leading order in $1/N$ [3].

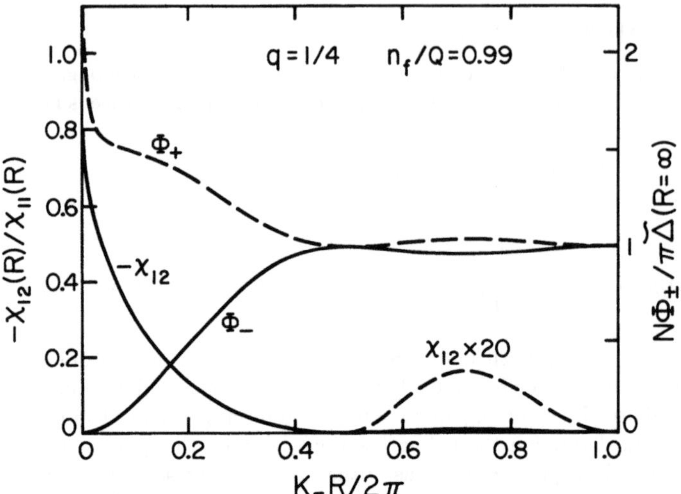

Fig. 3. showing (i) the ratio of the intersite susceptibility to the intrasite susceptibility χ_{12}/χ_{11} in the Kondo regime with $n_f/Q = 0.99$ and $q = \frac{1}{4}$; (ii) the direct (Φ_+) and exchange (Φ^-) interactions between the quasiparticles.

Fig. 3. shows the direct and exchange interactions as a function of separation for the case $q = \frac{1}{4}$. The rapid approach of the interactions to the the one impurity value as the separation is raised reflects a high degree of localization of the quasiparticles. In the case $q = \frac{1}{2}$ (not shown)in the Kondo regime, we actually find that the exchange term is larger than the direct term in the interaction, which poses the interesting question as to whether in a true spin half Kondo system, one might have ferromagnetic correlations between the spins at different sites.

Clearly, the gauge transformation of Read and Newns provides us with the link we sought between singular renormalization group behaviour and the well behaved nature of the Fermi liquid. Had we attempted to determine the interactions between the bare f-electron fields, we would always find that the Goldstone mode introduces a singularity in the low energy interactions. Read and Newns' transformation removes this singularity providing us with a natural low energy basis for describing the Fermi liquid properties. Indeed, we note that because the interactions between the gauge transformed fields are not singular at low energies, the full f-propagator, given by $\langle TX_{0m}(\tau)X_{m0}(0)\rangle$ is dominated at long times and low energies by the meanfield propagator , and we find in the leading fluctuations that

$$\langle TX_{0m}(\tau)X_{m0}(0)\rangle = \langle Tr(\tau)f_m(\tau)f^+_m(0)r(0)\rangle \to r_0^2\langle Tf_m(\tau)f^+_m(0)\rangle \qquad (6.9)$$

at long times or short frquencies, and that the high frequency charge fluctuations only affect the short time behaviour of the full f-propagator. We are able to identify r_0^2 as the quasiparticle pole strength.

6. Discussion.

Many aspects of the broken symmetry approach have not been covered here, and the interested reader is directed to more complete treatments currently in preperation [14]. We have made no attempt to take the renormalizations of the saddle point equations by the fluctuations into account. One of the principle affects of the fluctuations on the saddle point is to renormalize the scale invaraint f-level, which now becomes $\tilde{E}_f^{**} = E_f + (1 - 1/N)\Delta/\pi \ell n(D\pi/\Delta)$ due to the high frequency charge fluctuations. However, once we have taken these $1/N$renormalizations into account, our leading expressions for the correlations retain their current form in terms of renormalized expressions for the effective f-level width $\tilde{\Delta}$ and position $\tilde{E}_f$.

One of the interesting prospects offered by our approach is the possibility of extending the calculations to a lattice with a realistic hybridization between spin half band electrons and spin j- f electrons. It is straightforward to generalize the mixing matrix element entering into H_{mix} to accomodate more complicated mixing. Clearly, many interesting lattice properties in the Fermi liquid regime rely on the highly anistotropic narrow band structure that will accompany realistic mixing. Once the meanfield bandstructure can be determined, then the interactions and dynamic correlations can be modelled using the polarziation matrix Π. For instance, the dynamic spin susceptibility of the f-electrons is determined from the "angular stiffness" as in the one and two impurity models

$$\tilde{\chi}(\vec{q},\omega) = -\tilde{r}_0^2 \Pi_{\dot{\theta}\dot{\theta}}(\vec{q},\omega) \tag{7.1}$$

and at low frequencies we can expect significant $\vec{q}$ dependence in this quantity.

Another aspect of narrow band systems that seems likely to be closely related to the Bose fluctuations are the transport properties. It is tempting to attempt to ascribe the rapid rise in resistivity at low temperatures in the metallic mixed valence systems to the interaction between the band electrons and the phase fluctuations which are unconstrained at high energies and liable to lead to a great deal of scattering. Perhaps we can find a simple relationship between the angular fluctuations and the conductivity?

In conclusion then, we have presented a novel approach to dealing with strong coupling in mixed valence and heavy Fermion systems, which seems to provide a simple and fully consistent way of modelling the dynamics and interactions in these highly renormalized systems. Quite apart from the prospects it offers for modelling these systems, it offers the interesting new conceptual picture of strong coupling as a state of almost broken symmetry.

ACKNOWLEDGMENTS.

The author would particularly like to thank N. Andrei, A. Ruckenstein, D. J. Scalapino and J. R. Schrieffer for many useful discussions. This research was supported by the National Science Foundation Grant No. PHY77-27084.

References.

1. P. W. Anderson and G. Yuval, P.R.L. **23** , 89 (1969).
2. K. G. Wilson, Rev. Mod. Phys. **47** , 773 (1970).
3. P. Nozieres, J. Low Temp Phys. **17**, 31 (1974).

4. P. W. Anderson, in **Valence Fluctuations in Solids** , editors L. M. Falikov, W. Hanke and B. Maple (North Holland, Amsterdam, 1983).
5. S. Chakravarty, Proc. March Meeting American Physical Society, (1982).
6. D. Newns and N. Read, J. Phys C. **16**, 3273 (1983); J. Phys C. **29** L1055 (1983).
7. P. Nozieres and C. De Dominicis, Phys Rev **178** 1097 (1968).
8. P. W. Anderson, Phys Rev **86** ,894 (1952).
9. E. Witten, Nucl. Phys. B., **145**, 110 (1978).
10. N. Andrei and P. Coleman, Institute for Theoretical Physics Preprint, UCSB, CA 93106 (1985).
11. P. Coleman, Phys. Rev. B., **29** 3035 (1985); Proc. 5th Int. Conf. on Valence Fluctuations, J. Mag. Mat. , 1984.
12. A. Ruckenstein, private communication.
13. U. Shiba, Prog. Theor. Phys. **54** 967 (1975).
14. P. Coleman, Institute for Theoretical Physics Preprint, UCSB, CA 93106 (1985), to be published.
15. H. Keiter and J. Kimball, Int. J. Magn. **1** 233 (1971).
16. Y. Kuramoto, Zeit Phys. B. **53** 37 (1983).

Ground State of the Periodic Anderson Hamiltonian

K. Yamada
Research Institute for Fundamental Physics, Kyoto University
Kyoto 606, Japan

K. Yosida
Faculty of Science and Technology, Science University of Tokyo
Noda 278, Japan

Specific heat and magnetic susceptibility of the periodic Anderson Hamiltonian are discussed generally, on the basis of the microscopic Fermi liquid theory. It is pointed out that the Wilson ratio in the periodic system can take values over a different range from that of a single impurity system. This fact is related to coherent phenomena, such as decrease of the T-linear coefficient of the specific heat, observed in Kondo lattice systems. The enhancement factor of T-linear specific heat is calculated by the perturbation method with respect to Coulomb repulsion U. The local electronic structure of the paramagnetic ground state is shown to have a similar character to that of the singlet ground state of the dilute case by the discussion based on the Anderson orthogonality theorem.

1. Introduction

In our previous papers [1,2], we discussed the ground-state properties of the periodic Anderson Hamiltonian:

$$H = H_0 + H' \ , \tag{1.1}$$

$$H_0 = \sum_{k,\sigma} \varepsilon_k c^+_{k\sigma} c_{k\sigma} + \frac{1}{\sqrt{N}} \sum_{\substack{k,i \\ \sigma}} V_k (e^{-ik\cdot Ri} a^+_{i\sigma} c_{k\sigma} + e^{ik\cdot Ri} c^+_{k\sigma} a_{i\sigma}) \ , \tag{1.2}$$

$$H' = E_0 \sum_{i,\sigma} a^+_{i\sigma} a_{i\sigma} + U \sum_i a^+_{i\uparrow} a_{i\uparrow} a^+_{i\downarrow} a_{i\downarrow} \ , \tag{1.3}$$

where $c^+_{k\sigma}(c_{k\sigma})$ is the creation (annihilation) operator for the conduction electron with energy ε_k and momentum k. $a^+_{i\sigma}(a_{i\sigma})$ is that for the localized f-electron with energy E_0, interacting with each other by Coulomb repulsion U in the same atom. These two bands of electrons hybridize by mixing matrix element V_k. In the previous papers [1,2] we confined ourselves to the symmetric case with respect to electrons and holes, fixing the total electron number per site to 2. For this case, we concluded that the hybridization gap at the Fermi surface survives under the Coulomb repulsion.

According to recent developments [3,4], the Kondo temperature T_K in the system with two levels due to a crystal field is given by

$$k_B T_K = (D/\Delta)^{n_e/n_g} D \exp[-1/\rho J] \ , \ (k_B T_K \ll \Delta \ll D) \tag{1.4}$$

where n_e and n_g are the degeneracy of upper and lower levels separated by Δ, respectively. ρJ and D are an s-f exchange coupling constant and the width of conduction band, respectively. If we put $D/\Delta \simeq 10^2$ and $n_e/n_g=2$, the Kondo temperature increases by a factor of 10^4, because the binding energy between localized and conduction electrons can be gained by channeling through upper crystal levels. This fact is important to explain the reason why the Kondo temperature dominates RKKY interaction between localized spins at different sites.

The s-f exchange Hamiltonian can be derived from the periodic Anderson Hamiltonian by taking the limit, $U \to \infty$ and $E_0 \to -\infty$. In this limit, unless we include the orbital degeneracy, we would have a long-range order instead of the coherent dense Kondo state. In this paper we discuss the paramagnetic ground-state of the periodic Anderson Hamiltonian without orbital degeneracy. To avoid the long range order we assume $|E_0|$ and U are not so large and we discuss the ground-state properties on the basis of the perturbation theory with respect to Coulomb repulsion U. We believe that the paramagnetic state of our Hamiltonian contains essential properties of the dense Kondo and/or heavy fermion systems, on the basis of the continuation principle.

2. Specific Heat

A. General consideration

The Green's function of the f-electrons $g^f_{k\sigma}(\omega)$ and that of the conduction electrons $G_{k\sigma}$ are given by

$$g^f_{k\sigma}(z) = [z - \Sigma_{k\sigma}(z) - \frac{V_k^2}{z-\varepsilon_{k\sigma}}]^{-1} , \tag{2.1}$$

$$G_{k\sigma}(z) = [z - \varepsilon_{k\sigma} - \frac{V_k^2}{z-\Sigma_{k\sigma}(z)}]^{-1} , \tag{2.2}$$

where $\Sigma_{k\sigma}(z)$ is the true selfenergy part for f-electrons. The eigenvalues of quasi-partiles $E^*_{k\sigma}$ are given by the poles of (2.1) or (2.2). The density of states of f-electrons $\rho^f_\sigma(\omega)$ and that of conduction electrons $\rho^c_\sigma(\omega)$ are given as follows:

$$\rho^f_\sigma(\omega) = -\frac{1}{\pi}\sum_k \mathrm{Im}\, g^f_{k\sigma}(\omega + i\delta) = \sum_k \rho^f_k(\omega)$$

$$= \sum_k \alpha^f_k [1 - \alpha^f_k \frac{\partial \Sigma_k}{\partial \omega}]^{-1} \delta(\omega - E^*_k) , \tag{2.3}$$

where $$\alpha^f_k = [1 + V_k^2/(\omega - \varepsilon_k)^2]^{-1} . \tag{2.4}$$

$$\rho^c_\sigma(\omega) = -\frac{1}{\pi}\sum_k \mathrm{Im}\, G^c_{k\sigma}(\omega + i\delta)$$

$$= \sum_k \alpha^c_k [1 - \alpha^f_k \frac{\partial \Sigma_k}{\partial \omega}]^{-1} \delta(\omega - E^*_k) , \tag{2.5}$$

where $$\alpha^c_k = 1 - \alpha^f_k . \tag{2.6}$$

By combining (2.3) and (2.5), we obtain the following relation,

$$\rho^c(\omega) + \sum_k (1 - \frac{\partial \Sigma_k(\omega)}{\partial \omega}) \rho^f_k(\omega) = \sum_k \delta(\omega - E^*_k) . \tag{2.7}$$

Here we have used (2.4) and (2.6). The right hand side of (2.7) means the total density of states of quasi-particles and is given by the sum of ρ^c and ρ^f enhanced by the factor of $(1-\partial\Sigma_k(\omega)/\partial\omega)$. Eq.(2.7) gives the coefficient of the T-linear term of the specific heat as

$$\gamma = \rho^c(0) + \sum_k (1 - \frac{\partial\Sigma_k(\omega)}{\partial\omega})\rho_k^f(0) . \tag{2.8}$$

This is the rigorous result. If we neglect the momentum dependence of $\partial\Sigma_k(\omega)/\partial\omega$, γ is approximated by

$$\gamma \simeq \rho^c(0) + (1 - \frac{\partial\Sigma(\omega)}{\partial\omega})\rho^f(0) . \tag{2.9}$$

This was already given by Ohkawa [5] and is the same relation as that for a single impurity case [6]. The large enhancement of the γ-value in heavy fermion systems is originated from the second term of (2.8), because this term can be enhanced by the electron correlation effect.

B. Perturbation expansion with respect to U

First, we consider the following symmetric case for simplicity.

$$E_0 = -U/2 , \tag{2.10}$$

$$\varepsilon_k = \varepsilon_{-k} , \tag{2.11}$$

$$\varepsilon_{k+G} = -\varepsilon_k , \tag{2.12}$$

G being a half of a reciprocal lattice vector. Here we calculate selfenergy part $\Sigma_k(\omega)$ by using the perturbation method with respect to Coulomb repulsion U. We divide the Hamiltonian (1.1) into the unperturbed part H and the perturbation H' as [1,2]

$$H_0 = \sum_{k,\sigma} \varepsilon_k n_{k\sigma} + V\sum_{k\sigma} (a^+_{k\sigma}c_{k\sigma}+c^+_{k\sigma}a_{k\sigma}) - UN/4 , \tag{2.13}$$

$$H' = \frac{U}{N}\sum_{\substack{k,k' \\ q\neq 0}} a^+_{k+q\uparrow}a^+_{k'-q\downarrow}a_{k'\downarrow}a_{k\uparrow} . \tag{2.14}$$

The unperturbed Green's function of the f-electron is given by

$$g_k^0(i\omega) = [i\omega - \frac{V^2}{i\omega-\varepsilon_k}]^{-1} = -g^0_{k+G}(-i\omega) . \tag{2.15}$$

If the total electron number N_e is fixed to $N_e/N=2$, an insulating ground state survives in this symmetric case even under the Coulomb repulsion. This conclusion was shown as follows [1.2]. The selfenergy part of the f-electron satisfies the following relation by the use of symmetry of our system,

$$\Sigma_{k+G}(\omega) = -\Sigma_k(-\omega) . \tag{2.16}$$

On the other hand, we can show for every point k_0 on the Fermi surface of conduction band,

$$\Sigma_{k_0+G}(0)= \Sigma_{k_0}(0) , \tag{2-17}$$

which holds for one-dimensional lattice, square lattice, body-centered cubic lattice[1)] and probably simple cubic lattice, with nearest neighbor transfer. From (2.16) and (2.17), we obtain

$$\Sigma_{k_0}(0) = 0 \ . \tag{2.18}$$

This result means that $\Sigma_k(\omega=0)$ crosses the conduction band on the Fermi surface, and eigen-states split at every crossing point due to the hybridization. Therefore, the hybridization gap survives under the Coulomb repulsion, as far as the ground-state remains in a paramagnetic state. In this case the second-order term of the selfenergy $\Sigma_k^{(2)}(\omega+i\delta)$ is given by [1]

$$\Sigma_k^{(2)}(\omega+i\delta) = \left(\frac{U}{N}\right)^2 \sum_{q,k_1} \alpha_{k_1}^- \alpha_{k_1+q}^+ \left(\frac{\alpha_{k+q}^+}{\omega+i\delta+\mu+E_{k_1}^- - E_{k_1+q}^+ - E_{k+q}^+} + \frac{\alpha_{k+q}^-}{\omega+i\delta+\mu-E_{k+q}^- - E_{k_1}^- + E_{k_1+q}^+} \right) , \tag{2.19}$$

where

$$E_k^{\pm} = \frac{1}{2}(\varepsilon_k \pm \sqrt{\varepsilon_k^2+4V^2}) , \tag{2.20}$$

$$\alpha_k^{\pm} = \frac{1}{2}\left(1 \pm \frac{-\varepsilon_k}{\sqrt{\varepsilon_k^2+4V^2}}\right) . \tag{2.21}$$

Figs.1, 2 and 3 show $\Sigma_k^{(2)}(\omega)$ for one-dimensional, simple cubic and body-centered cubic lattices, respectively.

In order to obtain more general results and to compare with a single impurity case, we take a limit $V/D\to 0$. As shown in Figs.1, 2, 4, $\partial\Sigma_k^{(2)}/\partial\omega$ shows small momentum dependence and is scaled by V^4 in the region of small V. If we expand $-\partial\Sigma/\partial\omega$ with respect to $U/\pi\Delta$, as

$$-\frac{\partial\Sigma}{\partial\omega} = \sum_{n=0}^{\infty} a_n \left(\frac{U}{\pi\Delta}\right)^n , \tag{2.22}$$

where $a_n=0$ for n=odd and $\Delta=\pi\rho V^2$, ρ being the density of states of conduction band at the Fermi energy, we obtain

a) one dimensional lattice ($\varepsilon_k=-\cos k$)

$\rho = 1/\pi$ and $\Delta = V^2$,

$a_2 = 0.09$.

b) simple cubic lattice ($\varepsilon_k=-(\cos k_x+\cos k_y+\cos k_z)/3$)

---

1) kindly pointed out by D.M. Newns.

Fig. 1

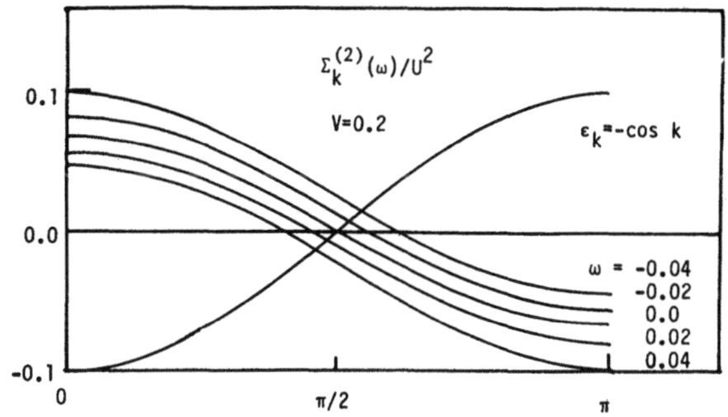

Fig. 2

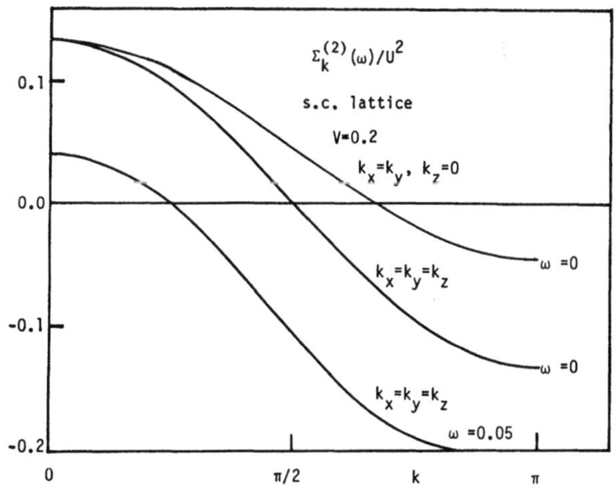

Fig. 3

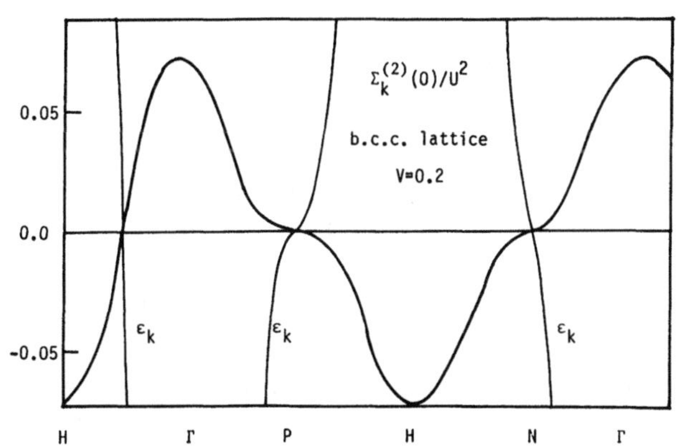

Fig.1 The second-order selfenergy $\mathrm{Re}\Sigma_k^{(2)}(\omega)$ for one dimensional lattice

Fig.2 The second-order selfenergy $\mathrm{Re}\Sigma_k^{(2)}(\omega)$ for simple cubic lattice, $\varepsilon_k=-(\cos k_x + \cos k_y + \cos k_z)/3$.

Fig.3 The second-order selfenergy $\mathrm{Re}\Sigma_k^{(2)}(0)$ for body-centered cubic lattice, $\varepsilon_k=-\cos k_x \cos k_y \cos k_z$.

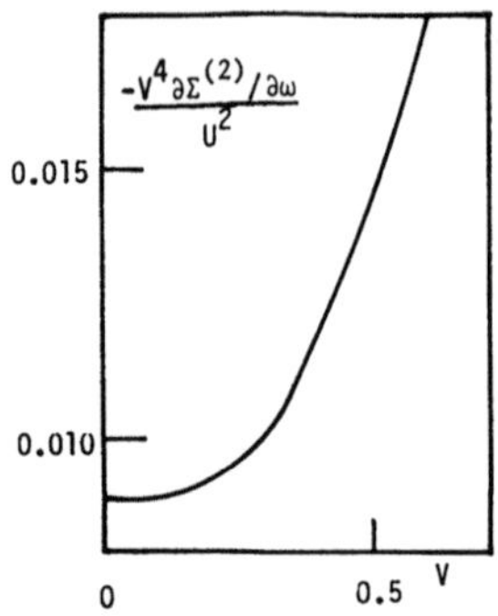

Fig.4 $\partial\Sigma^{(2)}/\partial\omega$ is scaled by V^4/U^2 for small V, ε_k = -cos k.

ρ = 0.856 and $\Delta = 0.856\ V^2$,

a_2 = 0.22 .

On the other hand, we have a_2=0.5326 for a single impurity Anderson model [6], where a constant density of states and infinite band width are assumed. In the present case, the state density of one dimensional conduction band is special because of singular behavior at the band edges, while that for cubic lattice is rather similar to constant one. In the periodic systems assumed here, an insulating gap exists and $\rho^f(\omega)$ vanishes at the Fermi energy. For metallic cases with nonzero $\rho^f(0)$, another term, $U^2\sum_q \rho^f_{k+q}(0)\chi^0_q$, is added to that considered here. Therefore, in usual metallic cases, values of a_2 are expected to be larger than in the insulating cases.

The ground-state energy also can be expanded for small V as

$$E_g/N = \pi\Delta\{4\log V - \frac{1}{4}(\frac{U}{\pi\Delta}) - b_2(\frac{U}{\pi\Delta})^2 + \cdots\}, \qquad (2.23)$$

where $b_2 \simeq 0.033$ for ε_k=-cosk in accordance with b_2=0.0369 for a single impurity.

3. Susceptibility and Wilson Ratio

When we apply a magnetic field to the present system, the Green's function for the localized f-electrons with spin σ is given by

$$g^f_{k\sigma}(i\omega) = [i\omega + \mu + H_\sigma - \Sigma_{k\sigma}(\omega) - \frac{V_k^2}{i\omega+\mu+H_\sigma-\varepsilon_k}]^{-1} , \qquad (3.1)$$

where $H_\sigma = g\mu_B H\sigma/2$ is defined by the Zeeman energy for σ spin, μ_B being the Bohr magneton. The g-values for f- and conduction electrons are assumed to be the same (g=2) for simplicity. Eigenvalue $E^*_{k\sigma}$ are given by the poles of (3.1). Total electron number N_e and magnetization M are given, respectively, by

$$N_e = \sum_{k,\sigma} \theta(\mu - E^*_{k\sigma}) , \qquad (3.2)$$

$$M = \mu_B \sum_{k,\sigma} \sigma\theta(\mu - E^*_{k\sigma}) . \qquad (3.3)$$

From (3.3), the spin susceptibility is given by

$$\chi_s = \lim_{H\to 0} \mu_B \sum_{k,\sigma} \sigma\delta(\mu - E^*_{k\sigma})(-\frac{\partial E^*_{k\sigma}}{\partial H})$$

$$= 2\mu_B \sum_k \delta(\mu - E^*_k)(-\frac{\partial E^*_{k\sigma}}{\partial H}) . \qquad (3.4)$$

By using eigen-value equation, we obtain

$$\frac{\partial E^*_{k\sigma}}{\partial H} = [1 - \frac{\partial \Sigma_k(\omega)}{\partial i\omega} + \frac{V_k^2}{(E^*_k-\varepsilon_k)^2}]^{-1}(-\mu_B\sigma)\{\tilde{\chi}_s(k) + \frac{V_k^2}{(E^*_k-\varepsilon_k)^2}\} , \qquad (3.5)$$

where $\tilde{\chi}_s(k)$ is given by $\tilde{\chi}_{\uparrow\uparrow}(k)+\tilde{\chi}_{\uparrow\downarrow}(k)$ and they are defined by

$$\tilde{\chi}_{\uparrow\uparrow}(k) = 1 - \frac{\partial \Sigma_{k\sigma}(0)}{\partial H_\sigma}\Big|_{H_\sigma=0} , \qquad (3.6)$$

$$\tilde{\chi}_{\uparrow\downarrow}(k) = \frac{\partial \Sigma_{k\sigma}(0)}{\partial H_{-\sigma}}\Big|_{H_{-\sigma}=0} . \qquad (3.7)$$

Thus, we obtain

$$\chi_s = 2\mu_B^2\{\sum_k \rho^f_k(0)\tilde{\chi}_s(k) + \rho^c(0)\} , \qquad (3.8)$$

where $\rho^f_k(0)$ and $\rho^c(0)$ were defined by (2.3) and (2.5), and

$$\rho^f_k(0) = \delta(\mu - E^*_k)[1 - \frac{\partial \Sigma_k(\omega)}{\partial i\omega} + \frac{V_k^2}{(E^*_k-\varepsilon_k)^2}]^{-1} , \qquad (3.9)$$

$$\rho^c(0) = \sum_k \rho^f_k(0)\frac{V_k^2}{(E^*_k-\varepsilon_k)^2} . \qquad (3.10)$$

It is noted in (3.8) that the susceptibility due to f-electrons is enhanced by $\tilde{\chi}_s(k)$, while that due to conduction electrons is not enhanced.

Now we discuss the parameter γ given by (2.8) and χ_s by (3.8) by using Ward identities. The coefficient γ of the specific heat is written as

$$\gamma = \rho^c(0) + \gamma^f , \qquad (3.11)$$

$$\gamma^f = \sum_k \rho^f_k \tilde{\gamma}(k) , \qquad (3.12)$$

$$\tilde{\gamma}(k) = 1 - \frac{\partial \Sigma_k(\omega)}{\partial i\omega} . \qquad (3.13)$$

By differentiating the unperturbed Green's function $g^0_k(i\omega)$ by $i\omega$, we obtain

$$[\frac{\partial}{\partial i\omega} g_k^0(i\omega)] = -[g_k^0(i\omega)]^2[1 + \frac{V_k^2}{(i\omega+\mu-\varepsilon_k)^2}] + \frac{\delta(\omega)}{i}\delta g_k^0 , \tag{3.14}$$

where δg_k^0 is given by putting $\Sigma_k(0)=0$ in δg_k and δg_k is given by

$$\delta g_k = g_k(0_+) - g_k(0_-) = -2\pi i\rho_k^f(0) . \tag{3.15}$$

On the other hand, $\partial g_k^0(\omega)/\partial\mu$ is given by

$$\frac{\partial g_k^0(i\omega)}{\partial\mu} = -[g_k^0(i\omega)]^2[1 + \frac{V_k^2}{(i\omega+\mu-\varepsilon_k)^2}] . \tag{3.16}$$

Using (3.14)~(3.16) and shifting the frequencies of every closed loop by external frequency ω, we obtain

$$\frac{\partial\Sigma_{k\sigma}(\omega)}{\partial i\omega} = \frac{\partial\Sigma_{k\sigma}(\omega)}{\partial\mu} + \sum_{k',\sigma'} \frac{\delta g_{k'}}{2\pi i}\Gamma_{\sigma\sigma';\sigma'\sigma}(k,k';\ k',k) . \tag{3.17}$$

By shifting the frequencies of every closed loop with only σ spin by ω,

$$\frac{\partial\Sigma_{k\sigma}(\omega)}{\partial i\omega} = \frac{\partial\Sigma_{k\sigma}(\omega)}{\partial H_\sigma} + \sum_{k'} \frac{\delta g_{k'}}{2\pi i}\Gamma_{\sigma\sigma;\sigma\sigma}(k,k';\ k',k) . \tag{3.18}$$

Thus, we obtain from (3.6), (3.13), (3.15) and (3.18),

$$\tilde{\gamma}(k) = \tilde{\chi}_{\uparrow\uparrow}(k) + \sum_{k'} \rho_{k'}^f(0)\Gamma_{\sigma\sigma;\sigma\sigma}(k,k';\ k',k) . \tag{3.19}$$

From (3.12) and (3.19),

$$\gamma^f = \sum_k \rho_k^f(0)\tilde{\gamma}(k) = \chi_{\uparrow\uparrow}^f + \delta_{\uparrow\uparrow}^f , \tag{3.20}$$

where $\chi_{\uparrow\uparrow}^f = \sum_k \rho_k^f(0)\tilde{\chi}_{\uparrow\uparrow}(k)$, (3.21)

and $\delta_{\uparrow\uparrow}^f = \sum_{k,k'} \rho_k^f(0)\Gamma_{\sigma\sigma;\sigma\sigma}(k,k';\ k',k)\rho_{k'}^f(0)$. (3.22)

Here we can assume the vertex between f-electrons with parallel spins, $\Gamma_{\uparrow\uparrow}=\Gamma_{\sigma\sigma;\sigma\sigma}$, is antisymmetrized. $\Gamma_{\uparrow\uparrow}$ vanishes identically for the single impurity case, but is nonzero for the periodic case because of the momentum conservation.

Eqs.(3.17) and (3.18) give the following equation,

$$\frac{\partial\Sigma_{k\sigma}(0)}{\partial H_{-\sigma}} = -\sum_{k'} \frac{\delta g_{k'}}{2\pi i}\Gamma_{\sigma-\sigma;-\sigma\sigma}(k,k';\ k',k) , \tag{3.23}$$

or

$$\tilde{\chi}_{\uparrow\downarrow}(k) = \sum_{k'} \rho^f_{k'}(0)\Gamma_{\sigma-\sigma;-\sigma\sigma}(k,k';\ k',k) \ . \tag{3.24}$$

Thus, we obtain

$$\chi^f_{\uparrow\downarrow} = \sum_k \rho^f_k(0)\tilde{\chi}_{\uparrow\downarrow}(k) = \sum_{kk'} \rho^f_k(0)\Gamma_{\uparrow\downarrow;\downarrow\uparrow}(k,k';\ k',k)\rho^f_{k'}(0) \ . \tag{3.25}$$

In a similar way, we can obtain the charge susceptibility as

$$\chi_c = \rho^c(0) + \chi^f_c \ , \tag{3.26}$$

$$\chi^f_c = \chi^f_{\uparrow\uparrow} - \chi^f_{\uparrow\downarrow} = \sum_k \rho^f_k(0)(\tilde{\chi}_{\uparrow\uparrow}(k) - \tilde{\chi}_{\uparrow\downarrow}(k)) \ . \tag{3.27}$$

The Wilson ratio R is given by

$$R = [\rho^c(0) + \chi^f_s]/[\rho^c(0) + \gamma^f] \ . \tag{3.28}$$

The ratio due to only f-electrons is given by

$$R^f = \chi^f_s/\gamma^f = [\chi^f_{\uparrow\uparrow} + \chi^f_{\uparrow\downarrow}]/[\chi^f_{\uparrow\uparrow} + \delta^f_{\uparrow\uparrow}]$$

$$= [1 + \chi^f_{\uparrow\downarrow}/\chi^f_{\uparrow\uparrow}]/[1 + \delta^f_{\uparrow\uparrow}/\chi^f_{\uparrow\uparrow}] \ . \quad (R^f \geq R) \ . \tag{3.29}$$

Thus, R^f depends on two parameters, $\chi^f_{\downarrow\uparrow}/\chi^f_{\uparrow\uparrow}$ and $\delta^f_{\uparrow\uparrow}/\chi^f_{\uparrow\uparrow}$, and $\chi^f_{\uparrow\downarrow}/\chi^f_{\uparrow\uparrow} \leq 1$. If we increase the Coulomb repulsion, it supresses the charge fluctuation,

$$\chi^f_c = \chi^f_{\uparrow\uparrow} - \chi^f_{\uparrow\downarrow} = 0 \qquad \text{for a large value of U.} \tag{3.30}$$

In this case we have

$$R^f = 2/(1 + \delta^f_{\uparrow\uparrow}/\chi^f_{\uparrow\uparrow}) \ . \tag{3.31}$$

Schlottmann discussed Wilson ratio for the case of a pair and a cluster of impurities [7]. In the periodic systems with an attractive interaction between parallel spins ($\Gamma_{\uparrow\uparrow}<0$), $\delta^f_{\uparrow\uparrow}$ is expected to be negative and R^f given by (3.29) can be larger than 2. As an examples, we show here the U^2-term of $\delta^f_{\uparrow\uparrow}$,

$$\delta^{(2)}_{\uparrow\uparrow} = U^2 \sum_{\substack{k\\q}} \rho^f_k(0)\rho^f_{k+q}(0)[\chi^0_q - \chi^0_{q=0}] \ . \tag{3.32}$$

It is noted that the momentum-dependence of the unperturbed susceptibility for f-electrons χ^0_q is important to determine the sign of $\delta^{(2)}_{\uparrow\uparrow}$. If f-electrons are localized at $U=U_c$ and behave as the Brinkman-Rice liquid, R^f is expected to be 4 and $\delta^f_{\uparrow\uparrow}$ tends to $-\tilde{\chi}_{\uparrow\uparrow}/2$. Thus, in principle, if we include correctly the momentum dependence, the range available for R^f is extended wider in contrast to the impurity case [6].

4. Application of Orthogonality Theorem to the Ground-State

When the ground-state has a long range order such as in magnetic or superconducting phases, it is easier to understand the nature of the ground-state. Here we consider the periodic Anderson Hamiltonian by using the Anderson orthogonality theorem. This theorem was generalized to the interacting electron systems by the present authors [8]. It should be noted that the expectation value of the mixing term of (1.1) in the paramagnetic ground-state Ψ_G should not vanish so as to gain the energy through the mixing term.

We consider first the unperturbed Hamiltonian given by (1.2). The energy gain comes from only the mixing term at the sacrifice of the kinetic energy of conduction electrons.[1)] Then, we introduce the perturbation (1.3) to this state. By this effect the configuration space available for localized electrons is reduced so as not to increase correlation energy. Though the correlation effect gives rise also to the reduction of the mixing term, the expectation value of the mixing term in the ground-state should not vanish in order to maintain the energy gain through the hybridization. As far as this situation holds in the paramagnetic ground-state, we can determine the local electronic structure of the ground-state by using the orthogonality theorem, as follows.

The wavefunction of the ground-state can be divided into the following parts depending on the occupation number of the localized electrons at arbitrary site i,

$$\Psi_G = A_0\Psi_i^{(0)} + A_\uparrow a^+_{i\uparrow}\Psi_i^{(\uparrow)} + A_\downarrow a^+_{i\downarrow}\Psi_i^{(\downarrow)} + A_2 a^+_{i\uparrow}a^+_{i\downarrow}\Psi_i^{(2)} , \tag{4.1}$$

where $|A_0|^2+|A_\uparrow|^2+|A_\downarrow|^2+|A_2|^2=1$. The average of the mixing term is given by

$$\begin{aligned}&\langle\Psi_G|V\sum_{i,\sigma}(c^+_{i\sigma}a_{i\sigma}+a^+_{i\sigma}c_{i\sigma})|\Psi_G\rangle\\ &= V\sum_{i,\sigma}\{A_0A_\sigma\langle\psi_i^{(0)}|c^+_{i\sigma}|\psi_i^{(\sigma)}\rangle + A_{-\sigma}A_2\langle\psi_i^{(-\sigma)}|c^+_{i\sigma}|\psi_i^{(2)}\rangle\\ &+ A_\sigma A_0\langle\psi_i^{(\sigma)}|c_{i\sigma}|\psi_i^{(0)}\rangle + A_2A_{-\sigma}\langle\psi_i^{(2)}|c_{i\sigma}|\psi_i^{(-\sigma)}\rangle\} .\end{aligned} \tag{4.2}$$

Here we define $n^{(\alpha)}_{i\sigma}$ as the number of local conduction electrons around site i with spin σ in the state of $\psi_i^{(\alpha)}$. For the matrix elements in (4.2) not to vanish, the local conduction electron numbers around site i in the left and right states should coincide with each other. This is the orthogonality

1) In the unperturbed state, $V\langle c_0^+f_0\rangle=-V^2\sum_k[(\varepsilon_k-E_0)^2+4V^2]^{-1/2}$. In general case, $V\langle c_0^+f_0\rangle=-V^2\sum_k\int_{-\infty}^{0}d\omega\ \rho^f_k(\omega)\frac{1}{\varepsilon_k-\omega}$.

theorem given by Anderson. Thus, we have the following relations;

$$n_{i\sigma}^{(0)} = n_{i\sigma}^{(\sigma)} + 1 , \tag{4.3}$$

$$n_{i-\sigma}^{(0)} = n_{i-\sigma}^{(\sigma)} , \tag{4.4}$$

$$n_{i\sigma}^{(-\sigma)} = n_{i\sigma}^{(2)} + 1 , \tag{4.5}$$

$$n_{i-\sigma}^{(-\sigma)} = n_{i-\sigma}^{(2)} . \tag{4.6}$$

In the paramagnetic state, the following relations also hold,

$$n_{i-\sigma}^{(2)} = n_{i\sigma}^{(2)} , \tag{4.7}$$

$$n_{i\sigma}^{(0)} = n_{i-\sigma}^{(0)} . \tag{4.8}$$

From (4.3),(4.4) and (4.8),

$$n_{i\sigma}^{(-\sigma)} = n_{i-\sigma}^{(\sigma)} = n_{i\sigma}^{(\sigma)} + 1 . \tag{4.9}$$

From (4.5),(4.6) and (4.7),

$$n_{i-\sigma}^{(\sigma)} = n_{i\sigma}^{(-\sigma)} = n_{i-\sigma}^{(-\sigma)} + 1 . \tag{4.10}$$

The relation of (4.9) or (4.10) is the same as that derived from the condition that the transverse component of the exchange interaction in the single impurity connects two components of the singlet ground-state. Therefore, Eq.(4.9) means that even in the periodic systems the ground state has the electronic structure of a local singlet state coupled with conduction electrons, as far as the ground-state remains in the paramagnetic state.

5. Summary and Remarks

In order to clarify the nature of the dense Kondo system at low temperatures or heavy fermion system, the ground-state of the periodic Anderson Hamiltonian is studied from the point of view of Landau-Luttinger Fermi liquid theory. For simplicity, orbital degeneracy of localized f-orbital is not taken into account. Corresponding to this simplification, the ground-state is assumed to remain in a paramagnetic state, not spin-ordered state.

In this paper, we investigated the following three things. First, the general expression for the specific heat has been derived which includes the ω-derivative of f-electron selfenergy. The selfenergy derivative with respect to ω can be expanded in a power series of $(U/\pi\Delta)$ and the coefficient of U^2-term was calculated and compared with a single impurity case. Then, the general expression for the magnetic susceptibility has also been derived in terms of the f-electron selfenergy and taking ratio to the coefficient of the T-linear specific heat, the Wilson ratio was derived. This Wilson ratio includes two independent quantities $\chi^f_{\uparrow\downarrow}/\chi^f_{\uparrow\uparrow}$ and $\delta^f_{\uparrow\uparrow}/\chi^f_{\uparrow\uparrow}$, the latter quantity is related to parallel spin vertex $\Gamma_{\uparrow\uparrow}$ which does not vanish in the periodic case. On the other hand, $\Gamma_{\uparrow\uparrow}$ vanishes in the single impurity case. This point is an important difference between periodic case and the impurity case. Because of the difference, Wilson ratio is able to attain, for negative $\delta^f_{\uparrow\uparrow}$, to a certain value larger than 2, the value in the impurity case, in the large U limit, although we could not calculate that value definitely.

Finally, the local electronic structure has been considered by the use of the Anderson orthogonality theorem. It has been proved that this local structure, as expressed by the relation between local numbers of conduction electrons with up and down spin, is the same as that of the singlet ground-state of the single impurity. This fact helps us to understand the reason why theories for the single impurity can reproduce qualitatively experimental results obtained in the dense Kondo systems as far as the local nature is concerned.

Although the local nature of the periodic system is similar to the impurity case, the coherent character is still important. The most striking effect of it appears in the non-zero value of $\Gamma_{\uparrow\uparrow}$. A negative value of $\delta_{\uparrow\uparrow}$ means that γ value decreases as degree of coherence grows up at low temperatures. On the other hand, the susceptibility does not change so much, because it does not include $\delta^f_{\uparrow\uparrow}$. Thus, the Wilson ratio is expected to become larger by increase of coherence as temperature is lowered. These two phenomena seem to be realized in dense Kondo systems such as $CeAl_3$ and $CeCu_2Si_2$. [9,10]. For positive value of $\Gamma_{\uparrow\uparrow}$, γ value increases and the Wilson ratio decreases with increase of coherence.

Acknowledgements

The authors would like to thank Professor F. Steglich, Dr. Y. Ōnuki, Dr. Y. Kitaoka for valuable comments concerning experimental results.

References

1. K. Yamada and K. Yosida: "Electron Correlation and Magnetism in Narrow Band Systems", ed. by T. Moriya (Springer-Verlag, Berlin, 1981) p.210.
2. K. Yamada and K. Yosida: J. Magn. Magn. Mater. 31-34, 461 (1983).
3. K. Yamada, K. Hanzawa and K. Yosida: Prog. Theor. Phys. 71, 450 (1984).
4. K. Hanzawa, K. Yamada and K. Yosida: J. Magn. Magn. Mater. 47-48, (1985) in press.
5. F.J. Ohkawa: J. Phys. Soc. Japan 53, 1389 (1984).
6. K. Yosida and K. Yamada: Prog. Theor. Phys. Suppl. No.46, 244 (1970). K. Yamada: Prog. Theor. Phys. 53, 970 (1975), K. Yosida and K. Yamada: Prog. Theor. Phys. 53, 1286 (1975). K. Yamada: Prog. Theor. Phys. 54, 316 (1975).
7. P. Schlottmann: Phys. Rev. B23, 3475 (1981), B24, 5394 (1981).
8. K. Yamada and K. Yosida: Prog. Theor. Phys. 68, 1504 (1982).
9. C.D. Bredl, S. Horn, F. Steglich, B. Lüthi and R.M. Martin: Phys. Rev. Lett. 52, 1982 (1984).
10. F.R. de Boer, J. Klaosse, J. Aarts, C.D. Bredl, W. Lieke, U. Rauchschwalke, F. Steglich, R. Felten, U. Umhofer and G. Weber: J. Magn. Magn. Mater. 47-48, (1985) in press.

Quasiparticles in Heavy Fermi Liquids and Their Interactions With Phonons

N. d'Ambrumenil
Department of Physics, University of Warwick, Great Britain

P. Fulde
Max-Planck-Institut für Festkörperforschung,
D-7000 Stuttgart 80, Fed. Rep. of Germany

We show how the low temperature properties of $CeCu_2Si_2$ and Kondo lattices in general can be predicted *semiquantitatively*. In particular we look at the density of quasiparticle excitations for energies much less than the Kondo temperature where the systems behave like heavy Fermi liquids. We also discuss the interactions of quasiparticles with phonons and how they may influence the properties of phonons. It is shown that they are sufficiently strong in order to result in a superconducting transition temperature in $CeCu_2Si_2$ of the observed size.

1. Introduction

We look at the nature of the quasiparticle excitations in Kondo lattices at energies and temperatures below the Kondo temperature, where the systems often behave like heavy Fermi liquids. The main emphasis will be on Ce compounds such as $CeCu_2Si_2$, $CeAl_3$, etc., but we will also comment on heavy fermion U compounds.

Our discussion is based on the Kondo-lattice Ansatz (KLA). The KLA was introduced in its simplest form by RAZAFIMANDIMBY et al. [1] to provide a phenomenological description of the heavy fermion properties found experimentally in systems like $CeCu_2Si_2$ [2]. It was developed further by D'AMBRUMENIL and FULDE [3], when the hope was expressed that the KLA could give a semiquantitative description of the quasiparticle dispersion, density of states and other Fermi surface properties of heavy fermion systems. We report here on the progress we have made in this direction and give a more general account of the theoretical basis of the KLA.

In the next section we describe the KLA making clear its basic assumptions. In Sect.3 we present the actual scheme of calculation using $CeCu_2Si_2$ as our working example. We also show some sample results. Section 4 tries to put our results in the context of the general body of work on the Anderson model.

The quasiparticle-phonon interaction and its influence on the phonon properties is discussed in Sect.5. This interaction is used in Sect.6 for an estimate of the superconducting transition temperature of $CeCu_2Si_2$. It is assumed in the following that the reader is familiar with the Anderson lattice model and with the concept of the compensated spin and orbital degrees of freedom at temperatures less than the spin fluctuation or Kondo temperature T_k. We will deal with the systems with compensated (singlet) ground states, e.g., $CeCu_2Si_2$ or $CeAl_3$. We will not touch on the systems which have magnetic ground states, e.g., $CeAl_2$ or CeB_6, where the RKKY interaction is thought to dominate the "Kondo compensation" [4].

2. The Kondo-Lattice Ansatz

The KLA describes the quasiparticle excitations of a Kondo lattice in the compensated temperature regime, $T \ll T_k$. It starts from two assumptions. The first makes use of the physical picture that at $T = 0$ the system is nonmagnetic and therefore should be describable in terms of scattering potentials or phase shifts at different sites like any other nonmagnetic metal. The assumption is therefore that such *local* scattering potentials can also be attributed to the sites of the individual Kondo ions.

These potentials are assumed well-behaved, i.e., that the energy-dependent phase-shifts at the sites i, $\delta_\ell^i(\varepsilon)$, can be represented by an expansion in $(\varepsilon - \varepsilon_F)$ and δn_ℓ^i. Here ε_F is the Fermi energy and δn_ℓ^i denotes the *local* departure from the ground state distribution function.

Secondly, it is assumed that the main scattering channel for the quasiparticles has the same local symmetry below the characteristic temperature, T_k, as the apparent "ground-state" symmetry of the individual Kondo ions above T_k. This ensures that the f-electron charge distribution around a Ce site is the same below and above T_k provided that T_k is much smaller than the crystalline electric field (CEF) splitting energies. The ground state of the 4f multiplet is known from the neutron scattering data of HORN et al. [5] to be a doublet. This is thought to have the form

$$|\Gamma_7{}^{1(2)}\rangle = c|\pm 5/2\rangle + d|\mp 3/2\rangle \tag{1}$$

where $c = 0.83$, $d = 0.56$ and the notation $|J_z\rangle = |J = 5/2, J_z\rangle$ has been used. (Recently a somewhat different form of the ground state doublet has been proposed [6]. For the purposes of our argument though we will assume that the original interpretation, Eq. (1), is correct. The problem has not yet been fully resolved).

These two assumptions allow us to develop our description of the Kondo-lattice. We write down an expansion for the local scattering potentials for energies close to ε_F and introduce a Kondo temperature T_k as the expansion parameter in energy, (see Sect.3). These local potentials do not imply a lattice of independent Kondo ions, NOZIERES [7]. We then require that the scattering of quasiparticles in the lattice is coherent; this leads to the well-known Korringa-Kohn-Rostoker (KKR) equations, which we solve numerically. The derivation of these KKR equations does not involve relating the quasiparticle picture, valid for $T \ll T_k$ to the "high-temperature parameters" – the exchange in integral J and density of states at the Fermi level – except in the one important aspect: we assume that the local symmetry of the scattering channel is that of the ground state of the individual Kondo ion.

Our approach bypasses the need for a detailed treatment of how the many-body interactions dress the low-lying excitations leading to their large effective masses.

Instead experimental information is used as input into the theory, which is within the spirit of the Landau approach to Fermi liquids. It is apparent that within the present phenomenological approach we cannot describe the Kondo lattices for temperatures $T \sim T_k$ or $T > T_k$ nor can we describe any high frequency spectral properties.

The strength of our approach is its generality. As such if experiments justify our second assumption in the actinides, the treatment can be extended to cover these systems as well. However, if there is no experimental information available on how the degeneracy of the 4f or 5f shell is lifted at low T (e.g., by the CEF) then the present approach loses its solid base. Of course, the total f-electron count is still related to the f-phase shifts via the Friedel sum rule:

$$n_f = \frac{2}{\pi} \sum_{\nu=1}^{7} \delta_{\ell=3,\nu}(\varepsilon_F) \quad , \tag{2}$$

but it is no longer clear how the phase shifts, $\delta_{\ell=3,\nu}(\varepsilon_F)$, in the different f-channels, ν, are to be weighted with respect to one another. If T_k is larger than the CEF splitting energies the prescription is again simple as to a good approximation all channels within the ground state J multiplet can be treated as equivalent.

3. Scheme of Calculation

We take as our working example the case of $CeCu_2Si_2$. We know that the ground state of the Ce ions is a doublet (see (1)). We define a pseudo spin, $\sigma = \pm 1$, to represent the occupation of one (+1) or the other (-1) of the Γ_7 states in the ground state doublet. The expansion for the phase shift at a site i is then given for the ground-state Γ_7 symmetry by

$$\delta_\sigma^i(\varepsilon) = \delta(\varepsilon_F) + \frac{(\varepsilon - \varepsilon_F)}{T_k} + \sum_{j\sigma'\varepsilon'} \phi_{\sigma\sigma'}^{i-j} \, \delta n_{\sigma'}^j(\varepsilon') \tag{3}$$

and for all other symmetries by

$$\delta_\ell^i(\varepsilon) = \delta_\ell^i(\varepsilon_F) \quad . \tag{4}$$

In (3) the energy scale, T_k, has been introduced as an expansion parameter. It is important to notice that, as in the one ion case, the expansion parameter, T_k, will almost certainly differ from the experimental estimate of the Kondo temperature, T_k^∞. For the one ion case $T_k^\infty / T_k = .324$ [7]. For the lattice the ratio is of course not known theoretically.

The parameters $\phi_{\sigma\sigma'}^{i-j}$ account for the two-body quasiparticle-quasiparticle interactions. They relate the change in phase shift at a site i to the local departure from equilibrium of the quasiparticle distribution $\delta n_{\sigma'}^j$ at site j. The phase shifts in (3) are defined for non-overlapping spheres around each ion. The requirement that scattering in the lattice is coherent leads directly to the KKR equations.

To proceed with a quantitative solution of the KKR equations we need values for the constants[1] in (4) and of the Fermi energy ε_F. Our scheme is then:

i) To calculate the constants and ε_F by performing a local density calculation for the system, but excluding the 4f electron channels on the Ce ions. n_f f-electrons should be treated as part of the core of each Ce ion to ensure charge neutrality. We start by setting $n_f = 1$.

ii) To insert the scattering potential (2) into the KKR equations. Start with a value of $\delta(\varepsilon_F) = \pi n_f N_f^{-1}$ consistent with the Friedel sum rule where n_f is chosen according what is known from experiments. N_f is the ground-state degeneracy of the f-level. In $CeCu_2Si_2$, N_f is 2.

iii) To solve the KKR equations.

iv) To check for consistency, e.g., does the value of n_f obtained by satisfying Luttinger's theorem agree with the value assumed on i)?

It is sufficient to solve the KKR equations in their simplified LMTO or ASW forms [8]. Because these calculational schemes evaluate structure constants in an orbital angular momenta basis it is necessary to make a basis transformation of the structure constants to a basis containing the Γ_7 doublet ground-state of the Ce 4f multiplet.

In Fig.1 we show the quasiparticle dispersion for $CeCu_2Si_2$. It differs slightly from the one shown in [3] which contained some inaccuracies. The most important feature is the presence of heavy as well as light quasiparticles at the Fermi surface. This is seen from the different slopes of the dispersion curves which inter-

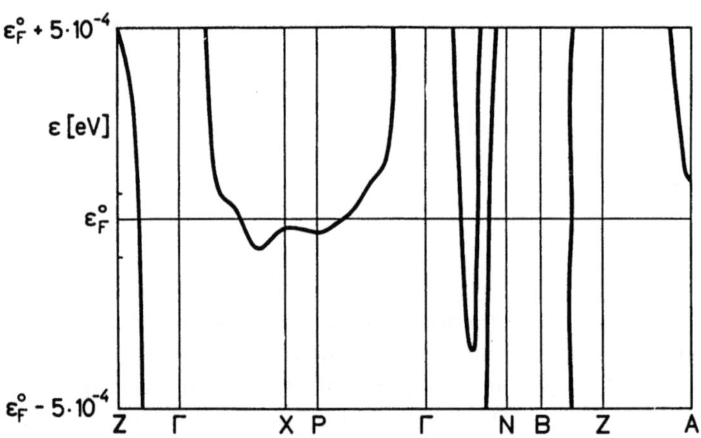

Fig.1. Quasiparticle bands near ε_F for $CeCu_2Si_2$ along selected directions in the Brillouin zone. $T_K = 15K \simeq 13\cdot10^{-4}$ eV

[1] In practice the phase shifts in (4) are taken as energy-dependent but they vary so slowly on a scale of T_k that in the region T_k about ε_F they are effectively constant

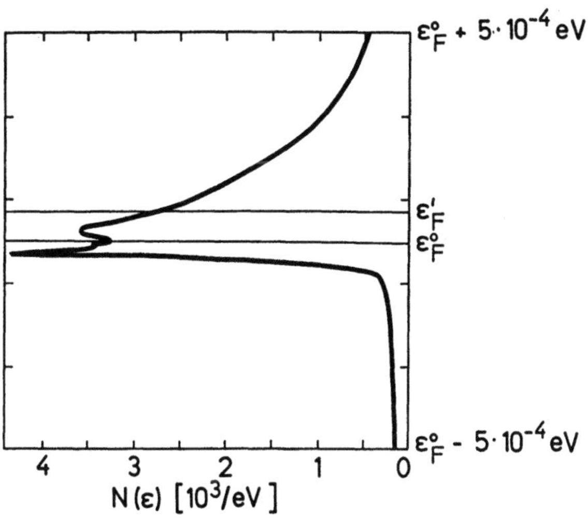

Fig.2. The density of quasiparticle excitations near ε_F for $CeCu_2Si_2$. On the basis of the single ion result of Nozières [7] the expansion parameter T_K would be expected to be about three times larger than T_K^∞, the temperature usually referred to as the Kondo temperature. However the case for the lattice is not clear (see text). $T_K = 15K$ was chosen

sect the Fermi energy. The light quasiparticles do not show up in the low temperature thermodynamic quantities but may become important in transport and Fermi surface properties. More details of the calculations as well as our results for $CeAl_3$ will be described elsewhere [9].

Figure 2 shows the corresponding quasiparticle density of states. Two horizontal lines are drawn to indicate the position of the Fermi energy. The lower value indicates the position of ε_F evaluated in step i). The upper value is the one obtained by requiring in step iv) that the number of occupied states equals the number of electrons (Luttingers theorem). In principle, we should repeat the calculation from step i) assuming a lower number for the f-occupancy than $n_f = 1$. Eventually after a few iterations the upper and lower values for ε_F should coincide. The calculations were done for two different values of T_k (i.e., 15 K and 30 K). Almost identical results were obtained when plotted in units of T_k. (If we assume $T_k^\infty / T_k = .324$ for the lattice these correspond to an experimental or high temperature value for the Kondo temperature, T_k^∞, of about 5 K and 10 K respectively.)

We note that the density of states at ε_F in Fig.2 is approximately a factor of seven larger than expected on the basis of specific heat measurements (1000 $mJmol^{-1} K^{-2} = 425\ eV^{-1}K^{-2}$ / formula unit [9]). If we assume that the scaling with T_k extends that far it would imply that a value of 100 K ($T_k^\infty \sim 35$ K) would have been more appropriate for T_k. We hope to report on this point later [10].

The important result from Fig.2 is that the DOS near ε_F shows structure and has exactly the form, quantitatively as well as qualitatively, as that proposed to account for experiments by BREDL et al. [11]. These results confirm the hypothesis of Bredl et al. that lattice coherence effects are responsible for the maximum observed in the heat capacity of $CeCu_2Si_2$ at about 2 K. Lattice coherence effects have also been demonstrated by measurements of the thermoelectric power [12]. We expect similar results for $CeAl_3$ [10].

4. Discussion

In this section we want to relate our Landau Fermi Liquid approach to other work on the Anderson lattice model.

The main body of work on the Anderson lattice has tended to follow one of two main approaches. The first looks for plausible low-energy expansions for the Green's function and relates the results directly to the low temperature behavior observed experimentally [13-17]. This work shares the same spirit as ours and, at least implicitly, makes the same two basis assumptions (see Sects.2,3).

The second approach has been to concentrate on giving a fairly rigorous account of the Anderson impurity model and to treat lattice effects subsequently in some perturbative scheme [18-24]. This approach is justified when the orbital degeneracy, N_f, is large, as interference effects between sites occur only in order $1/N_f$. The approach also has the advantage of being able to account for high energy (spectral) properties and here the agreement with experiment for the high T_k materials has been spectacular. The identification of $1/N_f$ as an expansion parameter [25] was also an important step conceptually, and we include here a brief account of the infinite N_f limit as it demonstrates how the Fermi liquid approach relates to the more standard approaches.

In the limit of large N_f the ground state and low-lying excited states of the Anderson impurity model are most easily written in the pictorial representation of GUNNARSSON and SCHOENHAMMER [26]. Figure 3 shows the ground state $|\psi_0\rangle$ and the excited (hole) states $|\psi^{ex}_{\mathbf{k}0}\rangle$ of the interacting system

$$
\begin{aligned}
|\psi_0\rangle &= A\Big(|FS\rangle + \sum_{\mathbf{p}\nu\sigma} g_{\nu\sigma}(\mathbf{p}) f^+_\nu \phi_{\mathbf{p}\sigma}|FS\rangle\Big) \quad . \\
|\psi^{ex}_{\mathbf{k}\sigma}\rangle &= A\Big(\phi_{\mathbf{k}\sigma}|FS\rangle + \sum_{\mathbf{p}\nu\sigma'} g_{\nu\sigma'}(\mathbf{p}) \phi_{\mathbf{k}\sigma} f^+_\nu \phi_{\mathbf{p}\sigma'}|FS\rangle\Big) \quad .
\end{aligned} \tag{5}
$$

The operators $\phi_{\mathbf{p}\sigma}$ and f_ν are annihilation operators for particles in the conduction band state $(\mathbf{p},\sigma)$ and in the localized f-level with spin and orbital quantum numbers denoted by ν. $|FS\rangle$ is the filled Fermi sea of the conduction electrons when the f-occupancy is zero. The function $g_{\nu\sigma}(\mathbf{p})$ can be calculated in the limit of large N_f using, for example, Brillouin-Wigner (BW) perturbation theory. The use of BW perturbation theory is only justified when N_f is large and for energies $|\varepsilon - \varepsilon_F| \ll T_k$. The constant A is just given by

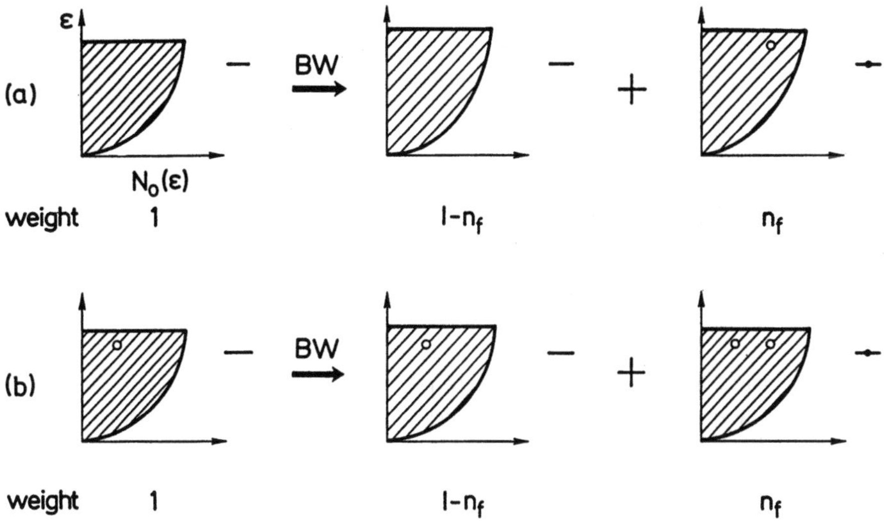

Fig.3. Effect of Brillouin-Wigner perturbation theory on the ground state (a), and on the 1-hole state (b). The weight of the bare states in the dressed states is $1-n_f$ and becomes very small for $n_f \to 1$ (after Ref. [18,27])

$$|A|^2 = 1 - n_f \tag{6}$$

where n_f is the f-occupancy of the interacting system as before. Figure 3 shows the one to one correspondence between the excitations of the interacting system and those of an electron gas. The excitation energy of the state $|\psi_k^{ex}\rangle$ is set in correspondence to that of a hole in an electron system with quantum numbers $(\mathbf{k},\sigma)$. GUNNARSSON and SCHÖNHAMMER [26] also treat higher order terms than those illustrated in Fig.3, but these couple in other conduction band states so that the one-to-one correspondence between quasiparticle excitations and unperturbed conduction band states is no longer transparent.

The quasiparticle of Fig.3b should be thought of as the dressed conduction band state. The 'dressing' or 'cloud' consists of the admixture at the impurity site of the localized f-state. The quasiparticle appears (for energies $|\varepsilon - \varepsilon_F| \ll T_k$) to have n_f f-electron weight at the impurity site and only $1 - n_f$ conduction weight.

The f-like weight measured *spectrally* at the Fermi energy e.g., via photoemission varies as $n_f(1 - n_f)$, where the added factor of $(1 - n_f)$ is from the matrix element $|\langle\psi_{k\sigma}^{ex}|f_\nu|\psi_0\rangle|^2 = |A|^2$ [19]. This factor is the well-known weighting factor of the so-called Kondo resonance.

The f-spectral weight is shown schematically in Fig.4. For small $T_k(n_f \sim 1)$ the f-spectral weight in the Kondo resonance vanishes as $(1 - n_f)$. At the same time the quasiparticle excitations near the Fermi energy appear to have almost completely f-like character. This is not a contradiction. The Kondo resonance is a many-body

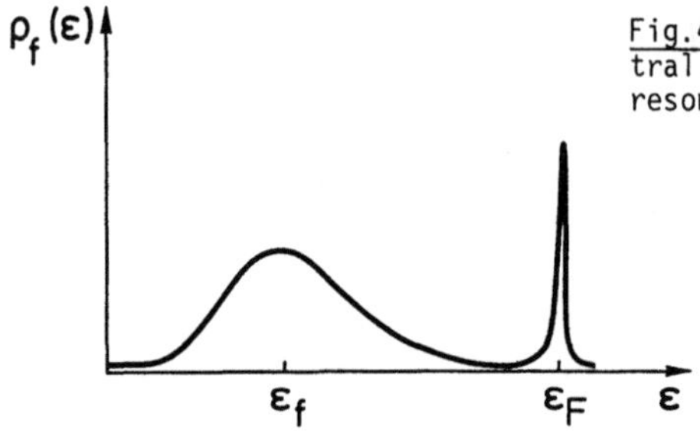

Fig.4. Schematic plot of the f-electron spectral density $\rho_f(\varepsilon)$. The contribution of the resonance at ε_F is of order $1-n_f$

resonance and should not be treated as single-particlelike. An exact calculation working with the Greens function would of course give the quasiparticle results.

These ideas are valid for the lattice in the case of large f-degeneracy N_f, as interference effects between virtual scattering events at different sites are of order $1/N_f$. The difference is that the conduction band state is 'dressed' at *each* site by the admixture of the localized state and not just at the one impurity site.

This discussion is not directly relevant to the heavy quasiparticles of $CeCu_2Si_2$ or $CeAl_3$ where $N_f = 2$. Its importance is conceptual. It demonstrates how the low-lying quasiparticle excitations in this simplified version of the Kondo lattice take on almost entirely f-like character. It also reemphasizes a point made in [3]. Although our model is similar to the virtual bound state model of ANDRES et al. [28], the narrow band in our calculation should not be thought of as an f-band. It is a band of single quasiparticle excitation energies, which do not correspond to a single-particle approximation for an f-electron wavefunction.

Finally we would like to mention briefly the role of the quasiparticle interaction parameters in (2). In the one impurity case NOZIÈRES [7] was able to determine their values from the requirement that the value of n_f, and hence of $\delta(\varepsilon_F)$, remains unchanged when the chemical potential changes. As mentioned in [3] only the sum rule affecting the electronic compressibility can be found this way for the lattice.

To give a full account of, say, magnetic susceptibility measurements we would need to know the values of all the interaction parameters, $\phi^j_{\sigma\sigma'}$. However, the measured value of the Sommerfeld ratio suggests that they may be small in $CeCu_2Si_2$. The Sommerfeld ratio between the susceptibility, χ, and the specific heat coefficient, γ, is given in our model by

$$R = (\pi k_B^2 / 3\mu_{eff}^2)(\chi/\gamma) = 1/[1 - \sum_j(\phi^j_{\uparrow\downarrow} - \phi^j_{\uparrow\uparrow})] \quad ; \qquad (7)$$

experimentally R is found in $CeCu_2Si_2$ to be slightly less than one, indicating that the interaction parameters are small for this compound. (This conclusion could change if it should turn out that (1) does not describe the ground state of the crystal field split 4f multiplet and that the correct ground state has a significantly smaller effective moment μ_{eff}.)

5. Interaction of Quasiparticles with Phonons

In metals the conventional interaction is between conduction electrons and phonons. As has been discussed in Sect.4 in heavy Fermi liquids the quasiparticles are mainly of 4f-character. The weight of the bare conduction electron states is very small for them. Therefore we must look for new coupling mechanisms between 4f-electrons and phonons. For $T \gg T_k$ the coupling mechanisms of phonons to 4f-electrons are well known and have been studied in detail [29]. The 4f-electrons can then be considered as localized and the interactions result from the changes of the CEF by lattice deformations. When $T \ll T_k$ the 4f-electrons participate in the formation of a heavy Fermi liquid. The phonons can then in addition cause quasiparticle-quasihole excitations. In order to derive that interaction it is not necessary to work with quasiparticle excitation bands which contain the full band structure details. Instead it suffices to limit oneself to their essential features. We shall assume in the following that we deal with scattering centers which have only one non-vanishing phase shift which is for $\ell = 0$. This phase shift is taken to be of a resonance form which reduces to (2) close to the Fermi surface, i.e.,

$$\cot g\, \delta_\sigma^i(\varepsilon) = \frac{(\varepsilon_F - \varepsilon)}{T_k} \quad . \tag{8}$$

As discussed in [1] the eigenvalues of the corresponding KKR equations are the same as those of the following effective two-band Hamiltonian

$$H = \sum_{\mathbf{k}\sigma} \left(\frac{k^2}{2m} - \mu\right) C^+_{\mathbf{k}\sigma} C_{\mathbf{k}\sigma} + \sum_{\mathbf{k}\sigma} E_f(\mathbf{k}) f^+_{\mathbf{k}\sigma} f_{\mathbf{k}\sigma} + \gamma_0 \sum_{\mathbf{k}\sigma} \left(f^+_{\mathbf{k}\sigma} C_{\mathbf{k}\sigma} + C^+_{\mathbf{k}\sigma} f_{\mathbf{k}\sigma}\right) \quad . \tag{9}$$

It contains two types of electrons, delocalized and quasilocalized ones with corresponding creation operators $C^+_{\mathbf{k}\sigma}$ and $f^+_{\mathbf{k}\sigma}$, respectively. The dispersion of the "f" electrons is that of a narrow band close to the Fermi surface

$$E_f(\mathbf{k}) = (\varepsilon_F - \mu) + a_1 T_k + a_2 T_k (k - k_F) / k_F \quad , \tag{10}$$

i.e., the band has a width of order T_k. a_1, and a_2 are of the order of unity, and μ is the chemical potential. The hybridization is $\gamma_0 \sim (T_k \varepsilon_F)^{1/2}$.

The narrow "f" band describes the energies of the low lying single quasiparticle excitations. As discussed in Sect.4, these excitations do not correspond to a single-particle approximation for an f-electron wave function. Therefore the Hamiltonian (9) must be considered as that of an equivalent one-electron system which has the same excitation spectrum as the Kondo lattice, a complex many-body system.

The Hamiltonian (9) does not contain the quasiparticle-quasiparticle interaction which are represented by the term $\sum_{\sigma' j \varepsilon'} \phi^{i-j}_{\sigma\sigma'} \delta n^j_{\sigma'}(\varepsilon')$ in (2). Their main effect is to ensure that the phase shift $\delta(\varepsilon_F) \simeq \pi/2$ remains unchanged as the chemical potential is changed (pinning to the Fermi surface). Strictly, when the electron density and hence the chemical potential change e.g., by applying pressure then also n_f will change slightly. Therefore, the pinning of the phase shift to the Fermi surface is *incomplete*. The quasiparticle-phonon interaction must be consistent with the nearly complete pinning of the phase shift to the Fermi surface. We show below that such an interaction results from a change of the Kondo temperature with volume. The Kondo temperature determines the width of the narrow "f" band and also n_f through the term a_1T_k in (10). An interaction based on the volume dependence of T_k does therefore not violate the requirements on the phase shift discussed above.

The Hamiltonian (9) is readily diagonalized by the operators

$$a^+_{\mathbf{k}\sigma} = x_{\mathbf{k}}c^+_{\mathbf{k}\sigma} + y_{\mathbf{k}}f^+_{\mathbf{k}\sigma} \tag{11}$$

which describe the quasiparticles. Their energies are given by the eigenvalue of H, or more realistically by the quasiparticle bands calculated before. The quasiparticle-phonon interaction is obtained from H by assuming a lattice deformation to be present which is characterized by small volume changes $d\Omega(\ell)$ of the different Wigner-Seitz cells ℓ. As a consequence the value of γ_0 and E_f vary from cell to cell. Those changes can be related to changes of the Kondo temperature with volume and result in the quasiparticle-phonon interaction we are looking for.

The result is (see [1])

$$H_{qp\text{-}ph} = \Omega \sum_{\mathbf{kq}} |\mathbf{q}| \, (2NM_0\omega_{\mathbf{q}})^{-1/2} \, [(\partial\gamma_0/\partial\Omega)(x_{\mathbf{k+q}}y_{\mathbf{k}} + y_{\mathbf{k+q}}x_{\mathbf{k}}) + 2(\partial E_f/\partial\Omega)$$

$$\times \, y_{\mathbf{k+q}}y_{\mathbf{k}}]a^+_{\mathbf{k+q}\sigma}a_{\mathbf{k}\sigma}(b_{\mathbf{q}} + b^+_{-\mathbf{q}}) \quad . \tag{12}$$

The $b^+_{\mathbf{q}}$, $b_{\mathbf{q}}$ are phonon creation and annihilation operators corresponding to compressional modes. N is the number of ions and Ω is the volume of a Wigner-Seitz cell. M_0 is the ion mass and $\omega_{\mathbf{q}}$ the phonon dispersion. $(\partial\gamma_0/\partial\Omega)$ and $(\partial E_f/\partial\Omega)$ can be related to changes of T_k with volume, i.e.,

$$\frac{\partial\gamma_0}{\partial\Omega} = \frac{\gamma_0}{2T_k}\frac{dT_k}{d\Omega} \qquad\qquad \frac{\partial E_f}{\partial\Omega} = \frac{E_f - \varepsilon_F + \mu}{T_k}\frac{dT_k}{d\Omega} \quad . \tag{13}$$

The direct dependence of γ_0 on Ω can be neglected. The change $dT_k/d\Omega$ can be directly measured by experiments. For that purpose one writes

$$\frac{dT_k}{d\Omega} = -\eta\frac{T_k}{\Omega} \quad . \tag{14}$$

TAKKE et al. [30] have shown that η can be considered as an electronic Grüneisen parameter. It has been determined experimentally for a number of heavy fermion systems by measuring the bulk modulus and the thermal expansion and by using thermodynamical relations which connect those quantities to the specific heat. Examples are $\eta = -200$ for $CeAl_3$ and $\eta = 20$-80 for $CeCu_2Si_2$. Alternatively, one can determine η by applying pressure to a sample and by measuring the changes in the Kondo temperature as they manifest themselves in the resitivity. For $CeCu_2Si_2$ one finds a value of $\eta \approx 80$ [1]. A similar result is found for Ce impurities in La [31]. We assume $\eta \approx 40$ to be a reasonable value for $CeCu_2Si_2$. With respect to the range of validity of the interaction (12) the following can be said. From Fig.1 it is seen that there are conduction electron branches which do not hybridize with the f electrons. For these branches the conventional conduction electron-phonon interaction holds. For the branches which give rise to the heavy mass quasiparticles an interaction of the form (12) holds as long as the energies of the initial and final states of the quasiparticle are within a range of k_BT_k of the Fermi energy. For higher excitation energies the quasiparticles change completely their character. They go over into conventional conduction electron excitations and CEF excitations, respectively. Therefore, one expects that for high frequency phonons ($h\omega \gg k_BT_k$) the interactions are of the form as in $LaCu_2Si_2$ (for the conduction electrons) and in CEF systems (for the 4f-electrons). Finally, although the interaction (12) was derived for longitudinal phonons we expect also a coupling of comparable size to transverse phonons due to the strongly anisotropic Fermi surface.

It is of interest to consider the influence of the interaction (12) on sound waves. For that purpose one can use the Hamiltonian

$$H = H_0 + H_{qp\text{-}ph}$$

$$H_0 = \sum_{\mathbf{k}\sigma} \left(\frac{k^2}{2m^*} - \mu\right) a^+_{\mathbf{k}\sigma} a_{\mathbf{k}\sigma} + \sum_{\mathbf{q}} \omega^{(0)}_{\mathbf{q}} b^+_{\mathbf{q}} b_{\mathbf{q}} \tag{15}$$

$$H_{qp\text{-}ph} = -\eta c T_k \sum_{\mathbf{k}\mathbf{k}'\sigma} |k - k'| \left(2NM_0 \omega^{(0)}_{\mathbf{k}-\mathbf{k}'}\right)^{-1/2} \cdot a^+_{\mathbf{k}\sigma} a_{\mathbf{k}'\sigma} \left(b^+_{-(\mathbf{k}-\mathbf{k}')} + b_{\mathbf{k}-\mathbf{k}'}\right) \quad .$$

A free-electron like dispersion of the heavy quasiparticles with mas m^* has been assumed. The quasiparticle-phonon interaction (12) has been written in a simplified form with c of order unity, and can be directly compared with the one postulated by LÜTHI [32] on the basis of the experimental findings. It has the form of a deformation potential with a coupling constant $C_1 = \eta c T_k$. The influence of $H_{qp\text{-}ph}$ on the sound velocity can be calculated by second order perturbation theory. One finds that

$$v_s = v_s^{(0)} - 2\frac{M^2_{\mathbf{q}}}{|\mathbf{q}|} \int d\varepsilon N(\varepsilon) \left(-\frac{\partial f}{\partial \varepsilon}\right) \quad , \tag{16}$$

where $v_s^{(0)}$ is the unrenormalized sound velocity. $N(\varepsilon)$ is the density of states of the quasiparticles for one spin direction. $f(\varepsilon)$ is the Fermi function. The matrix element squared $M^2_{\mathbf{q}}$ is given by

$$M_{\mathbf{q}}^2 = \frac{\eta^2 c^2 T_k^2 |\mathbf{q}|^2 \Omega}{2 M_0 \omega_{\mathbf{q}}^{(0)}} \tag{17}$$

where $\Omega = V/N$ (V = volume) has been assumed. A relation of the form of (16) has been used before by TAKKE et al. [30] on phenomenological grounds in order to explain their low temperature data on the elastic constants of heavy Fermi liquids. Equation (16) can be generalized to the case of an applied magnetic field [33]. The situation resembles the one in the A 15 compounds where a magnetic field dependence of the sound velocity has been observed due to the Zeeman splitting of a relatively large electronic density of states [34]. In the present case one obtains in the zero temperature limit

$$\omega_{\mathbf{q}}(H) = \omega_{\mathbf{q}}(0) + \frac{M_{\mathbf{q}}^2 R^2}{2} (\mu_{eff}\, H)^2 \left\{ -N''(\varepsilon_F) + \frac{[N'(\varepsilon_F)]^2}{N(\varepsilon_F)} \right\} + \Delta\omega_{\mathbf{q}}^{VV}(H) \quad . \tag{18}$$

Here $N^{(n)}(\varepsilon_F)$ is the n'th derivative of the quasiparticle density of states at the Fermi energy. The factor R^2 (see (7)) accounts for the quasiparticle interactions which enter into the susceptibility and therefore into the field dependent term. The sign of the field dependent term depends on the detailed shape of the quasiparticle density of states. A rough estimate shows that the relative change in sound velocity is of the order $(\mu_{eff}\, H / T_k)^2$. The contribution $\Delta\omega_{\mathbf{q}}^{VV}(H)$ takes the field dependent transitions to the higher CEF levels into account and is well known [35] (Van Vleck contribution).

6. Superconductivity

It is well known that $CeCu_2Si_2$ is a superconductor while $LaCu_2Si_2$ is not. Superconductivity must therefore result from the 4f-electrons and the question arises which are the relevant interactions which are responsible for that. There are only two possible sources for an effective quasiparticle-quasiparticle attraction. These are the interactions which result from the $\phi_{\sigma\sigma'}^{i-j}$, in (3) and the quasiparticle-phonon interaction (12). In [1] the proposal was made that in $CeCu_2Si_2$ superconductivity is due to the quasiparticle-phonon interaction (12). In the meantime high-resistance point-contact experiments by POPPE [36] have demonstrated the existence of a Josephson current between $CeCu_2Si_2$ and Al. From the existence of such a current and from the size of it one must conclude that the pairing in superconducting $CeCu_2Si_2$ is in a spin singlet state. That suggests strongly that the pairing results indeed from a phonon interaction. In order to obtain an estimate for T_c from the quasiparticle-phonon interaction one can write down the self-consistency equation for the superconducting order parameter $\Delta(\varepsilon)$, i.e.,

$$\Delta(\varepsilon) = \frac{2\tilde{\lambda}}{Z(\varepsilon)} \int_0^{T_k} \frac{d\varepsilon'}{\varepsilon'} \frac{N(\varepsilon')}{N(0)} \tanh \frac{\beta\varepsilon'}{2} \Delta(\varepsilon') \quad . \tag{19}$$

Here $N(\varepsilon)$ is the quasiparticle density of states. $Z(\varepsilon)$ is a renormalization function for the quasiparticles due to their interaction with phonons. Furthermore

$$\tilde{\lambda} = 2n^2c^2T_k^2 \langle\omega_{\mathbf{q}}\rangle^{-1} N(0) [\tilde{q}^2 / (2M_0\langle\omega_{\mathbf{q}}\rangle)] \quad , \tag{20}$$

where $\tilde{q}$ and $\langle\omega_{\mathbf{q}}\rangle$ are an average phonon frequency and momentum. M_0 is an ion mass. For $CeCu_2Si_2$ $\tilde{\lambda}$ is estimated to be $\tilde{\lambda} = 0.1$.

The cut-off of the integral at T_k results from the fact that for excitation energies larger than T_k the quasiparticles and the quasiparticle-phonon interaction change their character completely. For $\varepsilon' > T_k$ the 4f electron will act as a pair breaker in the same way as it would, for example, in $La_{1-x}Ce_xAl_2$. From (18) one obtains an estimated T_c of 1K for $CeCu_2Si_2$. This value may change due to quasiparticle repulsion which are not taken into account here. The purpose of this simple estimate is merely to demonstrate that the interaction (12) is sufficiently strong in order to account for a T_c of the observed size. For a more detailed investigation one must use a Hamiltonian which is more specific for a given material than the one (12) used here. Model calculations which take the quasiparticle repulsions into account and work explicitly with the distance dependence of quasiparticle-phonon interaction [37,38] came again to the conclusion that the attraction is strong enough to lead to spin singlet superconductivity. There is another interesting problem due to the presence of heavy *and* light quasiparticles at the Fermi surface. We expect the light quasiparticles to have essentially no energy gap because we expect that their superconducting properties are only induced by the heavy quasiparticles. It remains to be checked whether or not this might account for the nonexponential behavior of e.g., the specific heat in the low temperature limit. To the single-particle tunneling current heavy and light quasiparticles should contribute almost equally because the heavy quasiparticles leave their clouds behind when tunneling. Therefore, one would expect to see no energy gap in a tunneling experiment. Finally, we want to remark that it is possible that in the U compounds, in particular UPt_3 the dominating interacting mechanism for superconductivity is different from the one in $CeCuSi_2$. Considerable work remains to be done both, experimental and theoretical, before these open questions can be settled.

Acknowledgements. We thank Dipl. Phys. J. Sticht, Prof. J. Kübler and Dr. N. Christensen for the loan of some of their band structure programs and for help with the computations.

References

1 H. Razafimandimby, P. Fulde, J. Keller: Z. Phys. B **54**, 111 (1984)
2 F. Steglich, J. Aarts, C.D. Bredel, W. Lieke, D. Meschede, W. Franz, H. Schäfer: Phys. Rev. Lett. **43**, 1892 (1979)
3 N. d'Ambrumenil, P. Fulde: J. Magn. and Magn. Mat. **47-48**, 1 (1985)
4 S. Doniach: in *Valence Instab. Relat. Narrow Band Phenom.*, ed. by R.D. Parks (Plenum, New York 1977)

5 S. Horn, E. Holland-Moritz, M. Loewenhaupt, F. Steglich, H. Scheuer, A. Benoit, J. Flouquet: Phys. Rev. B **23**, 3171 (1981)
6 W. Hanzawa, S. Maekawa: private communications
7 P.A. Nozières: J. Low Temp. Phys. **17**, 31 (1974); *Proceedings of LT 14*, Vol.5 (North-Holland, Amsterdam 1975) p.339; J. de Physique **39**, 1117 (1978)
8 O.K. Andersen: Phys. Rev. B **12**, 3060 (1975)
A.R. Williams, J. Kübler, C.D. Gelatt: Phys. Rev. B **19**, 6094 (1979)
H.L. Skriver: *The LMTO Method*, Springer Ser. Solid-State Sci., Vol.41 (Springer, Berlin, Heidelberg 1984)
9 G.R. Stewart: Rev. Mod. Phys. **56**, 755 (1984)
10 N. d'Ambrumenil, J. Sticht, J. Kübler: to be published
11 C.D. Bredl, S. Horn, F. Steglich, B. Lüthi, R.M. Martin: Phys. Rev. Lett. **52**, 1982 (1984); for related measurements: J. Flouquet, J.C. Lasjaunias, J. Peyrard, M. Ribault: J. Appl. Phys. **53**, 2127 (1982)
12 F. Steglich, C.D. Bredl, W. Lieke, U. Rauchschwalbe, G. Sparn: Physica **126** B+C, 82 (1984)
13 H. Keiter, G. Morandi: Phys. Rep. **109**, 227 (1984)
14 N. Grewe: Solid State Comm. **50**, 19 (1984)
15 A. Yoshimori, H. Kasai: Journ. Magn. and Magn. Mat. **31-34**, 475 (1983)
M. Tachiki, S. Maekawa: Phys. Rev. B**29**, 2497 (1984)
16 F.J. Ohkawa: J. Phys. Soc. Japan **53**, 1389, 1828 (1984)
17 H. Jichu, T. Matsuura, Y. Kuroda: Prog. Theoret. Phys. **72**, 366 (1984)
18 O. Gunnarsson, K. Schönhammer: Phys. Rev. B**28**, 4315 (1983)
19 P. Coleman: Phys. Rev. B**29**, 3035 (1984)
20 F.C. Zhang, T.K. Lee: Phys. Rev. B**30**, 1556 (1984)
21 Y. Kuramoto, H. Kojima: Z. Phys. B **57**, 95 (1984)
22 N. Read, D. Newns: J. Phys. C **16**, 3273 (1983)
23 J.W. Rasul, A.C. Hewson: J. Phys. C **16**, L 933 (1983); **17**, 2555 (1984); **17**, 3337 (1984)
24 N.E. Bickers, D.L. Cox, J.W. Wilkins: Phys. Rev. Lett. **54**, 230 (1985)
25 T.V. Ramakrishnan: in *Valence Fluctuations in Solids*, ed. by L.M. Falicov, W. Hanke, B. Maple (North-Holland, Amsterdam 1981) and P.W. Anderson: same reference
26 O. Gunnarsson, K. Schönhammer: Phys. Rev. B **31**, 4815 (1985)
27 P.F. de Châtel: Solid State Comm. **41**, 853 (1982)
28 K. Andres, J.E. Graebner, H.R. Ott: Phys. Rev. Lett. **35**, 1779 (1975)
29 B. Lüthi: in *Dynamic Properties of Solids*, ed. by G.K. Horton, A.A. Maradudin, Vol.3 (North-Holland, Amsterdam 1980)
30 R. Takke, M. Niksch, W. Assmus, B. Lüthi, R. Pott, R. Schefzyk, D. Wohlleben: Z. Phys. B **44**, 33 (1981)
31 J.S. Schilling: Adv. Phys. **28**, 657 (1979)
32 B. Lüthi: in *Proceedings of the Vth International Conference on Crystalline Field and Anomalous Mixing Effects in f-Electron Systems*, Sendai 1985 (in preparation)
33 R.J. Wojciechowski, P. Fulde: to be published
34 J.N. Cheeke, H. Mallié, S. Roth, B. Seeber: Solid State Comm. **13**, 1567 (1973)
35 P. Fulde: "Crystal Fields" in *Handbook on the Physics and Chemistry of Rare Earths*, ed. by K.A. Gschneidner, L. Eyring (North-Holland, Amsterdam 1978) Chap.17
36 U. Poppe: in print
37 F.J. Ohkawa, H. Fukuyama: J. Phys. Soc. Japan **53**, 4344 (1984)
38 K. Miyake, T. Matsuura, H. Jichu, Y. Nagaoka: Progr. Theor. Phys. (Kyoto) **72**, 1063 (1984)

Z^{-1} Expansion in Dense Kondo Systems

H. Fukuyama

Institute for Solid State Physics, University of Tokyo
7-22-1, Roppongi, Minato-ku, Tokyo 106, Japan

It is emphasized that the coherent state of dense Kondo systems can be described conveniently by noting the existence of the small expansion parameter Z^{-1}, Z being the mass renormalization factor. This expansion naturally leads to the existence of the heavy quasi-particles at the fermi energy mixing with conduction electrons, but with the mixing matrix element renormalized by $Z^{-1/2}$. Such heavy quasi-particles form a band by themselves even without the direct mixing with conduction electrons and have residual repulsive interactions, whose coupling strength is of the order one. Based on this argument the effect of impurity scattering is also examined and it is indicated that the scattering rate due to potential fluctuations is reduced by Z^{-1} whereas that due to modifications of mixing matrix elements is not. Hence it is concluded that the superconductivity is more easily influenced by alloying which affects the latter.

1. Introduction

One of the most important and difficult problems in heavy electron systems [1] is the proper description of the fermi liquid properties of the coherent state realized at low temperatures in the dense Kondo systems [2]. Such systems are given by the periodic Anderson model [2],

$$H = \Sigma\, \xi_k c_{ks}^+ c_{ks} + \Sigma\, \varepsilon_f n_i + \Sigma U\, n_{i\uparrow} n_{i\downarrow} + \frac{V}{\sqrt{N}} \sum_i (e^{ikR_i} c_{ks}^+ a_{is} + \text{h.c}). \quad (1.1)$$

Here c_{ks} and a_{is} refer to conduction electrons and f-electrons, respectively, and each term in (1.1) has a conventional meaning. We are interested in the s-d limit of this model Hamiltonian. In this case, if the mixing matrix element, V, is absent, each f-state has a local moment and the excitation spectra of the whole system consist of those of conduction electrons and free localized spins as schematically shown in Fig.1(a). Here that of localized spins is N-fold degenerate at vanishingly small energy, N being the total number of lattice sites. In the presence of finite V this degeneracy should be removed and the excitation spectra of the combined system will be like that shown in Fig.1(b). Here and hereafter we assume the system stays in normal state and does not have magnetic ordering, i.e. the competition between magnetic and normal states is not discussed in this paper. In Fig.1(b) the characteristic energy scale of the removal of the degeneracy will be T_K, the Kondo temperature, and in this narrow energy region lie all of the excitations associated with fluctuations originating from the localized spins. These excitations will behave just as those of localized moments for high energy, ω, $\omega \gtrsim T_K$, or high temperature, T, $T \gtrsim T_K$, but once $T \lesssim T_K$ these will look like those

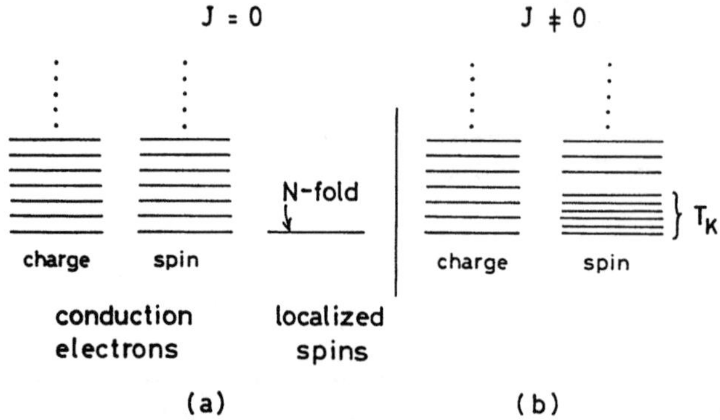

Fig.1 Schematic representation of excitation spectra of dense Kondo systems, without (a) and in the presence (b) of the coupling between conduction electrons and localized spins.

of fermi liquid because of the removal of the degeneracy. Due to the large density of states of such fluctuations associated with the spin degree of freedom, both the specific heat and the spin susceptibility should be enhanced in a similar way.

This will be the physical origin of the heavy electrons.

Then, how does one describe this on a microscopic ground?

We take unit of $\hbar=k_B=1$.

2. Z^{-1} Expansion

The f-electron Green function for (1.1), $G(k,i\varepsilon_n)$ ($\varepsilon_n=(2n+1)\pi T$), is formally given in terms of that of conduction electrons, $G_s(k,i\varepsilon_n)$, as

$$G^{-1}(k,i\varepsilon_n) = i\varepsilon_n - \varepsilon_f - V^2G_s(k,i\varepsilon_n) - \Sigma(k,i\varepsilon_n) \quad , \qquad (2.1)$$

where $\Sigma(k,i\varepsilon_n)$ takes into account interaction, U, between f-electrons and contains in principle all the information we need. Especially low temperature properties are correctly described by its k- and $i\varepsilon_n$- dependences near the fermi surface and in the limit of low energy. This is the fermi liquid theory, which was developed to the case of a single impurity by Noziere [3] some time ago and has recently be discussed for the lattice by Martin [4] and Ohkawa [5].

If the k-dependence of $\Sigma(k,i\varepsilon_n)$ is to be ignored, i.e. $\Sigma(k,i\varepsilon_n)\to\Sigma_0(i\varepsilon_n)$, one can determine $\Sigma_0(i\varepsilon_n)$ by placing the full scattering matrix of an impurity problem on every lattice site. This is the single site approximation. This approximation was originally proposed by Yoshimori and Kasai [6] in the discussion of the temperature-dependence of resistivity in dense Kondo systems, who, however, had to introduce ad hoc the finite band width of the f-electrons besides (1.1) because of the neglect of the k-dependence of $\Sigma(k,i\varepsilon_n)$ to be discussed in the following.

By the knowledge on a Kondo impurity the energy-dependence of $\Sigma_0(i\varepsilon_n)$ is given by [7,8]

$$\Sigma_0(i\varepsilon_n) = \Sigma_0(0) + i\varepsilon_n \left.\frac{\partial\Sigma_0(\varepsilon)}{\partial\varepsilon}\right|_{\varepsilon=0},$$

$$= \Sigma_0(0) - i\varepsilon_n(Z - 1) \quad , \qquad (2.2)$$

where $Z=\pi\Delta/4T_K \gg 1$ in the present case. Here $\Delta=\pi\rho V^2$, ρ being the density of states of conduction electrons. Diagrammatically $\Sigma_0(i\varepsilon_n)$ is given by the process shown in Fig.2, where the square, $\Gamma_{\downarrow\uparrow}$, implies all possible interaction processes at a given site, i, and lines are f-electron Green functions.

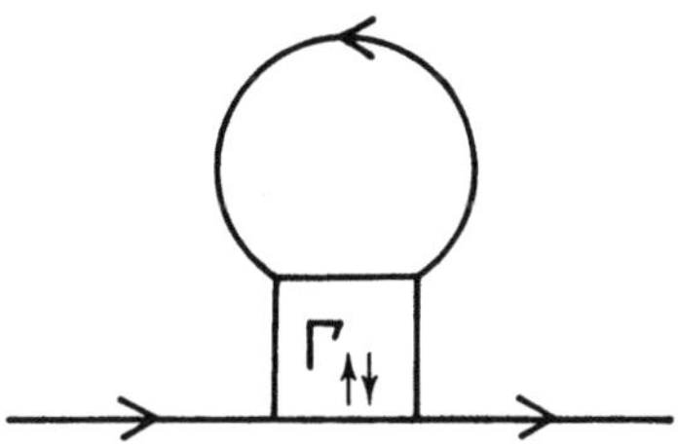

Fig.2 Self-energy in the single site approximation.

In general the k-dependent term of $\Sigma(k,i\varepsilon_n)$, defined as $\overline{\Sigma}(k,i\varepsilon_n)$, should be considered as well, i.e.

$$\Sigma(k,i\varepsilon_n) = \Sigma_0(i\varepsilon_n) + \overline{\Sigma}(k,i\varepsilon_n) \quad . \qquad (2.3)$$

A typical process contributing to $\overline{\Sigma}(k,i\varepsilon_n)$ is shown in Fig.3, where Γ_i is $\Gamma_{\uparrow\downarrow}$ at the i-th site. This process has been examined in detail by Jichu et al. [9] for the case where all of the energy variables of intermediate f-electron Green functions are small. In this case the resultant k- and $i\varepsilon_n$- dependences are shown to be weak, i.e. $\overline{\Sigma}(k,i\varepsilon_n)$ is of the order of $\rho\Delta\xi_k$ and ξ_k is the energy of conduction electrons and is without large factor Z. It turns out [10], however, that the contributions from the higher enegy region in the process of Fig.3 are also weakly dependent on k and $i\varepsilon_n$. Similarly all other possible processes can be shown to yield contributions, which have only weak k- and $i\varepsilon_n$- dependences, and we can approximate $\overline{\Sigma}(k,i\varepsilon_n)$ by some energy band, E_k, i.e.

$$\Sigma(k,i\varepsilon_n) \simeq \Sigma(0) - i\varepsilon_n(Z-1) + E_k \ , \qquad (2.4)$$

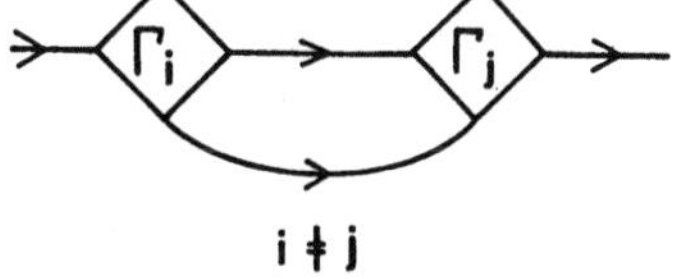

Fig.3 An example of processes leading to the k-dependence of the self-energy function.

Hence the assumption [5,9] of the single-site approximation for the lattice can be justified as far as Z>>1. Consequently we have

$$G(k,i\varepsilon_n)^{-1} = [i\varepsilon_n Z - E_k - V^2 G_s(k,i\varepsilon_n)],$$

$$= Z[i\varepsilon_n - E_k^* - V^{*2} G_s(k,i\varepsilon_n)] \quad , \tag{2.5}$$

where $E_k^*=E_k/Z$ and $V^*=V/\sqrt{Z}$.

Equation (2.5) implies that the renormalized f-electrons or quasi-particles form a coherent band energy, E_k^*, which is mixed with conduction electrons by the renormalized mixing matrix element, V^*. This conclusion is essentially in accordance with that of Razafimandimby et al. [11]. Note, however, these quasi-particles are not completely free but they have residual interactions. The effective strength of this interaction for quasi-particles is of the order one, since $Z^{-2}\Gamma_{\uparrow\downarrow}N(0)\sim O(1)$, $\Gamma_{\uparrow\downarrow}$ being defined in Fig.2 and $\Gamma_{\uparrow\downarrow}=\pi\Delta Z$ in the limit of low energy. Here N(0) is the density of states at the fermi energy, E_F, of quasi-particles defined by (ignoring the last term of eq.(2.5))

$$N(0) \equiv \sum_k \delta(E_F - E_k^*) \quad , \tag{2.6}$$

and Z^{-2} is due to the renormalization factor of f-electron Green functions.

It is straightforward to see that the specific heat is determined by N(0) and then enhanced by a factor Z, i.e. "heavy".

As regards the spin susceptibility, χ_s, one should note that $\Sigma_0(i\varepsilon_n)$ of a single impurity has a strong dependence on magnetic field, H, i.e. [7,8]

$$\frac{\partial}{\partial H}\Sigma_s = s\, g^*\, \mu_B \quad , \tag{2.7}$$

where s and μ_B are the spin index and the Bohr magneton. In (2.7) g^* is the effective g-factor given by $g^*=(1+\pi\Delta/2T_K)g_f$, g_f being the bare g-factor of f-electrons. From this fact, χ_s for the quasi-particles is seen to be [5] $\chi_s=(1/2)gg^*\mu_B^2N(0)/Z$ and the Wilson ratio is 2 as far as Z>>1 with possible fermi-liquid corrections.

3. Phase-Transition in Heavy Electron Systems

In the coherent state heavy electrons are interacting and there will be electronic phase-transitions driven by such mutual interactions. As mentioned in the previous section,heavy electrons have short-range repulsive interactions. The physical origin of the argument for the odd-parity superconductivity [12-14] will be based on such interaction. At the same time, however, one can think of a strong electron-phonon interaction [11, 15-18] modifying the mixing matrix element leading to the Kondo volume collapse [11,19]. In such circumstances there will be an appreciable effective interaction between neighboring f-sites [16,20]. In the presence of a repulsive on-site together with an attractive nearest neighbor interaction the order parameter of superconductivity can have a strong anisotropy even in the even-parity state, e.g. in a simple cubic crystal

$$\Delta(k) = \Delta_0 - \Delta_1 (\cos k_x a + \cos k_y a + \cos k_z a) \quad , \tag{3.1}$$

where a is a lattice constant and Δ_0 and Δ_1 are some constants dependent on relative strength of the repulsive and attractive interaction. If these interactions compete, we have $\Delta_0 \sim \Delta_1$, in which case it is possible to have $\Delta(k)=0$ on a line on the fermi surface. This is because the shape of the fermi surface is determined by E_k^* whose k-dependence is not necessarily equal to that of $\Delta(k)=0$. (In ref.16 the k-dependences of $\Delta(k)$ and E_k are assumed to be same and then $\Delta(k)=0$ is not possible on the fermi surface in this case.) (The details of such type of superconductivity are discussed by Ohkawa in this volume.)

Apart from superconductivity, an itinerant antiferromagnetism due to the fermi surface instability is also possible for heavy electrons because of the short-range repulsive interactions and depending on the actual shape of the fermi surface. U_2Zn_{17} [21] will belong to this category. In this system the specific heat jumps very sharply at the transition temperature,indicating the fermi surface instability instead of the alignment of localized moments, as the authors have stressed.

Hence,as far as the interaction processes are concerned, the heavy electrons behave quite similarly to ordinary fermi liquid.

4. Impurity Scattering

Depending on the types of randomness,there exists a drastic difference in the resultant effects, i.e. if these effects are phenomenologically represented in (2.5) by

$$G(k,i\varepsilon_n)^{-1} = Z[i\varepsilon_n - E_k^* - \frac{i}{2\tau^*}\,\mathrm{sgn}\varepsilon_n - V^{*2}G_s(k,i\varepsilon_n)] \quad , \tag{4.1}$$

τ^{*-1} will be of the order of Z^{-1} if either the potentials of f-sites, ε_f, or the conduction electrons are perturbed. On the other hand τ^{*-1} can be of the order of one when the mixing, V, and then Z is modulated locally, i.e.

$$\frac{1}{\tau^*} = \begin{cases} 1/\tau_0 Z & \text{(potential scattering)} \quad , & (4.2a) \\ 1/\tau_K & \text{(mixing scattering)} \quad , & (4.2b) \end{cases}$$

where τ_0 and τ_K are considered to be of the ordinary magnitude.

For (4.1) the diffusion propagator of f-electrons is given by

$$N_f(q,\omega) = \frac{1}{Z^2} N(0) \frac{D^*q^2}{D^*q^2 - i\omega} \equiv \frac{1}{Z^2} N(q,\omega) \quad , \tag{4.3}$$

where N(0) is given by (2.6) and $D^* = v_F^{*2}\tau^*/3$ with $v_F^* = \partial E_k^*/\partial k$ at the fermi surface and then $v_F^* \sim O(Z^{-1})$. In (4.3) N(q,ω) is the diffusion propagator of heavy quasi-particles.

The charge susceptibility of f-electrons is given by $N_f(0,0)$, which is of the order of Z^{-1} in accordance with the known results for a single impurity [7,8].

From (4.3) one sees that the ultrasonic attenuation coefficient [22], α_s, for sound waves modulating the mixing matrix elements and satisfying $D^*q>\omega$ is given by

$$\alpha_s \sim N(0)\ D^*q^2 / \omega \quad , \tag{4.4}$$

which will be of the order of ordinary metals if τ^* is due to potential scattering, (4.2a), since $N(0)D^* \sim O(1)$.

The Boltzmann conductivity, σ_0, due to heavy quasi-particles,is given by

$$\sigma_0 = e^2 D^* N(0) \quad , \tag{4.5}$$

which is of the order 1 or Z^{-1} for potential scattering or mixing scattering, respectively. The quantum corrections to the conductivity due to Anderson localization, σ' [23], are of the order of $\sigma'/\sigma_0 \sim \lambda^{d-1}$ for each dimensionality, d, where $\lambda=(2\pi\varepsilon_F^*\tau^*)^{-1}$ with $\varepsilon_F^*=v_F^*k_F$. Hence the corrections will be similar to ordinary dirty metals if the randomness is due to the potential scattering.but they are enhanced for mixing scattering.

Since the superconductivity with the anisotropy given by (3.1) is sensitive to impurity scattering [20] characterized by τ^*, it will be concluded that the impurities affecting the mixing matrix elements are strongly destructive to superconductivity, which appears to be the case in $CeCu_2Si_2$ [24].

5. Summary

It is emphasized that the characteristic features of heavy electrons achieved in the coherent state of dense Kondo systems can simply be extracted by noting the existence of small parameter Z^{-1}, Z being the mass renormalization factor. As far as $Z>>1$, Z is correctly given by that of the single-site approximation. These considerations make it straightforward to discuss various physical quantities associated with heavy electrons.including impurity scattering.

Acknowledgements

The author thanks F.J. Ohkawa and T. Matsuura for useful discussions.

References

1. For a recent review on experimental results see G.R. Stewart, Rev. Mod. Phys: **56**, 755 (1984).
2. See papers in Valence Fluctuations in Solids ed. L.M. Falicov, W. Hanke and M.P. Maple (North Holland, 1981).
3. P. Nozieres: J. Phys. (Paris) **39**, 1117 (1978).
4. R.M. Martin: Phys. Rev. Lett. **48**, 362 (1982).

5. F.J. Ohkawa: J. Phys. Soc. Jpn. **53**, 1389 (1984); **53**, 1828 (1984).
6. A. Yoshimori and H. Kasai: J. Magn. Magn. Mater. **31-34**, 475 (1983).
7. K. Yosida and K. Yamada: Prog. Theor. Phys. **53**, 1286 (1975).
8. K. Yamada: Prog. Theor. Phys. **53**, 970 (1975); **54**, 316 (1975); **55**, 1345 (1976).
9. H. Jichu, T. Matsuura and Y. Kuroda: Prog. Theor. Phys. **72**, 366 (1984).
10. H. Jichu, T. Matsuura and Y. Kuroda: ICCF5 (Sendai, 1985) to appear in J. Magn. Magn. Mater.
11. H. Razafimandimby, P. Fulde and J. Keller: Z. Phys. **B54**, 111 (1984).
12. C.M. Varma: Bull. Am. Phys. Soc. **29**, 404 (1984).
13. P.W. Anderson: Phys. Rev. **B30**, 1549 (1984).
14. H.R. Ott, H. Rudigier, T.M. Rice, K. Ueda, Z. Fisk and J.L. Smith: Phys. Rev. Lett. **52**, 1915 (1984).
15. M. Tachiki and S. Maekawa: Phys. Rev. **B29**, 2497 (1984).
16. T. Matsuura, K. Miyake, H. Jichu and Y. Kuroda: Prog. Theor. Phys. **72**, 402 (1984).
17. K. Miyake, T. Matsuura, H. Jichu and Y. Nagaoka: Prog. Theor. Phys. **72**, 1063 (1984).
18. K. Miyake, this volume.
19. J.W. Allen and R. Martin: Phys. Rev. Lett. **49**, 1106 (1982); J. Magn. Magn. Mater. **31-34**, 473 (1983).
20. F.J. Ohkawa and H. Fukuyama: J. Phys. Soc. Jpn. **53**, 4344 (1984).
21. H.R. Ott, H. Rudigier, P. Delsing and Z. Fisk: Phys. Rev. Lett. **52**, 1551 (1984).
22. D.J. Bishop, C.M. Varma, B. Batlogg, E. Bucher, Z. Fisk and J.L. Smith: Phys. Rev. Lett. **53**, 1009 (1984).
23. For review, e.g. Anderson Localization ed. Y. Nagaoka and H. Fukuyama (Springer Verlag, 1982).
24. F. Steglich, private communications.

Heavy Electron Fermi Liquids — Similarities and Differences to ^{3}He

T.M. Rice

Theoretische Physik, Eidgenössische Technische Hochschule, Hönggerberg CH-8093 Zürich, Switzerland

K. Ueda

Department of Applied Physics, The University of Tokyo, Tokyo 113, Japan

The starting point of the theory of heavy electron metals is the weak hybridization limit of the periodic Anderson model. A variational wavefunction for this limit is evaluated using a Gutzwiller approach and is related to the nearly half-filled Hubbard model in the limit of strong correlation. In the large degeneracy limit the binding energy of this coherent wavefunction has the Kondo form. ^{3}He has been described by an exactly half-filled Hubbard model also with strong correlations. Thus both systems can be viewed as examples of almost localized Fermi liquids.

1. Introduction

There is a clear underlying similarity between ^{3}He and the heavy electron metals. In both cases the onsite or short-range interactions are large and repulsive, and as a result charge (or density) fluctuations are strongly suppressed. Spin fluctuations on the other hand are not suppressed but enhanced, since local spin flip processes of the Fermions are low-energy processes when the charge (or density) fluctuations are restricted. The result is that the entropy associated with the spin degrees of freedom will appear already at low temperatures, and so if a Fermi liquid has formed, then the linear coefficient of the specific heat, which determines the effective mass of the quasiparticles, will be strongly enhanced, hence the name heavy electrons. This underlying similarity has been discussed in a qualitative way by a number of authors including ourselves, OTT et al [1], ANDERSON [2], VALLS and TESANOVIC [3], BEDELL and QUADER [4] and by RICE et al [5].

There has of course been an enormous amount of work on the normal state properties of ^{3}He and we will not attempt a review here but concentrate on one particular approach, namely that of an almost localized Fermi liquid. This model was suggested for ^{3}He by ANDERSON and BRINKMAN [6] and later VOLLHARDT [7] developed the model to describe the low-temperature Fermi liquid regime in detail. Recently the model has been extended to give a good qualitative description of the higher temperature thermodynamic properties of ^{3}He including the transition from a degenerate Fermi liquid to a non-degenerate liquid. Details of this extension will be published elsewhere [8]. The essence of this description is in terms of a half-filled single band Hubbard model with a large, but finite, onsite repulsion U which leads to a small density of doubly occupied and empty sites. The vast majority of sites are singly occupied, and it is the spin flip degree of freedom of these sites which gives rise to the entropy.

The heavy electron metals are generally believed to be described by the periodic Anderson model in the Kondo limit. In this model one has a broad conduction band which is coupled via hybridization to the narrow f-band, which in turn has very strong Coulomb interactions. In the Kondo limit, the f-level lies well below the Fermi energy, so that the number of electrons in the f-band per site would be integral in the absence of hybridization. Intuitively there should be a close relationship between the two models, and this is the question that we will focus on here. To this end we will use the Gutzwiller approximation [9] to describe the effect of the strong correlation on the f-band. There are many different approaches current in the literature to treat this Kondo lattice model, and we will not attempt here to compare with all of them. We hope, however, that the simplicity of our approach will make the underlying physics clear and clarify the relationship to the almost localized Fermion model of ^3He.

The origin and type of superconductivity in these systems will not be discussed here, but in view of the analogy to ^3He, p-wave superconductivity is clearly possible. The latter is investigated by us elsewhere in this volume.

The outline of the paper is as follows. First we review briefly the Gutzwiller approximation and then the application to ^3He. Section 4 is devoted to the Kondo lattice problem and the conclusions are in Section 5.

2. Almost Localized Fermions

In the noninteracting wavefunction, which simply minimizes the kinetic energy, there are no correlations between Fermions. In the presence of a strong onsite Coulomb repulsion the configurations with a larger number of doubly occupied sites will have high energies and will be suppressed. GUTZWILLER[9] proposed an approximate way to account for these correlations by introducing a reduction factor for the quantum mechanical kinetic energy. This reduction factor is calculated using a classical statistical method. A very readable account can be found in the work of VOLLHARDT [7].

We start by considering the single-band Hubbard Hamiltonian

$$H = \sum_{ij\sigma} t_{ij} c^{+}_{i\sigma} c_{j\sigma} + U \sum_{i} n_{i\uparrow} n_{i\downarrow} \tag{1}$$

where the first term is the kinetic energy due to hopping between sites i and j and the second term is the onsite repulsion energy, U. Gutzwiller proposed a variational wavefunction for the ground-state of the form

$$|\Psi\rangle = \prod_{i} [1-(1-g) n_{i\uparrow} n_{i\downarrow}] |\Psi_0\rangle \tag{2}$$

$$\equiv g^D |\Psi_0\rangle \tag{3}$$

with the variational parameter g controlling the number of doubly occupied sites D and $|\Psi_0\rangle$ is the ground-state of H when U=0. The effect of the projection operator was calculated by Gutzwiller by introducing a reduction factor q to take account of the reduced possibilities for hopping, and this factor was estimated using the classical statistical weighting of each configuration (see VOLLHARDT [7]). The result for the energy per site is

$$E_G = \sum_\sigma q_\sigma(d,n_\sigma,n_{-\sigma})\bar{\varepsilon}_\sigma + Ud \tag{4}$$

where d is the fraction of doubly occupied sites and $n_\sigma(n_{-\sigma})$ the concentration of $\sigma(-\sigma)$ spin electrons. The kinetic energy is the non-interacting value $\bar{\varepsilon}_\sigma$ times the reduction factor q_σ which is given by

$$q_\sigma = \left[\{(n_\sigma-d)(1-n_\sigma-n_{-\sigma}+d)\}^{\frac{1}{2}}+\{d(n_{-\sigma}-d)\}^{\frac{1}{2}}\right]^2 / n_\sigma(1-n_\sigma) \,. \tag{5}$$

The variational parameter g can be replaced by d, the fraction of doubly occupied sites. The almost localized limit occurs when the value of d which minimizes the energy, $d \to 0$ and was studied first BRINKMAN and RICE[10]. Two cases are of particular interest (i) a half-filled band with $n_\sigma = n_{-\sigma} = 1/2$; then

$$q_\sigma \equiv q(d) = 8d(1-2d) \tag{6}$$

(ii) a filling factor slightly less than 1; $n_\sigma + n_{-\sigma} = n \leq 1$ and very large U so that d=0; then

$$q_\sigma = (1-n)/(1-n_\sigma) \,. \tag{7}$$

The factor q which reduces the kinetic energy can be interpreted as a renormalization of the effective mass $m^*/m = 1/q$, and of the discontinuity in the Bloch state occupation from 1 to q and of the spectral weight in the quasiparticle pole in the one-particle Green's function also by q. In the almost localized limit $q \to 0$ and these renormalizations become very large.

It is straightforward to calculate the simplest parameters in the Landau Femi liquid theory from the energy expression. The linear coefficient of the specific heat is simply proportional to the effective mass m*. The magnetic and charge susceptibilities are enhanced over their non-interacting values by a factors of $(m^*/m)/(1+F_0^{a(s)})$ respectively. The spin asymmetric (symmetric) Landau parameters $F_0^{a(s)}$ in the limits (i) and (ii) take the forms

$$\text{(i)} \quad F_0^a = p\left(-1+\tfrac{1}{4}(1-2d)^{-2}\right) \xrightarrow[d\to 0]{} -3p/4 \tag{8}$$

$$F_0^s = p(-1-q^{-1}+1/2dq) \xrightarrow[d\to 0]{} \infty \tag{9}$$

where the factor p involves the product of the average kinetic energy and the density of states at the Fermi energy (incl. spin) N(0); $p = 2|\bar{\varepsilon}|N(0)$. For a parabolic band p=1.08. In the second case one finds

$$\text{(ii)} \quad F_0^a = -p \tag{10}$$

$$F_0^s = \tfrac{1}{2}(1-n)^{-1} \,. \tag{11}$$

Recently this approach has been generalized to finite temperatures [5] and the main results will be summarized here. The generalization proceeds by recognizing that excited state wavefunctions can be obtained by changing the occupation of the Bloch states $\{n(k,\sigma)\}$ from the choice, $n(k,\sigma)=1$ ($k<k_F$) and 0 ($k>k_F$) where k_F is the Fermi wavevector, which enters $|\Psi_0\rangle$. The internal energy can be calculated in the same way as (4). To obtain the free energy, we need an expression for the entropy. We must recognize that the set of excited states determined by changing $\{n(k,\sigma)\}$ are overcomplete because of the projection factor. It is easy to calculate the total number of independent states with a fixed value of d and in case (i) say this is a monotonically decreasing function of d from 4 per site with d=1/4 (corresponding to U=0) to 2 per site when $d\to 0$. However we know from Landau Fermi liquid theory that the low-energy excitations are independent quasiparticles whose entropy is determined by the free Fermion entropy formula. An ansatz consistent with this condition for the entropy is

$$S = -\sum_{\vec{k}\sigma} w(\vec{k},d,n)\{n(\vec{k},\sigma)\ln\big(n(\vec{k},\sigma)\big)+\big(1-n(\vec{k},\sigma)\big)\ln\big(1-n(\vec{k},\sigma)\big)\} \tag{12}$$

with the weighting function $w(k,d,n)$ chosen in such a way that $\lim w(k\to k_F,d,n) = 1$. The occupation number $n(k,\sigma)$ determined by minimizing the free energy has the form

$$n(\vec{k},\sigma) = \{\exp[q\varepsilon(\vec{k},\sigma)/w(\vec{k})T]+1\}^{-1}. \tag{13}$$

In the almost localized limit $q<<1$ and the occupation number $n(k,\sigma) = n/2$ as T increases beyond the renormalized Fermi energy qE_F. The entropy then saturates at a value determined by $\Sigma_k\ w(k,d,n)$. However, we know from the total number of degrees of freedom with fixed values of d and n the total possible entropy. In this way we get the average value of the weighting function $w(k,d,n)$ in the band and in the almost localized limit it approaches (1/2).

One can also calculate spin susceptibility $\chi_s(T)$ by including a coupling term $\mu_0 H\Sigma_k(n(k,\uparrow)-n(k,\downarrow))$ to the magnetic field H. In the low-temperature limit the Landau Fermi liquid results are recovered. In the higher temperature limit $T>>qE_F$, then a Curie law is obtained. For example, in the second limit one obtains

$$\chi_s = \mu_0^2\Big(\sum_{\vec{k}}(w(\vec{k}))^{-1}\Big)/2T. \tag{14}$$

In this limit, the number of magnetic moments should be simply the number of singly occupied sites, and this leads to an average value for $w^{-1}(k)$ in the band, namely 2n. A form for $w(k)$ can be chosen to comply with these conditions. The renormalized Fermi temperature (qE_F) is the crossover temperature between Fermi liquid behavior and non-degenerate behavior.

3. Normal State Properties of ^3He

We will review just briefly the application to ^3He. As discussed by ANDERSON and BRINKMAN [6] and especially VOLLHARDT [7] the low temperature or Fermi liquid behavior is well described if one takes the limit (i), namely a half-filled band with the value of U chosen as a parameter which depends on pressure. The assumption of an underlying lattice is of course an artificial one, but it can be interpreted as implying a high degree of short-range order in the liquid and the renormalized value of U as an effective energy describing the formation of an interstitial and a vacancy in the "lattice" determined by the short-range order. The big success of the model is that F_0^a goes to a constant value ($\simeq -3/4$) as $d \to 0$ while the effective mass diverges. This agrees well with the experimental behavior of ^3He. The effective mass determined by the specific heat rises with increasing pressure, reaching a value $\simeq 5$ at the solidification pressure. By contrast, the value of F_0^a is hardly pressure-dependent and is $\simeq -0.75$. These results are also rather different from the predictions of paramagnon theory, which presupposes an instability towards ferromagnetism (rather than localization) which requires $F_0^a \to -1$ and an effective mass proportional to $\ln(1+F^a)^{-1}$.

The extension to higher temperature has also been carried out. The results will be published in detail elsewhere [8]. The extension requires no extra parameters, but one must include the higher energy states since there is no upper limit in a liquid to the density of states. The simplest method is to assume that the band of states around the Fermi energy are strongly correlated and to be described by the lattice hypothesis, while the higher energy states are simply free and uncorrelated. The results give a good qualitative description of the entropy, specific heat and magnetic susceptibility as a function of temperature. For example, the results at a pressure of 30 bar show the clear tendency for the entropy to saturate at a value of Rln2 at a temperature $\simeq 0.5K$ and then increase more slowly. The transition from Pauli to Curie behavior in the magnetic susceptibility on a similar temperature-scale is also reproduced. Another thermodynamic property is the value of the pressure, P, as a function of the temperature at fixed volume. This shows a marked deviation from free Fermion behavior where P(T) is a monotonically increasing function of T corresponding to an increasing thermal expansion at fixed P. ^3He on the other hand shows a pronounced minimum at a temperature $\simeq 1.5K$ with a starting value $P(T=0)=30$ bar. Again this minimum is reproduced in the model. Its origin is in the strong volume dependence of the effective mass which governs the values of the entropy at low temperatures. At temperatures $T \gtrsim qE_F$, the spin fluctuation entropy saturates at Rln2 and is no longer volume-dependent, and then the free Fermion behavior of the high-energy states dominates.

The conclusion is that the almost localized model gives a good qualitative description of the thermodynamic properties of normal ^3He in the low-temperature regime, particularly the Fermi liquid behavior and the crossover to nondegenerate behavior at higher temperatures. So far the model has been limited to these properties, and an extension to describe the dynamics, transport properties, etc, of ^3He would require careful consideration.

4. Heavy Electron Metals

The starting point of a description of heavy electron metals has been the periodic Anderson model rather than the Hubbard model(1). This is an important difference and the relationship between the two models needs to be clarified. Qualitative discussions of this point were given in references [1-4]. In [5] we proposed that in the Kondo limit the periodic Anderson model was mapped onto the Hubbard model in the almost localized, almost half-filled, limit (ii). This proposal was made on the basis that in the single-site Kondo problem the number of f-electrons (or localized electrons) was very close to but not exactly integral, while the f-band was clearly in a large U limit. With this proposal the almost localized model was used, [5], to give a good description of the specific heat and entropy of UBe_{13} at temperatures up to ≈10K.

It is clear that a better justification for this proposal is required. The periodic Anderson Hamiltonian has the form

$$H_A = \sum_{\vec{k}\sigma} \varepsilon_k c^+_{\vec{k}\sigma} c_{\vec{k}\sigma} + V \sum_{i\ell\sigma} (c^+_{i\sigma} f_{i\ell\sigma} + f^+_{i\ell\sigma} c_{i\sigma})$$

$$+ E_f \sum_{i\ell\sigma} n_{i\ell\sigma} + \frac{1}{2} U \sum_{\substack{i \\ \ell\sigma\neq\ell'\sigma'}} n_{i\ell\sigma} n_{i\ell'\sigma'} \, . \tag{15}$$

The conduction electrons are described by Bloch states, energy $\varepsilon(k)$ with the band centre at zero-energy and width 2. The f-electrons (creation operator $f^+_{i\ell\sigma}$; number operator $n_{i\ell\sigma} = f^+_{i\ell\sigma} f_{i\ell\sigma}$) are dispersionless, but have a hybridization V with the conduction electrons. The Kondo limit is reached when $E_f \ll 0$ and $U \to \infty$. In this limit not more than one f-electron per site is allowed. The orbital degeneracy of the f-level is an important parameter in this limit, and we include it by summing over ℓ from 1 to L. The total degeneracy of the f-level is 2L in this model. Lastly there is the question of the number n of electrons per site. The case n=2 is a special case as we shall see, and we will consider values of $n < 2$ but $n > 1$.

We want to consider now the application of Gutzwiller methods to this Hamiltonian. In the limit $U \to \infty$ double occupation is totally forbidden, and the projection operator (2) on the wavefunction becomes simply a projection operator which forbids double occupancy, P. In the single-site Kondo problem Gunnarsson and Schonhammer [11] have shown that the spectral and other properties can be described by a variational wavefunction which in the lowest order takes the form

$$|\Psi_{ss}\rangle = (1 + N^{-1} \sum_{\substack{k<k_F \\ \ell\sigma}} a(\varepsilon) e^{i\vec{k}\cdot\vec{R}_i} c_{\vec{k}\sigma} f^+_{i\ell\sigma}) |\Psi_{00}\rangle \tag{16}$$

with $|\Psi_{00}\rangle$ as the state with all electrons in the lowest-energy Bloch states and the f-level empty. N is the number of sites. The function $a(\varepsilon)$ is determined variationally. This single-site wavefunction gives the correct value of the Kondo binding energy as $2L \to \infty$. An equivalent form of this wavefunction is

$$|\Psi_{ss}\rangle = P \prod_{\substack{k<k_F \\ \ell\sigma}} \left(1 + N^{-1} a(\varepsilon) e^{i\vec{k}\cdot\vec{R}_i} c_{\vec{k}\sigma} f^+_{i\ell\sigma}\right) |\Psi_{00}\rangle \tag{17}$$

where P is the projection operator which forbids double occupancy of the f-level.

In the many-site problem the wavefunction may be generalized by summing over site label i. Then the projection operator operates on a wavefunction $|\Psi_0>$ which is simply a one-electron hybridized band wavefunction i.e. hybridized conduction and f-bands. The key question is how to treat the projection operator. The Gutzwiller approximation was developed for the case of a single band with a fixed number of electrons. Therefore we must project first onto a fixed occupancy, n_f, of the f level and then remove the doubly occupied states leading to a many-site wavefunction

$$|\Psi_{ms}> = P_{n_f} P|\Psi_0> . \tag{18}$$

The projection operator P in the Gutzwiller approximation is treated by renormalizing all electron-hopping processes by a factor $q_\sigma(n_f, L)$ as in Eqn.(4). In the present problem a hopping process involves two factors of the hybridization matrix element V and so V itself is renormalized by a factor $q^{1/2}$. The result is an effective Hamiltonian for a fixed density of f-electrons, n_f

$$H_{eff}(n_f) = \sum_{\vec{k}\sigma} \{\varepsilon_k c^+_{\vec{k}\sigma} c_{\vec{k}\sigma} + \sum_\ell \tilde{V}_\sigma (c^+_{\vec{k}\sigma} f_{\vec{k}\ell\sigma} + f^+_{\vec{k}\ell\sigma} c_{\vec{k}\sigma}) + \sum_\ell E_f f^+_{\vec{k}\ell\sigma} f_{\vec{k}\ell\sigma}\} \tag{19}$$

with $\tilde{V}^2_\sigma = q_\sigma(n_f, L)V^2$. The density n_f is to be determined variationally. Instead of fixing n_f we can go over to the equivalent of a grand canonical ensemble, introducing a chemical potential μ (or selfenergy, $-\mu$) for the f-electrons with an effective Hamiltonian

$$K_{eff} = H_{eff} - \sum_{\vec{k}\ell\sigma} \mu_\sigma f^+_{\vec{k}\ell\sigma} f_{\vec{k}\ell\sigma} \tag{20}$$

which can be diagonalized at once to give a set of hybridized bands

$$\xi_{k\pm,\sigma} = \frac{1}{2}\left[\varepsilon_k + E_f - \mu_\sigma \pm \{(\varepsilon_k - E_f + \mu_\sigma)^2 + 4L\tilde{V}^2_\sigma\}^{\frac{1}{2}}\right] . \tag{21}$$

These bands are illustrated in Fig.1. In the special case that the total number of electrons $n = 2$ the system is insulating. We will consider the more general case, $1 < n < 2$, in which the lowest band is partially occupied, i.e. up to a Fermi momentum k_F and in this case the system is metallic. The density of f electrons is then given by

$$n_f = N^{-1} \sum_{k<k_F,\sigma} \frac{\varepsilon_k - E_f + \mu_\sigma}{2\left((\varepsilon_k - E_f + \mu_\sigma)^2 + 4L\tilde{V}^2_\sigma\right)^{\frac{1}{2}}} \tag{22}$$

and the energy per site is

$$N^{-1}<H_{eff}> = N^{-1} \sum_{k<k_F,\sigma} \xi_{k-,\sigma} + \sum_\sigma \mu_\sigma n_{f\sigma} . \tag{23}$$

We consider first the paramagnetic case and the weak hybridization limit, so that we can work to leading logarithmic accuracy. Then the energy can be evaluated as a function of n_f, using Eqn. (22) to eliminate μ and we get

$$N^{-1}\langle H_{eff}(n_f)\rangle = \varepsilon_0^2/2+(E_f-\varepsilon_0)n_f+n_f^2/2-1/2 - L\tilde{V}^2 \ell n[n_f(1+\varepsilon_0-n_f)/L\tilde{V}^2] \quad (24)$$

where $\varepsilon_0 = \varepsilon_{kF}$ (see Fig.1). The renormalization factor is calculated by taking the ratio of a f-electron (ℓ,σ) hopping from a filled to an empty site in the correlated and uncorrelated systems, so that $q_\sigma=(1-n_f)/(1-n_{f\sigma}/L)$. For a hopping process from i-site to j-site we can use the angular momentum around the axis connecting these sites as a label for the orbital degeneracy [12]. The label ℓ defined in this manner should be conserved in the hopping process. It implies that in actual situation there should be a weak k- and ℓ-dependence of the hybridization matrix element V, which we have neglected in Eqn.(15). Our form of q_σ is consistent with the conservation law mentioned. Another consequence of the k- and ℓ-dependence of V is that the state which can hybridize with the conduction band is not always the symmetric combination (see Fig.1) but an ℓ-dependent one the form of which is changing in the k-space. Minimizing the energy with respect to the f-band occupancy n_f leads to an optimum value

$$1-n_f = \frac{\varepsilon_0(2L-1)}{2L^2V^2}\exp\left(\frac{-(2L-1)(\varepsilon_0-E_f-1)}{2L^2V^2}\right). \quad (25)$$

This optimum value has an exponential form similar to that in the single-site Kondo problem. In leading logarithmic accuracy only the exponent can be compared. For a doubly degenerate band, L=1 the ex-

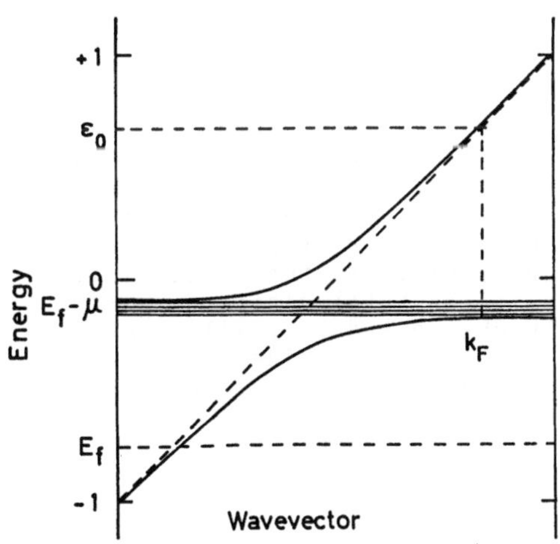

Fig.1. The energy-level structure for the case of L-degenerate bands and value of n=1.8. Only the symmetric combination of f-levels hybridizes with the conduction band. All f-levels are moved up by an energy ($-\mu$) so that the number of f-electrons is n_f which has a value $n_f \lesssim 1$.

ponent is a factor of 2 smaller than the single-site Kondo exponent. COLEMAN [12], KURAMOTO [13] and others have shown that in the large L limit the single-site Kondo results should be recovered. Taking this limit in Eqn.(25) leads to an exponent in agreement with the single-site Kondo exponent.

The condensation energy E_c defined as the energy difference per site between the hybridized and unhybridized problems can be evaluated to give to leading logarithmic accuracy

$$E_c = \{(1+E_f-\varepsilon_0)/(2L-1)-\tfrac{1}{2}\}(1-n_f)^2 . \tag{26}$$

This condensation energy is proportional to $(1-n_f)^2$. However, the characteristic energy to be compared to the Kondo temperature will be proportional to $(1-n_f)$, just as in a superconductor the characteristic energy is the energy-gap and not the condensation energy per site.

To summarize, we have found that there is a characteristic energy which has a Kondo form governing the coherent hybridization process in the paramagnetic state of the periodic Anderson model. The exponent agrees with the single-site Kondo result only as $L \to \infty$, indicating significant interaction effects when L is finite.

We turn now to a discussion of the simplest Landau Fermi liquid parameters of this paramagnetic metal. First, the specific heat is given by the density of quasiparticle states, which can be evaluated immediately from Eqn.(21) as

$$N_0 = 2\frac{(\varepsilon_0-E_f+\mu)^2}{qLV^2} = \frac{2}{qLV^2} . \tag{27}$$

Since it involves q^{-1} as expected, and also the ratio of the conduction band width to the hybridization energy, it can be very large and corresponds to an effective Fermi temperature which is of the order of the characteristic energy.

Next we consider the charge susceptibility, χ_c. A change in the total number of electrons in this problem is just a change in the filling parameter ε_0. If we calculate the second derivative of the total energy with respect to ε_0, the leading term comes from the change in the unhybridized term in the energy and the correction from the change in the condensation energy is small in the Kondo limit. In other words, when we make a change in the total electron density the number of f-electrons will hardly change and $1-n_f$ will remain small, and the hybridized quasiparticle bands will ride with the change in filling factor. As a result, the parameter F_0^s must be very large, since it must cancel the large effective mass term Eqn.(27), and neglecting the exponentially small corrections from the change in the characteristic energy we get

$$\chi_c = 1 \quad \text{and} \quad F_0^s \approx 2/qLV^2 . \tag{28}$$

This result emphasizes again the manybody character of the mass enhancement and the very strong residual interactions which occur between the quasiparticles. The positive (or repulsive) sign of F_0^s

also shows that it will tend to suppress s-wave superconductivity strongly.

Lastly we turn to the magnetic susceptibility χ_s. This reguires a careful treatment, since the renormalization factor q is explicitly dependent on the magnetizaion. Secondly, there is the question of the appropriate magnetic coupling scheme to use when L>1. In the single-site problem a j-coupling scheme may be appropriate and has been widely used (c.f. GUNNARSSON and SCHONHAMMER [11]), but in the periodic lattice problem it is not so obvious that the bands should be simply described in this manner. Therefore we will use simply a constant g-factor, and neglect any orbital magnetic moment.

The inverse of the spin susceptibility is given by the second derivative of the energy with respect to the magnetization, m. Taking the first derivative of the energy given by the expectation value of Eqn.(19) we get that the magnetic field is

$$B = \frac{\partial}{\partial m}\langle H_{eff}\rangle = \frac{1}{2}\left\{m-\frac{1}{2}(\mu_\uparrow-\mu_\downarrow) -\frac{1}{2}\left((\varepsilon_0+m-E_f+\mu_\uparrow)^2+4L\tilde{V}_\uparrow^2\right)^{\frac{1}{2}}+\frac{1}{2}\left((\varepsilon_0-m-E_f+\mu_\downarrow)^2+4L\tilde{V}_\downarrow^2\right)^{\frac{1}{2}}\right\}. \tag{29}$$

Note that although $\tilde{V}_\sigma$ has an important m-dependence through q_σ it gives no contribution to B since $\langle H_{eff}\rangle$ is stationary with respect to $n_{f\sigma}$ and μ_σ. However the m-dependence of $\tilde{V}_\sigma$ enters the second derivative leading to the result

$$\frac{1}{\chi} = \frac{1}{\chi_0} - \left(\frac{\partial\mu_\uparrow}{\partial m} + \frac{2qLV^2}{2L-1}\frac{\partial n_{f\uparrow}}{\partial m}\right) \tag{30}$$

with χ_0 determined by the quasiparticle density of states (Eqn.(27)) and $\chi_0 = N_0/2$. The term which includes the derivatives of the chemi cal potential (self-energy) and the f-electron number expresses the enhancement of the susceptibility over the density of states of quasiparticles and determines another Landau parameter $-F_0^a/\chi_0$. By differentiating the minimization conditions for $\mu_\uparrow$ and $n_{f\uparrow}$, we obtain for the inverse of the Wilson ratio in the Kondo limit,

$$\frac{\chi_0}{\chi_s} = 1 - \frac{4L-3}{(2L-1)^2} - \frac{2L}{(2L-1)^2}\ln\frac{\varepsilon_0}{qLV^2}. \tag{31}$$

First, we note that for the single orbital, L=1, χ_s is always negative, which means that the paramagnetic state is unstable towards a magnetic ordering in the Kondo limit. Our treatment is only for uniform magnetic states and the actual magnetic order will be determined by the detailed structure of the conduction band. For the single orbital case, the paramagnetic state is stable only in other limits such as in the limit of a small number of the f-electrons or in an extreme mixed valence regime. However, when the orbital degeneracy is big the paramagnetic state can be stable even in the Kondo regime. The condition is

$$2L \geqq \frac{\varepsilon_0-E_f-1}{LV^2} \tag{32}$$

where we have used Eqn.(25) to rewrite the logarithmic term in Eqn. (31). The quantity $\varepsilon_0 - E_f - 1$ is the depth of the bare f-level from the Fermi energy and $\pi L V^2/2$ is the width of the virtual level in the single site or impurity problem. This ratio governs the crossover between Kondo and mixed valence regimes. The present result that orbital degeneracy is essential to stabilize the paramagnetic state agrees with the conclusions of COLEMAN [12], READ, NEWNS and DONIACH [14], YAMADA, YOSIDA and HANZAWA [15].

5. Conclusions

In this paper we have emphasized the underlying similarity between the heavy electron metals and the Fermi liquid ^{3}He. We have put forward an almost localized description of these Fermi liquids in which spin flip processes are low energy processes and are responsible for the very large specific heat. Further we have tried to make this statement more precise by using a Gutzwiller approach to treat the very strong correlations in these almost localized Fermi liquids. Such an approach has been discussed for ^{3}He by ANDERSON and BRINKMAN [6] and VOLLHARDT [7]. In the case of a periodic Anderson model a numerical treatment has been given by VARMA [16] who found that the number of f-electrons remained nonintegral even as $U \to \infty$. Our results agree with this conclusion but the analytic treatment presented here shows that there is a formal similarity between the energetics of the Kondo effect and those of the coherent hybridization process in a lattice. Further we can treat the important question of the stability of the paramagnetic state against magnetic order. This gives us a criterion that the degeneracy factor 2L and the hybridization matrix elements cannot be too small, otherwise we must expect magnetic order at zero temperature. Lastly our method has an advantage in its simplicity in that the basic renormalizaions, q and μ, are conceptually simple and therefore there is the possibility that it can be generalized to treat realistic bandstructures.

There have been many schemes proposed to treat the Kondo limit of the Anderson model. Many of our conclusions agree with those reached by other methods, particularly the close relationship to single site Kondo problem in the large degeneracy limit. We will not discuss these here since many of them are covered elsewhere in this volume. Two papers which are closest in spirit to our approach are those by BRANDOW [17] and FAZEKAS [18]. Only the former treats the Kondo limit and uses a variational wavefunction like our Eqn.(18) which he treats by a diagrammatic technique. His results have similarities to ours in terms of the qualitative renormalizations but there are important differences especially in regard to the q-renormalization. He takes q to be independent of magnetization and finds a susceptibility even at L=1 which is not enhanced over the specific heat density of states. Our method reproduces the result that correlations are irrelevant for the completely polarized state at L=1.

References

1. H.R. Ott, H. Rudigier, T.M. Rice, K. Ueda, Z. Fisk and J.L. Smith, Phys. Rev. Lett. 52, 1915 (1984).
2. P.W. Anderson, Phys. Rev. B30, 1549 (1984).
3. O.T. Valls and Z. Tesanovic, Phys. Rev. Lett. 53, 1497 (1984).
4. K.S. Bedell and K.F. Quader (to be published).

5. T.M. Rice, K. Ueda, H.R. Ott and H. Rudigier, Phys. Rev. B31, 594 (1985).
6. P.W. Anderson and W.F. Brinkman in The Physics of Liquid and Solid Helium ed. by K. Bennemann and J.B. Ketterson (Wiley N.Y. 1978) p.177.
7. D. Vollhardt, Rev. Mod. Phys. 56, 99 (1984).
8. K. Seiler, C. Gros, T.M. Rice, K. Ueda and D. Vollhardt (to be published).
9. M.C. Gutzwiller, Phys. Rev. 137, A 1726 (1965).
10. W.F. Brinkman and T.M. Rice, Phys. Rev. B2, 4302 (1970).
11. O. Gunnarsson and K. Schonhammer, Phys. Rev. B28, 4315 (1983).
12. P. Coleman, Phys. Rev. B28, 5255 (1983).
13. Y. Kuramoto, in this volume.
14. N.B. Read, D.M. Newns and S. Doniach Phys. Rev. B30, 3841 (1984).
15. K. Yamada, K. Yosida and K. Hanzawa, Prog. Theor. Phys. 71, 450 (1984).
16. C.M. Varma in Proc. of the NATO A.S.I. Summer School, Vancouver Island, 1983 (in press).
17. B.H. Brandow (to be published).
18. P. Fazekas (to be published).

Part VI

Superconductivity in Heavy Fermion Systems

Effect of an External Magnetic Field on the Coherent Kondo State — Upper Critical Field in UBe_{13}

M. Tachiki, T. Koyama, and S. Takahashi

The Research Institute for Iron, Steel, and Other Metals
Tohoku University, Katahira, Sendai 980, Japan

1. Introduction

Intensive experimental and theoretical investigations have been done concerning the normal and superconducting state in the heavy fermion compounds [1] such as $CeCu_2Si_2$ [2] and UBe_{13} [3]. It has been shown in the previous paper [4] that the temperature dependence of the upper critical field in $CeCu_2Si_2$ [5] is well explained, if the singlet superconducting state is assumed and the saturation effect due to the Pauli paramagnetism is taken into account. However, in UBe_{13} the upper critical field at low temperatures could not be explained by any choice of the values of parameters. When we calculated the upper critical field of UBe_{13}, we did not take account of the following fact. The upper critical field H_{c2} of UBe_{13} is extremely high, although the superconducting transition temperature is rather low [6]. The upper critical field of UBe_{13} is about 9 Tesla at low temperatures. This intensity of the field is comparable to the Kondo temperature or the condensation energy of the coherent state in the compound. Therefore, the coherent state itself is affected by the external magnetic field. In this paper we take account of this effect in the following way. We assume the Anderson lattice model for the heavy fermion compounds. Then, we derive the Ward-Takahashi relations among the self-energy and vertex functions which originate from spin rotations. The field and frequency dependence of the self-energy and vertex functions are determined so as to fulfill the relations. Using these quantities in the normal state, we calculate the upper critical field in the superconducting state. It is shown that if proper values of the parameters are selected, the calculated upper critical field is in good agreement with the experimental values of the upper critical field in UBe_{13} obtained by Maple et al. [6] in all temperature range. In this calculation a singlet superconducting state is assumed.

2. Ward-Takahashi Relations in the Anderson Model

When we treat the Anderson lattice, we use a single site approximation for the self-energy and vertex functions of f electron [7,8]. Therefore, in this section we derive the Ward-Takahashi relations among the self-energy and vertex functions in the system of the single impurity Anderson model, using the path integral formulation of the quantum field theory [9]. In this section we confine ourselves to the case of absolute zero of temperature. We consider the following Lagrangian for the Anderson model

$$\mathcal{L}(t) = \sum_{\sigma}\int d^3x\psi_{\sigma}^{\dagger}(x)[i\partial_t - \varepsilon(-\nabla^2)]\psi_{\sigma}(x)$$

$$+ \sum_{\sigma}[f_{\sigma}^{\dagger}(t)\{i\partial_t - E_0\}f_{\sigma}(t) - V\{f_{\sigma}^{\dagger}(t)\psi_{\sigma}(\vec{0},t)$$

$$+ \psi_\sigma^\dagger(\vec{0},t)f_\sigma(t)\} - \frac{U}{2}f_\sigma^\dagger(t)f_{-\sigma}^\dagger(t)f_{-\sigma}(t)f_\sigma(t)] , \tag{2.1}$$

where $\psi_\sigma(x)$ and $f_\sigma(t)$ are respectively the field operator for the conduction electrons and annihilation operator for the f electron with spin σ at origin $\vec{0}$. In (2.1), $\varepsilon(\vec{p}^2)=\varepsilon_{\vec{p}}$ is the conduction band energy, and E_0, V, and U are the f-electron energy, the mixing interaction energy, and f-electron correlation energy, respectively. We define the causal Green's function $g_\sigma(t)$ for the f electron as

$$\begin{aligned} g_\sigma(t - t') &\equiv \langle f_\sigma(t)f_\sigma^\dagger(t')\rangle \\ &= \frac{i}{2\pi}\int d\omega g_\sigma(\omega)\exp[-i\omega(t - t')] , \end{aligned} \tag{2.2}$$

where $g_\sigma(\omega)$ is given by

$$g_\sigma(\omega) = [\omega - E_0 - V^2\int\frac{d^3p}{(2\pi)^3}\frac{1}{\omega - \varepsilon_{\vec{p}} + i\delta\mathrm{sgn}(\varepsilon_{\vec{p}})} - \Sigma_\sigma(\omega)]^{-1} . \tag{2.3}$$

The self-energy $\Sigma_\sigma(\omega)$ in (2.3) was introduced through the relation

$$\begin{aligned} &U\langle f_{-\sigma}^\dagger(t)f_{-\sigma}(t)f_\sigma(t)f_\sigma^\dagger(t')\rangle \\ &= \frac{i}{2\pi}\int d\omega\Sigma_\sigma(\omega)g_\sigma(\omega)\exp[-i\omega(t - t')] . \end{aligned} \tag{2.4}$$

Assuming the constant density of states for the conduction electrons ρ_s, we write the expression (2.3) in the form

$$g_\sigma(\omega) = [\omega - E_0 + i\pi V^2\rho_s\mathrm{sgn}(\omega) - \Sigma_\sigma(\omega)]^{-1} . \tag{2.5}$$

In deriving (2.5), the real part of the third term in the square brackets of (2.3) was assumed independent of ω and included in E_0.

We consider the following generating functional for the Green's functions.

$$\begin{aligned} W[\zeta,j] = \frac{1}{N}\int &\delta\psi_\sigma\delta\psi_\sigma^\dagger\delta f_\sigma\delta f_\sigma^\dagger\exp[i\int dt\{\mathcal{L}(t) + H_0\sigma_3^f(t) + \sum_\sigma(\bar{\zeta}_\sigma(t)f_\sigma(t) \\ &+ f_\sigma^\dagger(t)\zeta_\sigma(t)) + j_+(t)\sigma_-^f(t) + j_-(t)\sigma_+^f(t) + j_3(t)\sigma_3^f(t)\}] . \end{aligned} \tag{2.6}$$

In (2.6), $\zeta_\sigma(t)$ and $\bar{\zeta}_\sigma(t)$ are anticommuting fields, and $j_+(t)$, $j_-(t)$, and $j_3(t)$ are c-number source fields, and $\sigma_i^f(t)$ is defined by

$$\sigma_i^f(t) = f^\dagger(t)\sigma_i f(t) , \tag{2.7}$$

where σ_i is the Pauli spin matrix and $f^\dagger(t)$ is the spin doublet $(f_\uparrow^\dagger(t),$

$f^\dagger_\downarrow(t))$. The integrations by ψ_σ, $\psi^\dagger_\sigma$, f_σ and $f^\dagger_\sigma$in (2.6) are understood to be functional on the space of Grassman numbers. The term $H_0\sigma_3$ is to fix the magnetization to the third axis and H_0 is related to the external magnetic field H by $H_0=(g_f/2)\mu_B H$, g_f being the g-factor of f electron. N is the normalization factor such that W[0,0]=1. The Green's functions can be obtained by successive differentiations by the source fields. For example,

$$g_\sigma(t - t') = \langle f_\sigma(t)f^\dagger_\sigma(t')\rangle = -\delta^2 W[\zeta,j]/\delta\bar{\zeta}_\sigma(t)\delta\zeta_\sigma(t')\Big|_{\zeta=j=0} . \tag{2.8}$$

We introduce a rotation in the f-electron spin space:

$$f(t) \longrightarrow e^{i\theta(t)\sigma_+} f(t) . \tag{2.9}$$

Here, $\theta(t)$ is an arbitrary infinitesimal function depending on time. According to this rotaion, the Lagrangian is transformed as

$$\mathcal{L}(t) \to \mathcal{L}(t) - \dot{\theta}(t)\sigma^f_+(t) + i\theta(t)V\{f^\dagger_\uparrow(t)\psi_\downarrow(\vec{0},t) - \psi^\dagger_\uparrow(\vec{0},t)f_\downarrow(t)\} . \tag{2.10}$$

Since the transformation (2.9) is a change of integration variables in the generating functional (2.6), $W[\zeta,j]$ is invariant under the transformation. Therefore, we obtain

$$\begin{aligned}\delta W[\zeta,j] = i\int dt[-\dot{\theta}(t)\langle\sigma^f_+(t)\rangle_{\zeta,j} + i\theta(t)\{V\langle f^\dagger_\uparrow(t)\psi_\downarrow(\vec{0},t)\\ - \psi^\dagger_\uparrow(\vec{0},t)f_\downarrow(t)\rangle_{\zeta,j} + 2H_0\langle\sigma^f_+(t)\rangle_{\zeta,j} + \langle\bar{\zeta}_\uparrow(t)f_\downarrow(t)\\ - f^\dagger_\uparrow(t)\zeta_\downarrow(t) - j_+(t)\sigma^f_3(t) + 2j_3(t)\sigma^f_3(t)\rangle_{\zeta,j}\}] = 0 .\end{aligned} \tag{2.11}$$

Differentiating (2.11) by $j_+(t')$ and taking the limit of $\zeta,j\to 0$, we have

$$\begin{aligned}\int dt\theta(t)[i(\partial_t + 2iH_0)\langle\sigma^f_+(t)\sigma^f_-(t')\rangle - V\langle\{f^\dagger_\uparrow(t)\psi_\downarrow(\vec{0},t)\\ - \psi^\dagger_\uparrow(\vec{0},t)f_\downarrow(t)\}\sigma^f_-(t')\rangle - i\delta(t - t')\langle\sigma^f_3(t')\rangle] = 0 .\end{aligned} \tag{2.12}$$

Since $\theta(t)$ is an arbitrary function, (2.12) yields the relation

$$\begin{aligned}\delta(t - t')\langle\sigma^f_3(t')\rangle = (\partial_t + 2iH_0)\langle\sigma^f_+(t)\sigma^f_-(t')\rangle\\ + iV\langle\{f^\dagger_\uparrow(t)\psi_\downarrow(\vec{0},t) - \psi^\dagger_\uparrow(\vec{0},t)f_\downarrow(t)\}\sigma^f_-(t')\rangle .\end{aligned} \tag{2.13}$$

Note that

$$\langle f^\dagger_\uparrow(t)\psi_\downarrow(\vec{0},t)\sigma^f_-(t')\rangle = -iV\int d\tau G^0(\vec{0},t-\tau)\langle f^\dagger_\uparrow(t)f_\downarrow(\tau)\sigma^f_-(t')\rangle , \tag{2.14}$$

$$\langle\psi^\dagger_\uparrow(\vec{0},t)f_\downarrow(t)\sigma^f_-(t')\rangle = -iV\int d\tau G^0(\vec{0},\tau-t)\langle f^\dagger_\uparrow(\tau)f_\downarrow(t)\sigma^f_-(t')\rangle , \tag{2.15}$$

with

$$G^0(\vec{0},t) = \frac{i}{(2\pi)^4}\int d^4p \frac{e^{-ip_0t}}{p_0 - \varepsilon_{\vec{p}} + i\delta\,\mathrm{sgn}(\varepsilon_{\vec{p}})} . \tag{2.16}$$

We introduce the Green's function for the f-electron spin fluctuation

$$\langle\sigma_+^f(t)\sigma_-^f(t')\rangle = \frac{i}{2\pi}\int d\omega\,\Delta(\omega)\exp[-i\omega(t - t')] . \tag{2.17}$$

and the spin fluctuation-electron vertex function $\Gamma^{(3)}(\omega_1;\omega_2;\omega_3)$ by the relation

$$\begin{aligned}\langle f_\uparrow^\dagger(t)f_\downarrow(\tau)\sigma_-^f(t')\rangle &= \left(\frac{i}{2\pi}\right)^2\int d\omega_1 d\omega_2[-1 + \Delta(\omega_2 - \omega_1)\Gamma^{(3)}(\omega_2;\omega_2-\omega_1;\omega_1)] \\ &\times g_\uparrow^f(\omega_1)g_\downarrow^f(\omega_2)e^{i\omega_1 t - i\omega_2\tau + i(\omega_2 - \omega_1)t'} .\end{aligned} \tag{2.18}$$

Using (2.14), (2.15), and (2.18), we have

$$\begin{aligned}&\langle\{f_\uparrow^\dagger(t)\psi_\downarrow(\vec{0},t) - \psi_\uparrow^\dagger(\vec{0},t)f_\downarrow(t)\}\sigma_-^f(t')\rangle \\ &= -i\pi V\rho_s\left(\frac{i}{2\pi}\right)^2\int d\omega_1 d\omega_2\{\mathrm{sgn}(\omega_2) - \mathrm{sgn}(\omega_1)\} \\ &\times [-1 + \Delta(\omega_2 - \omega_1)\Gamma^{(3)}(\omega_2;\omega_2 - \omega_1;\omega_1)] \\ &\times g_\uparrow^f(\omega_1)g_\downarrow^f(\omega_2)e^{-i(\omega_2 - \omega_1)(t - t')} .\end{aligned} \tag{2.19}$$

Substituting (2.19) into (2.13) and making a Fourier transform of (2.13), we obtain the magnetization of f electron as

$$\begin{aligned}M_f &= (\nu - g_f\mu_B H)\Delta(\nu) - \pi V^2\rho_s\int\frac{d\omega}{2\pi}\{\mathrm{sgn}(\omega + \nu) - \mathrm{sgn}(\omega)\} \\ &\times [-1 + \Delta(\nu)\Gamma^{(3)}(\omega;\nu;\omega+\nu)]g_\uparrow(\omega)g_\downarrow(\omega+\nu) .\end{aligned} \tag{2.20}$$

If we apply the operation $\delta^2/\delta\bar{\zeta}_\uparrow(t_1)\delta\zeta_\downarrow(t_2)|_{\zeta,j=0}$ to (2.11) and follow a similar procedure to obtain (2.20), we have the following Ward-Takahashi relation

$$\begin{aligned}\Sigma_\downarrow(\omega+\nu) - \Sigma_\uparrow(\omega) &= (\nu - g_f\mu_B H)\Delta(\nu)\Gamma^{(3)}(\omega;\nu;\omega+\nu) - \frac{V^2\rho_s}{2}\int d\tilde{\omega}[\mathrm{sgn}(\tilde{\omega}+\nu) \\ &- \mathrm{sgn}(\tilde{\omega})\}g_\uparrow(\tilde{\omega})g_\downarrow(\tilde{\omega}+\nu)\Gamma_{\uparrow\downarrow}^{(4)}(\omega+\nu,\omega;\omega,\tilde{\omega}+\nu) .\end{aligned} \tag{2.21}$$

In (2.21) the four point vertex function is defined by

$$<f_\downarrow^\dagger(t)f_\downarrow(t')f_\uparrow(t_1)f_\uparrow^\dagger(t_2)>$$

$$= -(\frac{i}{2\pi})^2\int d\omega_1 d\omega_2[g_\downarrow(\omega_1)g_\uparrow(\omega_2)e^{-i\omega_1(t - t')}e^{-i\omega_2(t_1 - t_2)}$$

$$+ (\frac{i}{2\pi})^2\int d\omega_3 d\omega_4 g_\uparrow(\omega_1)g_\downarrow(\omega_2) \cdot 2\pi i\delta(\omega_1+\omega_2-\omega_3-\omega_4)$$

$$\times \Gamma^{(4)}_{\uparrow\downarrow}(\omega_1,\omega_2;\omega_3,\omega_4)g_\downarrow(\omega_3)g_\uparrow(\omega_4)$$

$$\times e^{i\omega_1 t_2 + i\omega_2 t - i\omega_3 t' - i\omega_4 t_1}] \ . \tag{2.22}$$

3. Magnetic Field-Dependence of the Self-energy and Vertex functions

When ν vanishes, the relation (2.20) is reduced to

$$M_f = -g_f\mu_B H\Delta(0) \ , \tag{3.1}$$

and the relation (2.21) is reduced to

$$\Sigma_\downarrow(\omega) - \Sigma_\uparrow(\omega) = -g_f\mu_B H\Delta(0)\Gamma^{(3)}(\omega;0;\omega) \tag{3.2}$$

As seen from (3.1), $-g_f\mu_B H\Delta(0)$ is the static susceptibility of f electron. Using (2.20) and (3.1), we have

$$g_f\mu_B H(\frac{\partial\Delta(\nu)}{\partial\nu})_{\nu=\pm 0} = \Delta(0) + V^2\rho_s[1 - \Delta(0)\Gamma^{(3)}(0;0;0)]g_\uparrow(0_\mp)g_\downarrow(0_\pm) \ . \tag{3.3}$$

Using (2.21) and (3.2), we obtain

$$\left(\frac{\partial\Sigma_\downarrow(\omega)}{\partial\omega}\right)_{\omega=0} = \Delta(0)\Gamma^{(3)}(0;0;0) - \frac{1}{2}\, g_f\mu_B H[\{(\frac{\partial\Delta(\nu)}{\partial\nu})_{\nu=+0} + (\frac{\partial\Delta(\nu)}{\partial\nu})_{\nu=-0}\}$$

$$\times \Gamma^{(3)}(0;0;0) + \Delta(0)\{(\frac{\partial}{\partial\nu}\,\Gamma^{(3)}(0;\nu;\nu)_{\nu=+0}$$

$$+ (\frac{\partial}{\partial\nu}\Gamma^{(3)}(0;\nu;\nu))_{\nu=-0}\}] - V^2\rho_s\mathrm{Re}[g_\uparrow(0_-)g_\downarrow(0_+)]$$

$$\times \Gamma^{(4)}_{\uparrow\downarrow}(0,0;0,0) \ . \tag{3.4}$$

From the relations (3.1), (3.2), (3.3), and (3.4), the quantities $\Sigma_\sigma(\omega)$, $\Delta(\nu)$, $\Gamma^{(3)}(0;\nu;\nu)$, and $\Gamma^{(4)}_{\uparrow\downarrow}(0,0;0,0)$ have the following forms of field-dependence up to third order in H

$$\Sigma_\sigma(\omega) = \Sigma_0^0 + \sigma(\Sigma_1 H + \Sigma_3 H^3) + \omega(\Sigma_0 + \Sigma_2 H^2) \ , \tag{3.5}$$

$$\Delta(\nu) = \Delta_0 + \Delta_2 H^2 + \Delta_1\nu H + i\Delta^I\nu\,\mathrm{sgn}(\nu) \ , \tag{3.6}$$

$$\Gamma^{(3)}(0;\nu;\nu) = \gamma_0 + \gamma_2 H^2 + \gamma_1 \nu H + i\gamma^I \nu \mathrm{sgn}(\nu) , \tag{3.7}$$

$$\Gamma^{(4)}_{\uparrow\downarrow}(0,0;0,0) = \Gamma_0 + \Gamma_2 H^2 . \tag{3.8}$$

Among the coefficient a_i, b_i, Δ_i, γ_i, and Γ_i, there are some relations which originate from the relations (3.1), (3.2), (3.3), and (3.4). For example, we can obtain

$$\Delta_0 = \rho_f(1 - 2\Sigma_1/g_f\mu_B) , \tag{3.9}$$

$$\Delta^I = \pi\rho_f^2(1 - 2\Sigma_1/g_f\mu_B)^2 , \tag{3.10}$$

$$\Sigma_0 = 2\Sigma_1/g_f\mu_B + \rho_f\Gamma_0 , \tag{3.11}$$

where $\rho_f=-V^2\rho_s[g_\uparrow(0_-)g_\downarrow(0_+)]_{H=0}$ is the density of states of f electron at the Fermi level. From these relations, it follows

$$\Delta^I = \pi\Delta_0^2 , \tag{3.12}$$

$$\Delta_0 = \rho_f(1 - \Sigma_0 + \rho_f\Gamma_0) . \tag{3.13}$$

The relations (3.12) and (3.13) give the Korringa relation [10] and the Wilson ratio obtained by Yosida and Yamada [11]. Since we are interested in low frequency phenomena, we retain the terms up to the first order in the frequency.

4. Model for the Heavy Fermion Systems

For the system being metallic, we tentatively use a two conduction bands model. However, the use of this special model is not essential in the following discussions. The conduction bands are expressed as ε_k+a and ε_k-a. The Hamiltonian for this Anderson lattice model is written as

$$\begin{aligned} H = & \sum_{k,\sigma} \varepsilon_{1k\sigma} c^\dagger_{1k\sigma} c_{1k\sigma} + \sum_{k,\sigma} \varepsilon_{2k\sigma} c^\dagger_{2k\sigma} c_{2k\sigma} + \sum_{k,\sigma} (E_0 + \tfrac{1}{2} g_f \mu_B \sigma H) f^\dagger_{k\sigma} f_{k\sigma} \\ & + \sum_{k\sigma} V(c^\dagger_{1k\sigma} f_{k\sigma} + f^\dagger_{k\sigma} c_{1k\sigma}) + \sum_{k\sigma} V(c^\dagger_{2k\sigma} f_{k\sigma} + f^\dagger_{k\sigma} c_{2k\sigma}) \\ & + \sum_i U f^\dagger_{i\uparrow} f_{i\uparrow} f^\dagger_{i\downarrow} f_{i\downarrow} , \text{ with} \end{aligned} \tag{4.1}$$

$$\varepsilon_{1k\sigma} = \varepsilon_k - a + \tfrac{1}{2} g_c \mu_B \sigma H , \tag{4.2}$$

$$\varepsilon_{1k\sigma} = \varepsilon_k + a + \tfrac{1}{2} g_c \mu_B \sigma H , \tag{4.3}$$

where k and i indicate the wave number and the lattice site of ion, respectively, and g_c is the g-factor of conduction electron. In (4.1), $c_{1k\sigma}$, $c_{2k\sigma}$, and $f_{k\sigma}$ are the electron operators for the two conduction bands and the f electron band, respectively, and V is the hybridization parameter which is assumed to be common to the two conduction bands and independent of k. From (4.1) the temperature Green's function of the f electron is given by

$$G_\sigma^f(k,\omega_n) = [i\omega_n - E_0 - \frac{1}{2}g_f\mu_B\sigma H - \Sigma_\sigma(k,i\omega_n)$$
$$- V^2(i\omega_n - \varepsilon_{1k\sigma})^{-1} - V^2(i\omega_n - \varepsilon_{2k\sigma})^{-1}]^{-1} , \quad (4.4)$$

with $\omega_n=\pi T(2n+1)$, n being integers. We use units such that $k_B=\hbar=1$. The quantity $\Sigma_\sigma(k,i\omega_n)$ in (4.4) is the self-energy due to the f electron correlation. Hereafter, we neglect the k-dependence of the self-energy and approximate it by that of the single ion (3.5). The frequency ω in (3.5) is replaced by $i\omega_n$. If the self-energy is substituted into (4.4), the Green's function (4.4) for a small value of $|\omega_n|$ is written as

$$G_\sigma^f(k,\omega_n) = (\tilde{\gamma} + \alpha)^{-1}[i\omega_n - \varepsilon_f - h\varepsilon_k - \frac{1}{2}g^*\mu_B\sigma H]^{-1} , \quad (4.5)$$

where

$$\alpha = (2V^2/a^2) , \quad (4.6)$$

$$\tilde{\gamma} = \tilde{\gamma}_0 - \Sigma_2H^2 , \quad (4.7)$$

$$\tilde{\gamma}_0 = 1 - \Sigma_0 , \quad (4.8)$$

$$h = \alpha/(\tilde{\gamma} + \alpha) , \quad (4.9)$$

$$\varepsilon_f = (E_0 + \Sigma_0^0)/(\tilde{\gamma} + \alpha) , \quad (4.10)$$

$$g^* = g_0^* + g_2^*H^2 , \quad (4.11)$$

$$g_0^* = [g_f + (2/\mu_B)\Sigma_1 + \alpha g_c]/(\tilde{\gamma}_0 + \alpha) , \quad (4.12)$$

$$g_2^* = [(2/\mu_B)\Sigma_3 + g_0^*\Sigma_2)]/(\tilde{\gamma}_0 + \alpha) . \quad (4.13)$$

Hereafter, we assume $\varepsilon_f=0$.

5. Scattering of the Quasifermions due to Impurities

It is understood in the expression (4.5) that $[i\omega_n-h\varepsilon_k-(1/2)g^*\mu_B\sigma H]^{-1}$ is the Green's function of the quasifermion and $(\tilde{\gamma}+\alpha)^{-1}$ is the Green's function renormalization factor. We introduce a fermion operator η_k where its Green's function is

$$G^{\eta}_{\sigma}(k,\omega_n) = (i\omega_n - h\varepsilon_k - \tfrac{1}{2}g^{*}\mu_B\sigma H)^{-1} \ . \tag{5.1}$$

The f electron operator is expanded in terms of η_k as

$$f_k = (\tilde{\gamma} + \alpha)^{-1/2}\eta_k + \text{higher order terms of } \eta_k \ . \tag{5.2}$$

In a similar way, the conduction electron operator is expanded as

$$c_k = \left[\frac{\alpha}{\tilde{\gamma} + \alpha}\right]^{1/2}\eta_k + \text{higher order terms of } \eta_k \ . \tag{5.3}$$

Since α is very small, the renormalization coefficient $[\alpha/(\tilde{\gamma}+\alpha)]^{1/2}$ in (5.3) is much smaller than $(\tilde{\gamma}+\alpha)^{-1/2}$ in (5.2). This fact indicates that quasifermions are mainly composed of the f electrons in this model. Therefore, the scattering of the quasifermions is mostly caused by the impurity potential acting on the f electrons. The impurity potential is written as [12]

$$H_{imp} = \sum_j \sum_{k,k'} \sum_{\beta\beta'} V_{\beta\beta'}(k,k') f^{\dagger}_{k\beta} f_{k'\beta'} \exp[i(k - k')\cdot r_j] \ , \tag{5.4}$$

with

$$V_{\beta\beta'}(k,k') = v\delta_{\beta\beta'} + i(v_{so}/k_F^2)(k \times k') \cdot \sigma_{\beta\beta'} \ . \tag{5.5}$$

In (5.4) and (5.5), β expresses the spin state, r_j the position of impurity, and $\sigma_{\beta\beta'}$ a matrix element of the Pauli spin. The first term at the right hand side of (5.5) is the normal impurity potential and the second term is the spin-orbit interaction due to impurities. If (5.2) is used and the higher order terms of η_k are neglected, the Hamiltonian (5.4) is rewritten as

$$H_{imp} = \sum_j \sum_{k,k'} \sum_{\beta\beta'} \tilde{V}_{\beta\beta'}(k,k') \eta^{\dagger}_{k\beta} \eta_{k'\beta} \exp[i(k - k') \cdot r_j] \ , \tag{5.6}$$

with

$$\tilde{V}_{\beta\beta'}(k,k') = (\tilde{\gamma} + \alpha)^{-1} V_{\beta\beta'}(k,k') \ . \tag{5.7}$$

From (5.7) we see that the scattering potential for the quasifermions is reduced from the bare potential by the renormalization effect.

The scattering potential (5.6) gives an extra contribution to the self-energy of the quasifermion. The Green's function which includes this self-energy is expressed as

$$G^{\eta}_{\sigma}(k,\omega_n) = (i\omega_{\sigma} - h\varepsilon_k)^{-1} \ , \quad \text{with} \tag{5.8}$$

$$i\omega_{\sigma} = i\omega_n + \frac{1}{2\tau}\cdot\frac{i\omega_n}{|\omega_n|} - I\sigma \ . \tag{5.9}$$

The quantities in (5.9) are defined by

$$I = \frac{1}{2} g^* \mu_B H \,, \tag{5.10}$$

$$\frac{1}{\tau} = \frac{1}{\tau_1} + \frac{1}{\tau_{s,o}} \,, \tag{5.11}$$

$$\frac{1}{\tau_1} = \frac{1}{\alpha(\tilde{\gamma} + \alpha)} \cdot \frac{1}{\tau_1^0} \quad \text{with} \quad \frac{1}{\tau_1^0} = 2\pi n_i \rho V^2 \,, \tag{5.12}$$

$$\frac{1}{\tau_{s,o}} = \frac{1}{\alpha(\tilde{\gamma} + \alpha)} \cdot \frac{1}{\tau_{s,o}^0} \quad \text{with} \quad \frac{1}{\tau_{s,o}^0} = \frac{4\pi}{3} n_i \rho v_{s,o}^2 \,, \tag{5.13}$$

where n_i is the concentration of impurities. The life times τ_1^0 in (5.12) and $\tau_{s,o}^0$ in (5.13) are those for bare f-electrons. In the above calculation, the position average of each impurity was made over the volume of the sample [12].

6. Upper Critical Field

There is a controversy whether the superconducting state in the heavy fermion systems is a singlet state or a triplet state. In this paper it is assumed that even in the heavy fermion compounds the phonon-mediated attractive interaction dominates and the singlet superconducting state is realized. Under this assumption, we calculate the upper critical field using the Green's function (5.8) and the vertex function (3.8). If the standard theory of superconductivity is applied, the equation to determine the upper critical field is given in the dirty limit by [13-16]

$$\ln\left[\frac{T}{T_c}\right] + \frac{1}{2} \Big[\Big\{1 + \frac{b}{\sqrt{b^2 - I^2}}\Big\}\psi\left(\frac{1}{2} + \rho_-\right) + \Big\{1 - \frac{b}{\sqrt{b^2 - I^2}}\Big\}\psi\left(\frac{1}{2} + \rho_+\right)\Big] - \psi\left(\frac{1}{2}\right) = 0 \,, \text{ where} \tag{6.1}$$

$$\rho_\pm = \frac{1}{2\pi T}\left[a \pm \sqrt{b^2 - I^2}\,\right] \,, \tag{6.2}$$

$$a = \frac{1}{3\tau_{so}} + \frac{1}{6}\tau v_F^2 h^2 \left(\frac{2eH}{c}\right) \,, \tag{6.3}$$

$$b = \frac{1}{3\tau_{so}} \,, \tag{6.4}$$

$$I = \frac{1}{2} g^* \mu_B H \,. \tag{6.5}$$

In (6.1) $\psi(z)$ is the di-gamma functions. If the magnetic field-dependence of $\tilde{\gamma}$ and g^* is taken into account, a, b, and I are written as

$$a = a_0 / \left(1 - \frac{\Sigma_2}{\tilde{\gamma} + \alpha} H^2\right) \,, \tag{6.6}$$

$$b = b_0/(1 - \frac{\Sigma_2}{\tilde{\gamma} + \alpha} H^2) \ , \tag{6.7}$$

$$I = I_0(1 + \frac{g_2^*}{g_0^*} H^2) \ , \tag{6.8}$$

where a_0, b_0, and I_0 are the values when the field-dependence of $\tilde{\gamma}$ and g^* is neglected. The superconducting transition temperature T_c in (6.1) is given by [8]

$$T_c = 1.134 T_K \exp[- \frac{\alpha(\tilde{\gamma} + \alpha)}{\rho(V_{BCS} - \Gamma_{\uparrow\downarrow})}] \ , \tag{6.9}$$

where V_{BCS} is the attractive interaction constant and $\Gamma_{\uparrow\downarrow}$ is given by (3.8). In (6.9), we introduced the cutoff energy T_K for the frequency summation, which is comparable to the Kondo temperature. Using (6.6), (6.7), (6.8), and (6.9), the equation (6.1) is rewritten as

$$\ln \frac{T}{T_{co} e^{AH^2}} + \frac{1}{2} [(1 + \frac{b_0/(1 - BH^2)}{\sqrt{\{b_0/(1 - BH^2)\}^2 - \{I_0(1 + CH^2)\}^2}}) \psi(\frac{1}{2} + \rho_-)$$

$$+ (1 - \frac{b_0/(1 - BH^2)}{\sqrt{\{b_0/(1 - BH^2)\}^2 - \{I_0(1+CH^2)\}^2}}) \psi(\frac{1}{2} + \rho_+)] - \psi(\frac{1}{2}) = 0 \ . \tag{6.10}$$

where

$$\rho_\pm = \frac{1}{2\pi T} [\frac{a_0}{1 - BH^2} \pm \sqrt{(\frac{b_0}{1 - BH^2})^2 - \{I_0(1 + CH^2)\}^2}] \ , \tag{6.11}$$

$$A = \frac{\alpha(\tilde{\gamma}_0 + \alpha)}{\rho(V_{BCS} - \Gamma_0)} (\frac{\Sigma_2}{\tilde{\gamma}_0 + \alpha} - \frac{\Gamma_2}{V_{BCS} - \Gamma_0}) \ , \tag{6.12}$$

$$B = \frac{\Sigma_2}{\tilde{\gamma}_0 + \alpha} \ , \tag{6.13}$$

$$C = \frac{g_2^*}{g_0^*} \ . \tag{6.14}$$

The temperature T_{co} is the superconducting-transition temperature at zero field.

In the numerical calculation of the upper critical field, A,B, and C are considered to be adjustable parameters. The upper critical field in UBe_{13} is calculated from (6.10) in the following way. We use the experimental value for T_{co}. We assume $g_0^*=2$ and $\lambda_{so}=2\hbar/3\pi k_B T_{co}\tau_{so}^0 = 9$. The values of a_0 and b_0 are determined by the experimental value of $(\partial H_{co}/\partial T)_{T=T_c}$ and the value of λ_{so}, respectively. The calculated result is shown in Fig. 1. The solid circles indicate the experimental values obtained by Maple

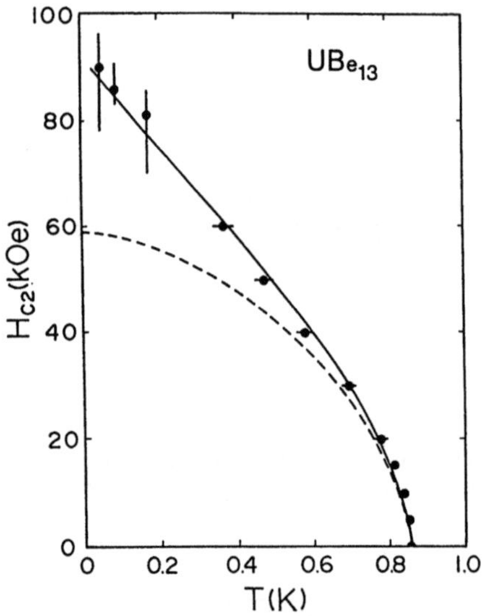

Fig. 1 Upper critical field H_{c2} as a function of temperature. The Solid curve denotes H_{c2} when the effect of external magnetic field on the coherent Kondo state is included. The parameter values used are $g_0^*=2$, $\lambda_{so}=9$, $A=2\times10^{-5}(KOe)^{-2}$, $B=2.5\times10^{-5}(KOe)^{-2}$, and $C=-2.64\times10^{-5}(KOe)^{-2}$. The dashed curve shows H_{c2} when A=B=C=0. The Solid circles denote the experimental data on UBe_{13} obtained by Maple et al. [6]

et al. The solid curve shows the calculated values when $A=2\times10^{-5}(KOe)^{-2}$, $B=2.5\times10^{-5}(KOe)^{-2}$, and $C=-2.64\times10^{-5}(KOe)^{-2}$ are used. In Fig. 1, the values of H_{c2} for A=B=C=0 are also plotted by the dashed curve. As seen from Fig. 1 the introduction of the field-dependence of the parameters much improves the agreement between theory and experiment.

This improvement is caused in the following way. The term of AH^2 in (6.10) works to increase T_c. As seen in (6.10), the term of BH^2 has the effect of increasing the spin-orbit interaction, since b_0 is proportional to the spin-orbit interaction. The term of CH^2 acts to decrease the g-factor, as seen in (4.11). Therefore, all the effects work to increase H_{c2}. Although the values of AH^2, BH^2, and $-CH^2$ are much smaller than unity, these effects work cooperatively to cause the enormous increase in H_{c2} at low temperatures. The values of the parameters which were used for the calculation should be justified by other experimental results or a calculation from first principle. Scharnber and Klemm [17] calculated the upper critical field in p wave superconducors. According to their theory, the H_{c2} curve in p wave superconductors does not show the convex curvature near T_c which was observed in UBe_{13}. It is a future problem that if the same field effect to the coherent state is included in the theory of p wave superconductors, how the temperature-dependence of H_{c2} is modified.

In this paper, we presented a phenomenological theory for the coherent Kondo state when the intensity of an external magnetic field is comparable to the condensation energy of the coherent state. Using the Ward-Takahashi relations, we obtained the field-dependence of the self-energy and vertex functions. If the proper values of the parameters included in the self-energy and vertex functions are used, the peculiar temperature-dependence of H_{c2} in UBe_{13} is explained. Therefore, the possibility of the singlet superconductivity in UBe_{13} can not be eliminated.

Acknowledgments

The authors would like to thank Professor H. Fukuyama and Professor S. Maekawa for valuable discussions.

References

1. G.R. Stewart, Rev. Mod. Phys. 56, 755 (1984) and reference therein.
2. F. Steglich, J. Aarts, C.D. Bredl, W. Leike, D. Meschede, W. Franz, and J. Schäfer, Phys. Rev. Lett. 43, 1892 (1979).
3. H.R. Ott, H. Rudigier, P. Delsing, and Z. Fisk, Phys. Rev. Lett, 52, 1551 (1984).
4. M. Tachiki, S. Maekawa, and S. Takahashi, Phys. Rev. 31, 228 (1985).
5. W. Assums, M. Herrman, U. Rauchschwalbe, S. Riegel, W. Lieke, H. Spille, S. Horn, G. Weber, F. Steglich, and G. Cordier, Phys. Rev. Lett. 52, 469 (1984).
6. M.B. Maple, J.W. Chen, S.E. Lambert, Z. Fisk, J.L. Smith, H.R. Ott, J.S. Brooks, and M.J. Naughton, Phys. Rev. Lett, 54, 477 (1985).
7. A. Yoshimori and H. Kasai, J. Magn. Magn. Mater. 31-34, 475 (1983).
8. M. Tachiki and S. Maekawa, Phys. Rev. 29, 2497 (1984).
9. T. Koyama and M. Tachiki, to be appear in Prog. Theor. Phys. Suppl. No. 80.
10. H. Shiba, Prog. Theor. Phys. 54, 967 (1975).
11. K. Yosida and K. Yamada, Prog. Theor. Phys. 53, 1286 (1975); K. Yamada, Prog. Theor. Phys. 54, 316 (1975).
12. A.A. Abrikosov and L.P. Gor'kov, Zh. Exsp. Teor. Fiz. 42, 1088 (1962) [Sov. Phys. -JETP 15, 752 (1962)].
13. P. Fulde and K. Maki, Phys. Rev. 141, 275 (1966).
14. K. Maki, Phys. Rev. 148, 362 (1966).
15. N.R. Werthermer, E. Herfrand, and P.C. Hohenberg, Phys. Rev. 147, 295 (1966).
16. P. Fulde, Adv. Phys. 148, 667 (1973).
17. K. Scharnberg and R.A. Klemm, (unpublished).

Superconductivity in Kondo Lattices

*F.J. Ohkawa**

Institute for Solid State Physics, University of Tokyo
Minato-ku, Tokyo 106, Japan

The Fermi liquid theory of Anderson lattices is developed to describe coherent states of Kondo lattices at low temperatures. The origin of heavy electrons is the spin fluctuation nearly confined within each single f site; the situation is similar to that in dilute Kondo alloys. On the basis of the Fermi liquid theory, one can conclude that an attractive interaction between adjacent f electrons is indispensable to realize heavy-electron superconductivity. Consequently the attractive interaction should be of finite range, so that even s-symmetry Cooper pairs can be anisotropic; the energy gap, for example, can vanish along lines on the Fermi surface. It is also discussed that phonon-mediated attractive interaction between adjacent f electrons depends on the lattice structure. For example, it is expected in CsCℓ type lattices, while it should be absent in NaCℓ type lattices. Typical heavy-electron superconductors such as $CeCu_2Si_2$ and UBe_{13} can be actually regarded as a kind of CsCℓ type.

Section 1. Introduction

Recently exotic superconductivity has been found in $CeCu_2Si_2$ [1,2], UBe_{13} [3] and UPt_3 [4]. Both the specific heat, C, and its jump, ΔC, at the transition temperature, T_c, are $10^2 \sim 10^3$ times as large as those of usual superconductors. The electronic specific heat coefficient, γ, is of the order of $1 J/K^2 \cdot mol$. The slope of the upper critical field, H_{c2}, at T_c is also as large as $10 \sim 10^2 T/K$. Because these large values of γ and H_{c2} suggest that heavy quasi-particles are condensed into a superconducting state, they are called heavy-electron superconductors or heavy-fermion superconductors. More remarkable experimental results are the temperature dependences of the specific heat [2,5], the nuclear magnetic relaxation time [6,7] and the ultrasonic attenuation [8], which imply gapless superconductors.

Based on the exotic experimental results mentioned above, it is reasonable to argue that a new type of superconductivity might be realized, for example, triplet p-symmetry superconductivity caused by non-phonon attractive interaction as in superfluid ^3He [9,10]. One of the most interesting problems is to clarify the character of Cooper pairs actually realized in heavy-electron superconductors.

Because the density of states obtained from the specific heat coefficient is very large, one can argue that high T_c or high H_{c2} superconductors might be found in heavy-electron systems. Another problem is to explain the roles of the large density of states of quasi-particles.

---

*) Present address: Department of Physics, Faculty of Science, Hokkaido University, Sapporo 060, Japan

Therefore the normal Fermi liquid state should be discussed first to clarify the origin of the large γ value, i.e., heavy electrons.

Section 2. Model

Even in crystals, *3d*, *5f* and *4f* electrons are well localized and retain their atomic character. Among them, the degree of localization is the strongest for *4f* electrons in lanthanides, while it is the weakest for *3d* electrons in transition metals. It is intermediate for *5f* electrons in actinides. The following Hamiltonian is one of the simplest models to describe the essential feature of such localized electrons in crystals.

$$\mathcal{H}_0 = \sum_{np\sigma} \varepsilon_n(p) a^\dagger_{np\sigma} a_{np\sigma} + \sum_{i\sigma} E_f f^\dagger_{i\sigma} f_{i\sigma} + \sum_{ij\sigma} t_{ij} f^\dagger_{i\sigma} f_{j\sigma}$$

$$+ (1/\sqrt{N}) \sum_{np} \sum_{\sigma i} [V_n(p)\exp(-ip\cdot R_i) a^\dagger_{np\sigma} f_{i\sigma} + \text{h.c.}]$$

$$+ \frac{1}{2} U \sum_{i} \sum_{\sigma\tau} f^\dagger_{i\sigma} f^\dagger_{i\tau} f_{i\tau} f_{i\sigma} . \quad (1)$$

Here localized electrons such as *3d*, *5f* and *4f* are tentatively called f electrons, and creation and annihilation operators of f electrons with spin σ at site i are denoted by $f^\dagger_{i\sigma}$ and $f_{i\sigma}$, respectively. Those of conduction electrons are denoted by $a^\dagger_{np\sigma}$ and $a_{np\sigma}$, where n is the band index and p is wave vector.

There are two mechanisms to make f electrons itinerant. One is the transfer integral between adjacent f orbits, and the other is the hybridization between f orbits and conduction bands. The contribution of the transfer integral to the f band width is approximately given by

$$W \sim 2z|t| , \quad (2)$$

where z is the coordination number of the lattice and t is the transfer integral between a pair of the nearest neighbor f orbits. On the other hand, the contribution of the hybridization is approximately estimated as

$$\Gamma = \frac{\pi}{N} \sum_{np} V_n(p) V^*_n(p) \delta(\mu - \varepsilon_n(p)) \sim \pi |V|^2 \rho_c(\mu) . \quad (3)$$

Here $\rho_c(\mu)$ is the density of states of conduction electrons at the Fermi level, μ, and $\rho_c(\mu) \sim 1/2D$ where $2D$ is the conduction band width; V is a typical hybridization energy, i.e., the transfer integral between f orbits and their adjacent s, p, *4d* and *5d* orbits.

In transition metals, various transfer integrals between adjacent sites are of the same order, i.e., $|t| \sim |V| \ll D$. Therefore the hybridization between f orbits and conduction bands can be ignored, because $W \gg \Gamma$. Then the Hubbard model is used for *3d* systems. On the other hand, $|t| \ll |V|$ in lanthanides because of strong atomic character of *4f* orbits. Therefore the transfer integral, t, can be ignored, unless the distance between the nearest neighbor f sites is extremely small. In this case, eq.(1) is reduced to an Anderson lattice, i.e., a periodic Anderson model, which is a primary concern in the present paper. Experimentally, $CeCu_2Si_2$ is a typical *4f* system, and UBe_{13} is rather close to *4f* systems because of the large nearest neighbor U-U distance (5.13A)[11]. The transfer integral between f orbits can be ignored in $CeCu_2Si_2$, and approximately in UBe_{13}. However it is not certain that the transfer integral can be ignored in UPt_3.

Section 3. Fermi Liquid Theory

Let us confine ourselves to the case where U can be considered to be infinitely large in eq.(1). Then the number of occupied f electrons per site, n_f, is one at largest, i.e., $0<n_f<1$. When $U \to \infty$ and $n_f \sim 1$, f electrons are almost localized even if they are still itinerant. What will be discussed here is the normal Fermi liquid behavior or a coherent behavior at low temperatures of such almost localized liquids.*)

We will follow the theoretical framework used by Luttinger[12-15]. Because a coherent behavior of f electrons can be included simply by taking this theoretical framework, the problem is how to treat the strong correlation between f electrons. When $U = 0$, the system described by eq.(1) is clearly a normal Fermi liquid. Here two parallel conduction bands are assumed in order to make the system metallic even in the case of $t_{ij} = 0$, and half-filled electron number. For finite U, a single particle Green function of f electrons can be generally written as

$$G_{ff}(i\varepsilon,\boldsymbol{p}) = \frac{1}{i\varepsilon+\mu-E_f(\boldsymbol{p})-\Sigma(i\varepsilon,\boldsymbol{p})-M(i\varepsilon,\boldsymbol{p})}, \quad \text{where} \tag{4}$$

$$E_f(\boldsymbol{p}) = E_f + \sum_j t_{ij}\exp(-i\boldsymbol{p}\cdot R_{ij}), \tag{5}$$

$$M(i\varepsilon,\boldsymbol{p}) = \sum_n \frac{|V_n(\boldsymbol{p})|^2}{i\varepsilon+\mu-\varepsilon_n(\boldsymbol{p})}, \tag{6}$$

and the self-energy $\Sigma(i\varepsilon,\boldsymbol{p})$ can be calculated perturbatively in terms of U, as long as no order parameters exist. Various rigorous Fermi liquid relations such as the Fermi surface sum rule can be proved nearly in parallel to Luttinger's discussion[14,15]. What will be used in the following section is the relation between the electronic specific heat coefficient, γ, and the density of states of quasi-particles;

$$\gamma = \frac{2\pi^2}{3} k_B{}^2\rho^*(\mu), \quad \text{where} \tag{7}$$

$$\rho^*(\mu) = \frac{1}{N}\sum_p \delta(\mu-E^*(p)), \tag{8}$$

is the density of states of quasi-particles at the Fermi level, and $E^*(p)$ is the spectrum of quasi-particles defined by the poles of eq.(4).

For the sake of simplicity, it will be assumed in the following part that there exists only a single Fermi surface around the Γ point in the Brillouin zone, and that every physical quantity is isotropic, although the

---

*) It should be discussed whether the normal Fermi liquid is possible even in the large U limit. In case of Anderson lattices at least, the normal Fermi liquid is possible as discussed later.

generalization is straightforward. Let us expand $\Sigma(i\varepsilon,p)$ around the Fermi surface as

$$\Sigma(i\varepsilon,p) = \Sigma(+i0,p_F) + (1-\chi)i\varepsilon + \frac{\hbar^2}{m}p_F(p-p_F) + \cdots\cdots , \tag{9}$$

in order to discuss the quasi-particles in the neighborhood of the Fermi level. Then we obtain

$$E^*(p) = \mu + \frac{\hbar^2}{m^*}p_F(p-p_F) + \cdots\cdots , \quad \text{with} \tag{10}$$

$$\frac{1}{m^*} = \frac{1}{Z}\left[\frac{1}{m} + \frac{1}{m_0} + \sum_n \frac{|c_n|^2}{m_n}\right] . \tag{11}$$

Here p_F is the Fermi wave vector, m_0 is the mass of f electrons in the absence of hybridization defined at $p = p_F$ and $m_n (n \neq 0)$ is the mass of conduction electrons defined in the same way; the momentum dependence of $|V_n(p_F)|^2$ has been ignored;

$$Z = \chi + \sum_n |c_n|^2 , \quad \text{and} \tag{12}$$

$$c_n = \frac{V_n(p_F)}{\mu - \varepsilon_n(p_F)} . \tag{13}$$

In the following discussion, $\Sigma(+i0,p_F)$ does not appear explicitly, because it satisfies

$$\mu - E_f(p_F) - \Sigma(+i0,p_F) - M(+i0,p_F) = 0 . \tag{14}$$

Then single-particle Green functions can be approximately written as

$$G_{ln}(i\varepsilon,p) = \frac{c_l^\dagger c_n}{Z} \frac{1}{i\varepsilon + \mu - E^*(p)} , \tag{15}$$

where l and n are band indexes; band with n=0 for f band and c_0=1, while band with $n \neq 0$ for conduction band.

Because we are interested in $4f$ and $5f$ systems, it can be assumed that $|m_0| \gg |m_n|$ $(n \neq 0)$. Furthermore it can be reasonably assumed that $|m| \gg |m_n|$, because the dispersion of $\Sigma(+i0,p)$ can not be so large as that of conduction electrons. Then it can be easily seen from eq.(11) that the effective mass of quasi-particles m^* depends on $|c_n|^2$. When m^* is large, $|c_n|^2 \ll 1$ for $n \neq 0$. This means that heavy quasi-particles should be mainly composed of f electrons.

Two problems are still left to be settled in the present Fermi liquid theory. One is to show that the normal Fermi liquid is actually possible even in the large U limit, and the other is to estimate two Fermi liquid parameters of χ and m. Then, let us discuss such problems in Anderson lattices, i.e., in case of $t_{ij} = 0$ [16,17]. When f levels are so shallow that n_f might be almost equal to zero, the effects of U can be ignored. Therefore the ground-state is a normal Fermi liquid in the limit of $n_f \to 0$. When f levels are so deep that $n_f \sim 1$, f electrons are almost localized because of large U. However conduction electrons can be polarized, and the RKKY exchange interaction can work between f electrons. It is clear that the ground-state is magnetic in the limit of $n_f \to 1$. Therefore there should exist a critical n_{cr} where the first order transition occurs from a nonmagnetic state to a magnetic state, when n_f increases from 0 to 1.

Let us estimate n_{cr} in terms of the competition between the Kondo effect and the exchange interaction. By using a single f-site Anderson Hamiltonian instead of Anderson lattices, and by following Yosida [18] and Inagaki [19], the Kondo temperature is calculated straightforwardly as [16,17]

$$T_K = 2\frac{\Gamma}{\pi}\frac{1-n_f}{n_f}, \tag{16}$$

as a function of n_f. Here it is assumed that the ground multiplet of f levels is a doublet, and that the crystal−field splitting and the spin-orbit splitting are so large that the number of occupied f electrons in excited multiplets can be ignored. On the other hand, the energy gain per site by the exchange interaction is calculated as [16]

$$J = \frac{z\Gamma}{4\pi}\frac{|F(2p_F a)|}{\ln(D/\Gamma)}, \tag{17}$$

where a is the lattice constant, and $F(x)$ is an RKKY-like oscillating function. Although it is difficult to calculate $F(x)$ for arbitrary x, it is easy to calculate $F(0)$, i.e., $F(0)=1$. Here it is assumed that f levels are so shallow that J might compete with T_K. Let us estimate n_{cr} roughly by the condition,

$$T_K = J. \tag{18}$$

Because $D/\Gamma \gg 1$, and $|F(2p_F a)| \ll 1$, one can conclude that n_{cr} is very close to one, for example, 0.9 or 0.99.

According to Haldane's scaling relation [20] or the density function theorem discussed by Yosida and Sakurai [21], and Ohkawa [16,17], there is a unique one-to-one relation between the depth of f levels and n_f, which is given by

$$E_1 - \mu + \frac{\Gamma}{\pi}\ln\frac{D}{\Gamma} + \sum_{\nu}^{\text{excited}} 2\frac{\Gamma}{\pi}\ln\frac{D}{E_\nu - E_1} = 2\frac{\Gamma}{\pi}\left[\frac{1-n_f}{n_f} + \ln\left\{\frac{2}{\pi}\frac{1-n_f}{n_f}\right\}\right]. \tag{19}$$

Here E_1 is the depth of the ground f doublets, while E_ν ($\nu \neq 1$) are those of excited doublets. In the limit of $n_f \to 1$, the Kondo temperature can be expressed as

$$T_K = D\sqrt{\Gamma/D}\,\exp\left[\frac{\pi}{2\Gamma}(E_1-\mu)\right]\prod_{\nu}^{\text{excited}}\frac{D}{E_\nu - E_1}, \tag{20}$$

as a function of the depth of f levels. This expression agrees with the result in the next divergent approximation of the Coqblin-Schrieffer model. Here the multiplication over ν is made over nearly empty excited doublets. If T_K is expressed as a function of the depth of f levels instead of n_f, there are two mechanisms making the Kondo temperature high. One is the depth of f levels. The Kondo temperature depends on them exponentially. The other is the existence of nearly empty excited multiplets [22]. When n_f is significantly smaller than one, i.e., $n_f<0.5\sim0.8$, on the other hand, it is given by

$$T_K = E_1 - \mu + \frac{\Gamma}{\pi}\,\ell n\frac{D}{\Gamma} + \sum_{\nu}^{\text{excited}} 2\,\frac{\Gamma}{\pi}\,\ell n\frac{D}{E_\nu - E_1} \,. \tag{21}$$

This result can be easily understood from Haldane's scaling discussion [20]. The energy gain is due to the hybridization rather than the Kondo effect.

The crossover as a function of the depth of f levels can be easily understood as follows:

(i) Mixed or intermediate valence states; When f levels are shallow, charge fluctuations are as important as spin fluctuations. One-body hybridization makes the ground-state singlet and paramagnetic. A large value of γ is difficult to be realized in this regime.

(ii) Kondo lattices; When f levels are so deep that $n_f\sim1$, Anderson lattices are reduced to Kondo lattices. In this regime, charge fluctuations are not so important as spin fluctuations. The Kondo effect makes the ground-state paramagnetic, and a large γ value is easily realized by spin fluctuations. Once $n_f > n_{cr}$, however, the exchange interaction overcomes the Kondo effect, and the ground-state is magnetic.

Because we are interested in nonmagnetic Kondo lattices, the Fermi liquid parameters can be approximately replaced by those of dilute Kondo alloys because spin fluctuations are similar to those in dilute Kondo alloys [23]. Then

$$\chi = \frac{\pi\Gamma}{4T_K} \sim \frac{\pi^2}{8(1-n_f)} \,, \tag{22}$$

and $m \sim \infty$. It should be noticed that Z is much larger than unity in Kondo lattices. If we put $n_f \sim 0.95$, for example, $\chi \sim 20$ and $Z \sim 20$. The specific heat coefficient is calculated as

$$\gamma = 54.7\;\left(\frac{\pi\varepsilon_0}{4D}\right)^2 \frac{1\text{K}}{T_K}\;[\text{J/K}^2\cdot\text{mol}] \;\propto\; O(Z) \,, \tag{23}$$

where $2\varepsilon_0$ is the separation between the two conduction bands. If we put T_K=10K and $\varepsilon_0=D/2$, γ is as large as 0.8J/K·mol. It should be mentioned here that γ is of the order of Z, and that the origin of large γ value, i.e., heavy-electrons is the spin fluctuation similar to that in dilute Kondo alloys. Figure 1 shows a schematical band structure and a schematical density of states according to our model [24].

Finally, let us discuss the effective repulsive interaction between quasi-particles. Because f electrons can hybridize between conduction bands whose band width is as large as $2D$, the retardation effect should be included. By including such a retardation effect approximately, the reducible vertex function obtained by Yamada and Yosida [25,26] will be used as the irreducible vertex functions of Cooper channels. This treatment is the same procedure as that of obtaining μ^* from μ in Morel and

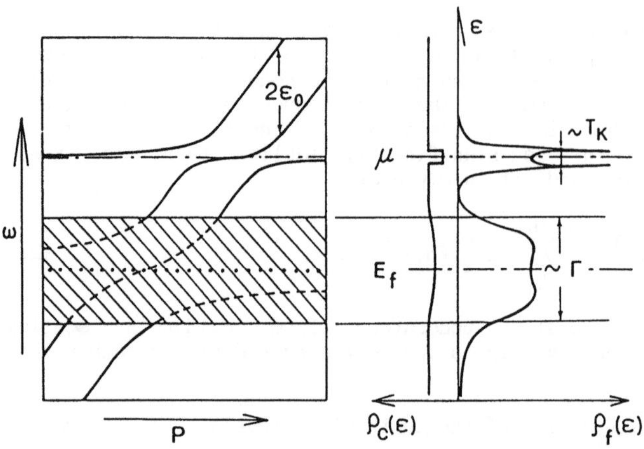

Fig.1 Model of the normal state [24]. The density of states around the Fermi level can be obtained from the Fermi liquid theory, while that around bare f levels can be calculated in the alloy analogy of the CPA. It should be noted that the present model system is always metallic as long as no order parameters exist, and that the density of states has a double-peak structure in the neighborhood of the Fermi level.

Anderson's discussion [27]. In our approximation, the repulsive interaction is of short range.

$$\Gamma_{\uparrow\downarrow} = \frac{\pi^2\Gamma^2}{4T_K} \propto O(Z) \ , \quad \Gamma_{\sigma\sigma} = 0 \ . \tag{24}$$

Section 4. Attractive Interaction between Adjacent f Electrons

Because heavy quasi-particles should be mainly composed of f electrons as discussed in the previous section, an attractive interaction between conduction electrons, or between conduction electrons and f electrons can play only a minor role even if it exists. Therefore it is the most probable that an attractive interaction exists between f electrons.

In usual superconductors, an attractive interaction is of short range as well as the repulsive Coulomb interaction. It is generally accepted that the retardation effect is important in order that the weak phonon-mediated attractive interaction might win the strong Coulomb repulsion. In heavy-electron systems, however, the on-site part of the attractive interaction can never overcome the repulsive interaction because of the narrow f band and strong Coulomb repulsion. Therefore the mutual attractive interaction between neighboring f electrons is indispensable to realize heavy-electron superconductivity; it is indispensable, more precisely, when $|c_n|^2$ defined by eq.(13) is much smaller than unity for conduction bands ($n \neq 0$) [28].

In the previous section, we have argued that spin fluctuations are confined nearly within each single site as in dilute Kondo alloys. Therefore it is rather difficult to expect that paramagnon might play an important role in nonmagnetic Kondo lattices except in the regime very close to magnetic Kondo lattices.

The present author [29] has proposed an attractive interaction between adjacent f electrons due to one-phonon exchange, which is investigated by treating vibrations of ions or ligands in the Einstein model. The main results of his theory are as follows: When an angle made by f site-ligand (or f site)-f site is smaller than $\pi/2$, one-phonon exchange results in an attractive interaction between neighboring f electrons. Typical examples where an attractive interaction can exist are a pair of f electrons between the nearest neighbor sites of fcc lattices and a pair between the next nearest neighbor sites of bcc lattices. For example, an attractive interaction can be expected in CsCℓ type lattices, while it can not in NaCℓ type lattices. Although this prediction disagrees with the observation of usual superconductors, it agrees with the observations of uranium compounds such as UCo and UC; UCo has a distorted CsCℓ structure and is a superconductor, while UC has an NaCℓ structure and is normal [30]. Actually typical heavy-electron superconductors such as $CeCu_2Si_2$ and UBe_{13} satisfy the proposed criterion; they can be regarded as a kind of CsCℓ type. If f ions occupy Cu-sites of $AuCu_3$ type lattices, an attractive interaction can be expected. However it can not be expected, if only Au-sites are occupied by f ions. It is interesting to examine in detail the correlation between the lattice structure and superconductivity in a large number of heavy-electron systems.

Section 5. Anisotropic Superconductivity

A possibility of the s-wave superconductivity has been argued by several groups [31-33,28]. Because the attractive interaction should be of finite range, however, even the s-symmetry Cooper pairs can be anisotropic in Kondo lattices [34,35]. For the sake of simplicity, it is assumed that f sites form a simple cubic lattice, and that an attractive interaction exists between their nearest neighbor sites, for example, an attractive interaction due to phonon exchange. Our model Hamiltonian is $\mathcal{H} = \mathcal{H}_0 + \mathcal{H}_g$ where $\mathcal{H}_0$ is defined by eq.(1) and

$$\mathcal{H}_g = -\frac{1}{2} g_0 \sum_{(ij)} \sum_{\sigma\tau} f^\dagger_{i\sigma} f^\dagger_{j\tau} f_{j\tau} f_{i\sigma} . \quad (25)$$

Here (ij) stands for the nearest neighbor sites, and the coupling constant, g_0 is of the order of Z^0.

First let us introduce an effective Hamiltonian to make our discussion simple. Within the mean field approximation, the result is the same as the straightforward treatment without introducing any effective Hamiltonian [35].

$$\mathcal{H}_{eff} = \sum_{p\sigma} E^*(\boldsymbol{p}) c^\dagger_{p\sigma} c_{p\sigma} + \frac{1}{2N} \sum_{kpq} \sum_{\sigma\tau} I(\boldsymbol{q})\, c^\dagger_{p+q\sigma} c^\dagger_{k-q\tau} c_{k\tau} c_{p\sigma}$$

$$+ \frac{w}{N} \sum_{(j)} \sum_{pq} \exp[-i(\boldsymbol{p}-\boldsymbol{q})\cdot \boldsymbol{R}_j]\, c^\dagger_{p\sigma} c_{q\sigma} , \quad \text{where} \quad (26)$$

$$I(\boldsymbol{q}) = u - \sqrt{6} f_s(\boldsymbol{q}) v , \quad \text{with} \quad (27)$$

$$u = \Gamma_{\uparrow\downarrow}/Z^2 \propto O(1/Z) , \quad (28)$$

$$v = g_0/Z^2 \propto O(1/Z^2) , \quad (29)$$

$$w = W/Z \propto O(1/Z) \ , \quad \text{and} \tag{30}$$

$$f_s(q) = \sqrt{2/3}\,[\cos(q_x a) + \cos(q_y a) + \cos(q_z a)] \ . \tag{31}$$

Here factors $1/Z^2$ in eqs.(28) and (29), and factor $1/Z$ in eq.(30) are due to the residue of the single particle Green function of f electrons as shown in eq.(15); W is the strength of nonmagnetic impurities.

By assuming the broken symmetry of superconductivity, the effective Hamiltonian is reduced to the BCS-type Hamiltonian as

$$\mathcal{H}_{BCS} = \sum_{p\sigma} E^*(p) c^\dagger_{p\sigma} c_{p\sigma} - \frac{1}{2}\sum_{p}\sum_{\sigma\tau} [\Delta_{\tau\sigma}(p) f^\dagger_{p\sigma} f^\dagger_{-p-\sigma} + \text{h.c.}] \ . \tag{32}$$

Here the impurity term is ignored, and $\Delta_{\tau\sigma}(p)$ is the order parameter, which is divided into two parts as

$$\Delta_{\tau\sigma}(p) = \Delta^{(1)}_{\tau\sigma} + \Delta^{(2)}_{\tau\sigma}(p) \ , \quad \text{where} \tag{33}$$

$$\Delta^{(1)}_{\tau\sigma} = -\frac{u}{N}\sum_q \langle f_{-q\tau} f_{q\sigma} \rangle_s \ , \tag{34}$$

is due to the on-site repulsive interaction, while

$$\Delta^{(2)}_{\tau\sigma}(p) = \frac{v}{N}\sum_q \sqrt{6} f_s(p-q) \langle f_{-q\tau} f_{q\sigma} \rangle_s \ , \tag{35}$$

is due to the attractive interaction between neighboring sites; the average $\langle \cdots\cdots \rangle_s$ is taken in the superconducting state.

Because the attractive interaction works only between the nearest neighbors of the simple cubic lattice, there are six degrees of freedom, which can be classified according to the symmetry.

(i) even parity (s- or $d\gamma$-symmetry)

$$\Delta^{(2)}_{\uparrow\downarrow}(p) = \sum_\mu A_\mu \cos(p_\mu a)\delta \ , \tag{36}$$

and $\Delta^{(2)}_{\sigma\sigma}(p) = 0$. Here $A_x = A_y = A_z$ for the s-symmetry, while $A_x + A_y + A_z = 0$ for the $d\gamma$-symmetry. Here $\Delta^{(1)}_{\sigma\tau}$ is determined uniquely from $\Delta^{(2)}_{\sigma\tau}$ through the gap equation.

(ii) odd parity (p-symmetry)

$$\Delta^{(2)}_{\sigma\tau}(p) = \sum_\mu B_{\sigma\tau;\mu} \sin(p_\mu a)\delta \ , \qquad \Delta^{(1)}_{\sigma\tau} = 0 \ . \tag{37}$$

The anisotropic feature of the s-symmetry superconductivity will be argued in the following part.

The transition temperature in the absence of impurities is calculated as [36,35]

$$T_c = 1.14 T_K \exp(-1/\lambda_s) \ , \tag{38}$$

with

$$\lambda_s = \frac{2\lambda_0(\lambda_2-\lambda_1)}{-(\lambda_2-\lambda_0)+[(\lambda_2-\lambda_0)^2+4\lambda_0(\lambda_2-\lambda_1)]^{1/2}} . \quad (39)$$

Here the dimensionless coupling constants are defined by

$$\lambda_0 = \Gamma_{\uparrow\downarrow}\rho^*(\mu)/Z^2 \propto O(Z^0) , \quad (40)$$

$$\lambda_1 = g_0\langle f_s\rangle^2\rho^*(\mu)/Z^2 \propto O(1/Z) , \text{ and} \quad (41)$$

$$\lambda_2 = g_0\langle f_s^2\rangle\rho^*(\mu)/Z^2 \propto O(1/Z) . \quad (42)$$

The coupling constant, λ_s, can be simply expressed in the following limiting cases; $\lambda_s = \lambda_2-\lambda_0$ for $\lambda_2 \gg \lambda_0$, while $\lambda_s = \lambda_2-\lambda_1$ for $\lambda_2 \ll \lambda_0$. As shown in Fig.2, T_c is a decreasing function of λ_0, but it is still finite even for infinitely large λ_0 because of $\langle f_s^2\rangle \geqq \langle f_s\rangle^2$ which leads to $\lambda_2 \geqq \lambda_1$. Although Z is much larger than unity in heavy-electron systems, λ_1 can compete with λ_0 under reasonable conditions [28,29]. One can easily see from the dependence of $\Gamma_{\uparrow\downarrow}$ on T_K that T_c is an increasing function of T_K [28].

The specific heat jump at the transition temperature is of the order of γT_c, i.e., $\Delta C \propto O(Z)$, but it is smaller than the value, $1.42\gamma T_c$, predicted by the BCS theory in general [37,35].

The energy gap can vanish along lines on the Fermi surface in this anisotropic s-symmetry superconductivity. The energy gap $\Delta_{sp}(\boldsymbol{p},T)$ of the single-particle Green function is given by [35]

$$\Delta_{sp}(\boldsymbol{p},T) \propto [f_s(\boldsymbol{p}) - \langle f_s\rangle R] \left[\frac{T_c-T}{T_c}\right]^{1/2} , \quad (43)$$

near the transition temperature with

$$R = \lambda_0 \ln(1.14T_K/T_c)/[1+\lambda_0 \ln(1.14T_K/T_c)] . \quad (44)$$

The Fermi surface exists in the Brillouin zone. Another surface is defined by

$$\Delta_{sp}(\boldsymbol{p},T) = 0 . \quad (45)$$

If the two surfaces intersect, the energy gap vanishes along lines. The average of eq.(43) over the Fermi surface identically vanishes for $R = 1$. If there exists only a single Fermi surface, therefore, the energy gap always vanishes along lines in the limit of $R \to 1$, i.e., in the limit of strong correlation. Although the qualitative feature does not depend on the anisotropy of the dispersion relation of quasi-particles, the quantitative feature depends on it. According to the Fermi liquid theory developed here, the dispersion relation depends on the hybridization as well as the lattice structure of f sites. Miyake et al. [34] have considered a similar anisotropic character of the s-symmetry superconductivity in Kondo lattices. However their model is special; the anisotropy of the dispersion relation of quasi-particles is the same as that of the attractive interaction. If one treats superconductivity in the approximation used here, one can easily see that the gap should vanish over the whole Fermi surface in the limit of $R \to 1$. Taking the approximation

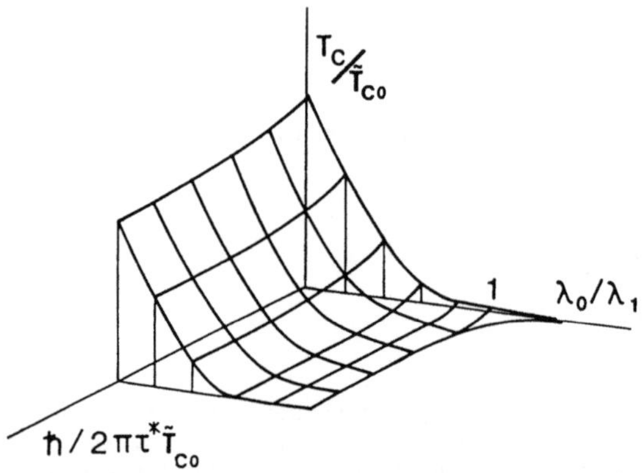

Fig.2 The transition temperature, T_c, as a function of λ_0/λ_1 and $\hbar/2\pi\tau^*\tilde{T}_{c0}$; τ^* is the relaxation time of quasi-particles and $\tilde{T}_{c0}$ is the transition temperature in the absence of both the repulsive interaction and impurities [35] .

into account, however, one can conclude that the gap can be small, but that the gap does not vanish in general.

In the presence of nonmagnetic impurities, the transition temperature is reduced [38,39,35] . However the reduction of the transition temperature is different in two regimes; in the case of $\lambda_0 \gg \lambda_1$, a large suppression of T_c is expected, and it is similar to the Abrikosov-Gorkov theory for the isotropic s-wave superconductivity with magnetic impurities [40] . Therefore T_c vanishes in this case, once $\hbar/\tau^* T_{c0} \sim 1$; T_{c0} is the transition temperature in the absence of impurities, and τ^* is the relaxation time of quasi-particles. It is calculated as

$$\frac{\hbar}{2\tau^*} = n_i \pi \rho^*(\mu) \, | w |^2 \propto O(1/Z) , \tag{46}$$

in the Born approximation, where n_i is the concentration of impurities. When some magnetic ions such as Ce ions are replaced by nonmagnetic ions such as La ions, they work as impurities. They can not be treated, however, in the Born approximation. Such a case has been investigated by the present author [24] , and the life-time is still of the order of $1/Z$. When $\lambda_1 \gg \lambda_0$, on the other hand, T_c never vanishes as shown in Fig.2. The reduction of T_c in this regime has already been investigated by Tsuneto [38] , and Markovitz and Kadanoff [39] in the different theoretical context. However the reduction is enhanced by the existence of the strong repulsive interaction in Kondo lattices.

The slope of the upper critical field is calculated as [28]

$$- \left(\frac{\hbar e}{m_e c k_B}\right) \left[\frac{dH_{c2}}{dT_c}\right]_{H_{c2}\to 0} = \frac{4}{3\pi} \frac{\hbar}{\tau^* T_K} \frac{m^*}{m_e} \propto O(Z) , \tag{47}$$

in the dirty limit, where m_e is the free electron mass. The dimensionless number given by eq.(47) is nearly equal to the number of the slope of the upper critical field in units of Tesla and Kelvin. If the residual resistivity is as high as 100$\mu\Omega$cm as in UBe_{13}, then $\hbar/\tau^* T_K$ is estimated to be of the order of unity.

Finally, let us discuss the nuclear magnetic relaxation rate. The relaxation rate in the superconducting state relative to that in the normal state is calculated as

$$\frac{W_s}{W_n} = \int \frac{d\Omega}{S} \int \frac{d\Omega'}{S} \int d\varepsilon \left(-\frac{df}{d\varepsilon}\right) \frac{\varepsilon^2 + \Delta_{sp}(\Omega)\Delta_{sp}(\Omega')}{[\varepsilon^2 - \Delta_{sp}^2(\Omega)]^{1/2}[\varepsilon^2 - \Delta_{sp}^2(\Omega')]^{1/2}} \ . \tag{48}$$

Here the integrations over Ω and Ω' mean those over the Fermi surface, and S is the area of the Fermi surface; the integration over the energy should be made over the region of $|\varepsilon| > \mathrm{Max}(|\Delta_{sp}(\Omega)|, |\Delta_{sp}(\Omega')|)$. The energy gap is approximately a function of the direction on the Fermi surface as

$$\Delta_{sp}(\Omega) = \Delta_{sp}(\boldsymbol{p}) \ . \tag{49}$$

In case of odd parity Cooper pairs, the second term in the nominator of the integrand in eq.(48) does not contribute to the integration due to the symmetry. In case of gapless s-symmetry Cooper pairs, the energy gap, $\Delta_{sp}(\boldsymbol{p})$, changes its signs on lines of zero energy gap on the Fermi surface. Therefore the contribution of the second term should be small, although it does not vanish in general. Then the nuclear magnetic relaxation rate near the transition temperature can be different from that of ordinary BCS s-wave superconductivity. It can be similar to that of the polar type p-symmetry superconductivity. However, the quantitative feature depends on the anisotropy of the Fermi surface as well as the energy gap.

Section 6. Summary and Discussion

Superconductivity in Kondo lattices has been examined near the transition temperature. On the basis of the Fermi liquid theory, one can conclude that the large value of the specific heat coefficient, i.e., the large effective mass of quasi-particles is due to spin fluctuations similar to those in dilute Kondo alloys, but that there remains only a small degree of freedom of charge fluctuations in heavy-electron; the spin fluctuation is of the order of Z, while the charge fluctuation is of the order of $1/Z$. Therefore the transition temperature can not be high in Kondo lattices because only a small degree of freedom is left in the charge fluctuation. Theoretically this can be seen from eqs.(40)-(42) where the dimensionless coupling constants are divided by Z^2; Z^2 is of the order of $10^2 \sim 10^3$. However the spin fluctuation contributes to the specific heat jump as well as the specific heat itself, and to the slope of the upper critical field; they are all of the order of Z.

The s-symmetry superconductivity investigated here has novel anisotropic features in the presence of strong correlation, which are different from the ordinary isotropic BCS s-wave superconductivity. For example, the specific heat jump is reduced from the BCS value, and the reduction of the transition temperature by nonmagnetic impurities can also be expected. This is not inconsistent with the Anderson theorem on the transition temperature of dirty s-wave superconductors, which is valid only when the interaction is of short range[41]. The most remarkable feature of this anisotropic s-symmetry superconductivity is that the energy gap can vanish along lines on the Fermi surface. Although our discussion has been confined to superconductivity in Kondo lattices such as $CeCu_2Si_2$ and UBe_{13}, the qualitative anisotropic feature can also be applied to another class of heavy-electron superconductors such as UPt_3.

Both the nuclear magnetic relaxation and the ultrasonic attenuation imply that the energy gap vanishes along lines down to low temperatures [6-8]. These experimental results do not exclude a possibility of the anisotropic s-symmetry superconductivity. Furthermore they can be considered to be evidences of anisotropic s-symmetry Cooper pairs rather than the polar type p-symmetry Cooper pairs: There is an internal degree of freedom in p-symmetry pairing. Such a degree of freedom should be determined in order that the free energy might be as low as possible, i.e., that the energy gap might be opened as widely as possible. If the mutual interaction is independent of spin, the BW-like state should be stabilized; the gap is opened over the whole Fermi surface. If the feed back effect of paramagnons [42] is strong, on the other hand, the ABM-like state is presumably stabilized; the gap is opened except at points on the Fermi surface. A new unknown mechanism seems to be required in order to stabilize such a polar state. So far the spin orbit interaction is assumed to be weak. If the spin-orbit interaction is strong, the gap never vanishes along lines on the Fermi surface for p-symmetry Cooper pairs [43]. In conclusion, the anisotropic s-symmetry superconductivity examined here should be taken into account for the identification of the symmetry realized in heavy-electron superconductors.

Acknowledgement

The author is grateful to Professor K. Yosida, Professor H. Fukuyama and Professor H. Shiba for useful discussions. He is also thankful to Dr. Y. Hasegawa for discussion about superfluid ^{3}He.

References

[1] F. Steglich, J. Aarts, C. D. Bredl, W. Lieke, D. Mescheded, W. Franz and H. Schäfer, Phys. Rev. Lett. **43** (1979) 1892.
[2] F. Steglich, C. D. Bredl, W. Lieke, U. Rauchschwalbe and G. Sparn, Physica **126** B+C (1984) 82.
[3] H. R. Ott, H. Rudigier, Z. Fisk and J. L. Smith, Phy. Rev. Lett. **50** (1983) 1595.
[4] G. R. Stewart, Z. Fisk, J. O. Willis and J. L. Smith, Phy. Rev. Lett. **52** (1984) 679.
[5] H. R. Ott, H. Rudigier, T. M. Rice, K. Ueda, Z. Fisk and J. L. Smith, Phys. Rev. Lett. **52** (1984) 1915.
[6] Y. Kitaoka, K. Ueda, T. Kohara and K. Asayama, Solid State Commun. **51** (1984) 461.
[7] D. E. MacLaughlin, Cheng Tien, W. G. Clark, M. D. Lan, Z. Fisk, J. L. Smith and H. R. Ott, Phys. Rev. Lett. **53** (1984) 1833.
[8] D. J. Bishop, C. M. Varma, B. Batlogg, E. Bucher, Z. Fisk and J. L. Smith, Phys. Rev. Lett. **53** (1984) 1009.
[9] C. M. Varma, Bull. American Phys. Soc. **29** (1984) 404.
[10] P. W. Anderson, Phys. Rev. **B30** (1984) 1549.
[11] N. C. Baezinger and R. E. Rundle, Acta. Cryst. **2** (1949) 258.
[12] J. M. Luttinger and J. C. Ward, Phys. Rev. **118** (1960) 1417.
[13] J. M. Luttinger, Phys. Rev. **119** (1960) 1153.
[14] R. M. Martin, Phys. Rev. Lett. **48** (1982) 362.
[15] F. J. Ohkawa, J. Phys. Soc. Jpn. **53** (1984) 1389.
[16] F. J. Ohkawa, J. Phys. Soc. Jpn. **52** (1983) 3886.
[17] F. J. Ohkawa, J. Phys. Soc. Jpn. **53** (1984) 2697.

[18] K. Yosida, Phys. Rev. 147 (1966) 223.
[19] S. Inagaki, Prog. Theor. Phys. 62 (1979) 1441, and Prog. Theor. Phys. Suppl. 69 (1980) 232.
[20] F. D. M. Haldane, Phys. Rev. Lett. 40 (1978) 416.
[21] K. Yosida and A. Sakurai, Prog. Theor. Phys. 61 (1979) 1597, and J. Magn. & Magn. Mater. 15-18 (1980) 949.
[22] K. Yamada, K. Yosida and K. Hanzawa, Prog. Theor. Phys. 71 (1984) 450.
[23] A. Yoshimori and H. Kasai, J. Magn. & Magn. Mater. 31-34 (1983) 475. They assumed finite transfer integral between adjacent f orbits in order to make systems metallic. Therefore their model is not an Anderson lattice in the sense used in the present paper.
[24] F. J. Ohkawa, J. Phys. Soc. Jpn. 53 (1984) 1828.
[25] K. Yamada, Prog. Theor. Phys. 53 (1975) 970, *ibid.* 54 (1975) 316, and *ibid.* 55 (1976) 1345.
[26] K. Yosida and K. Yamada: Prog. Theor. Phys. 53 (1975) 1286.
[27] P. Morel and P. W. Anderson, Phys. Rev. 125 (1962) 1263.
[28] F. J. Ohkawa, J. Phys. Soc. Jpn. 53 (1984) 3577.
[29] F. J. Ohkawa, J. Phys. Soc. Jpn. 53 (1984) 3568.
[30] B. S. Chandrasekhar and J. K. Hulm: J. Phys. Chem. Solid 7 (1958) 259.
[31] M. Tachiki and S. Maekawa, Phys. Rev. B29 (1984) 2497.
[32] H. Razafimandimby, P. Fulde and J. Keller, Z. Phys. B54 (1984) 111.
[33] T. Matsuura, K. Miyake, H. Jichu and Y. Kuroda, Prog. Theor. Phys. 72 (1984) 402.
[34] K. Miyake, T. Matsuura and H. Jichu, Prog. Theor. Phys. 72 (1984) 652.
[35] F. J. Ohkawa and H. Fukuyama, J. Phys. Soc. Jpn. 53 (1984) 4344.
[36] M. D. Whitmore and J. P. Carbotte, Phys. Rev. B23 (1981) 5782.
[37] V. L. Pokrovskii and M. S. Ryvkin, Sov. Phys.-JETP 16 (1963) 67.
[38] T. Tsuneto, Prog. Theor. Phys. 28 (1962) 857.
[39] D. Markowitz and L. P. Kadanoff, Phys. Rev. 131 (1963) 563.
[40] A. A. Abrikosov and L. P. Gorkov: Soviet Physics JETP 12 (1961) 1243.
[41] P. W. Anderson, J. Phys. Chem. Solids 11 (1959) 26.
[42] P. W. Anderson and W. F. Brinkman: *The Helium Liquids, Proc. 15th Scottish Universities Summer School in Physics*, 1974, ed. J. G. M. Armitage and I. E. Farquhar (Academic Press, New York 1975) p.315.
[43] K. Ueda and T. M. Rice: preprint.

Some Aspects of Heavy Fermion Superconductivity

K. Miyake

Department of Physics, Nagoya University, Nagoya 464, Japan

1 Introduction

Some compounds containing Cerium and Uranium, such as $CeCu_2Si_2$ [1], UBe_{13} [2] and UPt_3 [3], which have recently been found to show superconductivity, are called heavy fermion systems, because they have very large linear specific heat γT, with γ being of order ~Joule/mole$\cdot K^2$, in their normal phase. It has now turned out that various properties of their superconducting states are quite different from those of well investigated states, such as BCS state or triplet p-states of ^{3}He.

In this paper the following three points are discussed. (i) A tight-binding picture of quasiparticle of Kondo lattice system, in which both the on-site repulsive interaction and the inter-site attractive one are the order of T_k, the Kondo temperature. (ii) Formation of Cooper pair for such system in which the interaction potential is expressible in terms of superposition of separable forms each of which has s-, d-, p-, or f-like symmetry. (iii) Practical irrelevancy of frozen spin degrees of freedom, due to the strong spin-orbit coupling and crystal field around heavy ions, such as U and Ce, to the odd parity Cooper pairing.

2 Tight-Binding Model of Kondo Lattice System

2.1 Backgrounds

Since the band width of heavy fermion Λ~10K is much smaller than the Debye frequency ω_D~10^2-10^3K, phonon mediated interaction is instantaneous.
So, the effective interaction $V(q,\omega)$ including the Coulomb repulsion is expressible in terms of the static dielectric constant $\varepsilon(q,0)$: $V(q,\omega)\simeq V(q,0)=4\pi e^2/q^2\varepsilon(q,0)$. It is believed that $\varepsilon(q,0)>0$ for all q under the usual condition, and so $V(q,0)>0$ for all q [4]. This would be one of the reasons why the superconducting state of heavy fermion system is expected to be the triplet (odd parity) state, assisted by the paramagnon mechanism analogous to superfluid ^{3}He [5,6]. However, MATSUURA et al [7] and MIYAKE et al [8] have recently proposed counter examples. They showed that the singlet s-pairing is possible in principle if the q-dependence of $V(q,0)$ or the r-dependence of $V(r,0)\equiv\Sigma_q\exp(iq\cdot r)V(q,0)$ has a form like Fig.1.
Essential point of their discussions is that two quasiparticles can form the s-pair using the attractive part of $V(r,0)$ and avoiding each other so as not to feel the repulsion. Though their conclusion is due to the

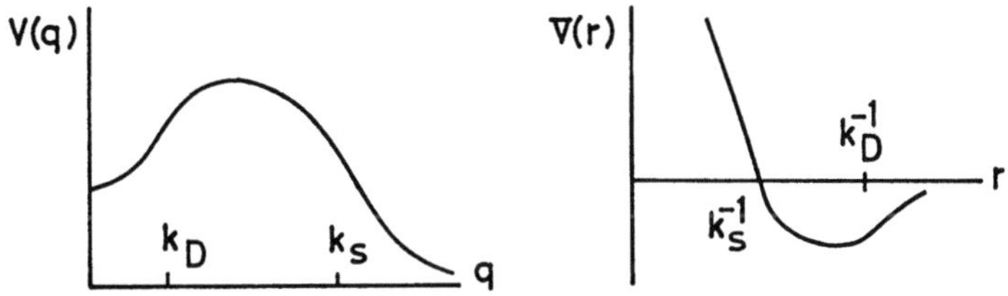

Fig.1 A model potential of $V(q)>0$ making the s-pairing possible

phenomenological model, a microscopic basis can be given to them on the basis of JICHU-MATSUURA-KURODA's (JMK) [9] theory for the Kondo lattice and the idea of Kondo volume collapse (KVC) which has been used by RAZAFIMANDIMBY, FULDE and KELLER [10] to explain the on-site attraction.

2.2 Jichu-Matsuura-Kuroda's Theory for Kondo Lattice [9]

JMK treat the periodic Anderson model in the so-called "sd-limit". First of all, they assume that single impurity problem has already been solved [11]; i.e., the on-site selfenergy M_{ii} and the on-site repulson $\Gamma_{\uparrow\downarrow}$ for the low energy excitation of f-electron are assumed to be given by

$$\left.\begin{aligned} M_{ii}(i\omega_n) &= -i\omega_n(\tilde{\chi}-1) + M_{ii}(0) \ , \\ \Gamma_{\uparrow\downarrow} &= \pi\tilde{\chi}\Gamma \ , \end{aligned}\right\} \tag{1}$$

where $\Gamma\equiv\pi\rho V^2$ with the density of states ρ of the conduction electron and the mixing potential V, and the dimensionless susceptibility $\tilde{\chi}$ for single impurity is defined as

$$\tilde{\chi} \equiv \pi\Gamma/4T_k \ . \tag{2}$$

On the basis of these results for the single impurity problem, JMK take into account the inter-site selfenergy of a type like Fig.2 and obtain the f-electron Green function as follows:

$$G_{ij} = G_i[\delta_{ij}+M_{ij}G_j+\sum_{\ell\neq i,j} M_{i\ell}G_\ell M_{\ell j}G_j+\cdots] \ , \tag{3}$$

where the on-site Green function G_i is given by

$$G_i^{-1}(i\omega_n) = i\omega_n-\varepsilon_{fi}-M_{ii}(i\omega_n) = \tilde{\chi}_i(i\omega_n-\tilde{\chi}_i^{-1}\varepsilon'_{fi}) \ . \tag{4}$$

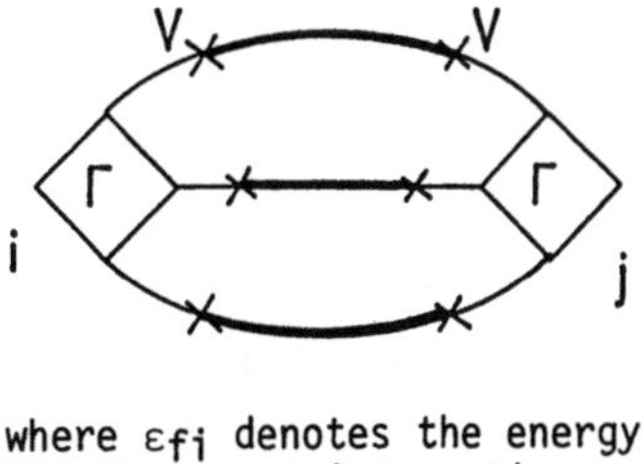

Fig.2

A diagram for $M_{ij}(i\neq j)$. Light and heavy lines denote the Green function for single site f-electrons and conduction electron, respectively. Crosses stand for the hybridization V

where ε_{fi} denotes the energy of localized f-electron at i-th site. The frequency integrations concerning internal lines of the selfenergy $M_{ij}(i\omega_n)$ for $|\omega_n|<T_k$ are divided into three regions; (a) all of them are smaller than T_k, (b) one of them is smaller and two of them are larger

than T_K, and (c) all of them are larger than T_K. Contribution to M_{ij} from the region (a), (b), and (c) is the order of Γ, $\Gamma(\Gamma\rho)/\tilde{\chi}\sim 10^{-3}\Gamma$, and $\Gamma(\Gamma\rho)^2 \sim 10^{-4}\Gamma$, respectively; so, the dominant contribution comes from the region (a). The Green function (3) is written in the Fourier transformed form like

$$G_k^{-1}(i\omega_n) = \tilde{\chi}(i\omega_n - \tilde{\chi}^{-1}\varepsilon_f' - \tilde{\chi}^{-1}M_k) \ , \tag{5}$$

where M_K, the Fourier transform of M_{ij}, is at most of order $\tilde{\chi}T_K$ and has k-dependence. Therefore, localized f-electrons have gotten coherence and turned to the heavy fermions with narrow bandwidth of order T_K.

Then, in order to turn to the quasiparticle picture, we renormalize the wave function, the selfenergy, and the interaction vertex:

$$G_{ij} \to \tilde{G}_{ij} = \tilde{\chi}_i^{1/2}\, G_{ij}\tilde{\chi}_j^{1/2}, \quad M_{ij} \to \tilde{M}_{ij} = \tilde{\chi}_i^{-1/2} M_{ij}\tilde{\chi}_j^{-1/2} \ ,$$

and

$$\Gamma_{\uparrow\downarrow} \to \tilde{\Gamma}_{\uparrow\downarrow} = \tilde{\chi}^{-2}\Gamma_{\uparrow\downarrow} \sim \pi\Gamma/\tilde{\chi} = 4T_K \ .$$

Thus we obtain the effective Hamiltonian describing the system of heavy fermions and conduction electrons as follows:

$$\begin{aligned} H = & \sum_{ij}\sum_{\sigma}(\tilde{M}_{ij}a_{i\sigma}^{+}a_{j\sigma} + \mathrm{h.c.}) + \sum_{i}\sum_{\sigma}\tilde{\varepsilon}_{fi}n_{i\sigma} + \sum_{i}\tilde{\Gamma}_{\uparrow\downarrow}n_{i\uparrow}n_{i\downarrow} \\ & + \sum_{i}\sum_{\sigma}\tilde{V}_i(a_{i\sigma}^{+}c_{i\sigma} + \mathrm{h.c.}) + \sum_{k}\sum_{\sigma}\xi_k c_{k\sigma}^{+}c_{k\sigma} \ , \end{aligned} \tag{6}$$

where $a_{i\sigma}$ and $c_{k\sigma}$ are annihilation operators of heavy fermion and conduction electron, respectively, and $n_{i\sigma} \equiv a_{i\sigma}^{+}a_{i\sigma}$, $\tilde{\varepsilon}_{fi} \equiv \tilde{\chi}_i^{-1}\varepsilon_{fi}'$, $\tilde{V}_i = \tilde{\chi}_i^{-1/2}V_i$, and ξ_k is the kinetic energy of conduction electron. The subscript σ denotes z-component of spin for both fermions. However, as will be discussed below, σ should be read as two components of quasi-spin for the Kramers doublet in such cases where the spin-orbit coupling and the crystal field are strong.

2.3 Heavy Fermion-Phonon Coupling [12,13]

The basic idea is that parameters in the Hamiltonian (6) is modulated by lattice vibrations and the volume dependence of T_K (so $\tilde{\chi}$) is very large. (KVC [10]) Indeed, on the basis of the Anderson model, the volume (Ω) dependence of T_K (which is measured by $\eta \equiv -d\ell nT_K/d\ell n\Omega$) is estimated as

$$\eta \sim \frac{U}{V^2\rho}\frac{d\ell n\rho}{d\ell n\Omega} \sim \frac{U}{V^2\rho} \sim (10\text{-}10^2) \ , \tag{7}$$

which is quite large compared to that for the density of states of conduction band $d\ell n\rho/d\ell n\Omega$.

Since $T_K \ll \omega_D$, instantaneous interaction is obtained after the phonon degrees of freedom have been eliminated as usual: Its general form is

$$-\sum K_{\ell m,rs}a_{\ell\sigma}^{+}a_{m\sigma}a_{r\sigma'}^{+}a_{s\sigma'} \quad , \tag{8}$$

where ℓm, rs denote the lattice points. If we restrict the terms in (8) to those relevant to the same or nearest neighbor sites, we obtain the indirect interaction potential

$$V_{indirect} = -g_0 \sum_i n_{i\uparrow}n_{i\downarrow} - g_1 \sum_{<i,j>\sigma,\sigma'} a^+_{i\sigma}a_{j\sigma}a^+_{j\sigma'}a_{i\sigma'}$$
$$-g_2 \sum_{<i,j>\sigma\sigma'} n_{i\sigma}n_{j\sigma'} - g_3 \sum_{<i,j>\sigma} n_{i\sigma}(a^+_{i,-\sigma}a_{j,-\sigma}+\text{h.c.}) \ , \qquad (9)$$

The coupling constants $g_m(m=0\sim3)$ are positive and their orders of magnitude are estimated as

$$g_m \sim \frac{T_K}{Ms^2}\,\eta^2 T_K \sim 10^{-3}\eta^2 T_K \ , \qquad (10)$$

where M is the ion mass, s is the sound velocity, and it has been assumed that $Ms^2 \sim \varepsilon_F \sim 10^4$K and $T_K \sim 10$K. By (7), amounts to 10-10^2; so that g_m can be of the order of T_K in general.

Adding the indirect interaction (9) to (6), we obtain the following Hamiltonian in the k-representation:

$$H = \sum_{k\sigma} \xi_k c^+_{k\sigma}c_{k\sigma} + \tilde{V}\sum_{k\sigma}(c^+_{k\sigma}a_{k\sigma}+\text{h.c.}) + \sum_{k\sigma} \tilde{\varepsilon}_{k\sigma}a^+_{k\sigma}a_{k\sigma}$$
$$+ \frac{1}{N}\sum_{k,k'}[(\tilde{\Gamma}_{\uparrow\downarrow}-g_0)-2g_1\gamma_{k+k'}-2g_2\gamma_{k-k'}-2g_3(\gamma_k+\gamma_{k'})]a^+_{k\uparrow}a^+_{-k\downarrow}a_{-k'\downarrow}a_{k'\uparrow}$$
$$- \frac{1}{N}\sum_{k,k'}\sum_{\sigma}(g_1\gamma_{k+k'}+g_2\gamma_{k-k'})a^+_{k\sigma}a^+_{-k\sigma}a_{-k'\sigma}a_{k'\sigma} \ , \qquad (11)$$

where $\gamma_k \equiv \sum_\delta \exp(ik\cdot R_\delta)$.

3 Cooper Pairing in the System with On-Site Repulsion and Inter-Site Attraction

3.1 Pairing Potential and the Gap Structure [12,13]

First of all, we analyze the properties of $\gamma_{k\pm k'}$ appearing in (11). For clear presentation, we assume the simple cubic structure for the lattice. (Similar discussions can be done for other types of crystal structure.) Then, γ_k is given by

$$\gamma_k = 2(\cos k_x a+\cos k_y a+\cos k_z a) \ , \qquad (12)$$

where a is the lattice constant. Hereafter, we use the abbreviations C^x_k and S^x_k for cos $k_x a$ and sin $k_x a$, respectively. After some rearrangements, $\gamma_{k\pm k'}$ is expressible in terms of the superposition of separable interactions:

$$\gamma_{k\pm k'} = \frac{1}{6}(\gamma_k\gamma_{k'}+\eta_k\eta_{k'}+\zeta_k\zeta_{k'}) \pm 2(S^x_kS^x_{k'}+S^y_kS^y_{k'}+S^z_kS^z_{k'}) \ , \qquad (13)$$

where η and ζ are defined as

$$\left.\begin{aligned} \eta_k &= \sqrt{6}\,(C^x_k - C^y_k) \ , \\ \zeta_k &= \sqrt{2}\,(C^x_k + C^y_k - 2C^z_k) \ . \end{aligned}\right\} \qquad (14)$$

The symmetry of γ_k, (η_k,ζ_k), and $S^\alpha_k(\alpha=x,y,z)$ under the lattice rotation is s-, dγ-, and p-like, respectively.

Then, the interaction part of (11), H_{int}, is transformed to the following form:

$$H_{int} = \frac{1}{N}\sum_{kk'}\sum_{\sigma\sigma'} V_{kk'}^{\sigma\sigma'} a_{k\sigma}^{+} a_{-k\sigma'}^{+} a_{-k'\sigma} a_{k'\sigma} \ , \tag{15}$$

where

$$V_{kk'}^{\sigma\sigma'} = \frac{1}{2}(V_{kk'}^{s}+V_{kk'}^{d})\delta_{\sigma,-\sigma'} + V_{kk'}^{p}(\delta_{\sigma,\sigma'}+\delta_{\sigma,-\sigma'}) \ , \tag{16}$$

with

$$\left.\begin{aligned} V_{kk'}^{s} &= (\tilde{\Gamma}_{\uparrow\downarrow}-g_0) - 2g_3(\gamma_k+\gamma_{k'}) - \frac{1}{3} g_s\gamma_k\gamma_{k'} \ , \\ V_{kk'}^{d} &= -\frac{1}{3} g_s(\eta_k\eta_{k'}+\zeta_k\zeta_{k'}) \ , \\ V_{kk'}^{p} &= -2g_t \sum_{\alpha=x,y,z} S_k^{\alpha}S_{k'}^{\alpha} \ . \end{aligned}\right\} \tag{17}$$

Here the abbreviations $g_s=g_2+g_1$ and $g_t=g_2-g_1$ have been used. In case the crystal symmetry has inversion, the gap equations for both the singlet and the triplet unitary pairings are written in the form of

$$\Delta_k^{\sigma\sigma'} = -\frac{1}{N}\sum_{k'} V_{kk'}^{\sigma\sigma'} \sum_{\tau=\pm1} (\Delta_{k'}^{\sigma\sigma'}/2E_{k'}^{\tau})\,\mathrm{th}(E_{k'}^{\tau}/2T) \ , \tag{18}$$

where the excitation energy E_k^{τ} is defined by

$$E_k^{\tau} \equiv [(\varepsilon_k^{\tau})^2+\hat{\Delta}_k^{+}\hat{\Delta}_k]^{1/2} \tag{19}$$

with the hybridized band energy,

$$\varepsilon_k^{\tau} \equiv \frac{1}{2}\{\tilde{\varepsilon}_k+\xi_k+\tau[(\tilde{\varepsilon}_k-\xi_k)^2+4\tilde{V}^2]^{1/2}\} \ . \tag{20}$$

As usual the gap $\hat{\Delta}_k$ in (19) is proportional to $i\hat{\sigma}_y$ for the singlet pairing and $i(\hat{\sigma}_y\hat{\sigma}_j)d_j$ for the triplet pairing.

From (16~18), one can see that the gap Δ_k for the singlet pairing has the form

$$\Delta_k = \Delta_1 + \Delta_\gamma\gamma_k + \Delta_\eta\eta_k + \Delta_\zeta\zeta_k \ , \tag{21}$$

where the coefficients Δ_1~Δ_ζ are complex numbers in general. The structure of the d-vector for the triplet pairing is more complicated.

3.2 Comparison of the Transition Temperatures [12,13]

The transition temperature T_c ($\equiv 1/\beta_c$) is determined separately for s-, d-, and p-pairing:

For s-pairing,

$$\begin{vmatrix} 1+(\tilde{\Gamma}_{\uparrow\downarrow}-g_0)\Phi_1(\beta_c) & -2g_3\Phi_\gamma(\beta_c) \\ -2g_3\Phi_1(\beta_c) & 1-\frac{1}{3}g_s\Phi_\gamma(\beta_c) \end{vmatrix} = 0 \ . \tag{22}$$

For d-pairing,

$$\frac{1}{g_s} = \frac{1}{3}\Phi\ (\beta\ ) \ . \tag{23}$$

For p-pairing,

$$\frac{1}{g_t} = 4\Phi_p(\beta_c) \ . \tag{24}$$

The functions Φ's are difined as follows: $\Phi_1 \equiv 1/N\Sigma F_k$, $\Phi_\gamma \equiv 1/N\Sigma\gamma_k^2 F_k$, $\Phi_d \equiv 1/N\Sigma\eta_k^2 F_k = 1/N\Sigma\zeta_k^2 F_k$, and $\Phi_p \equiv 1/N\Sigma(S_k^\alpha)^2 F_k$, with $F_k = \Sigma_\tau \mathrm{th}(\beta E_k^\tau/2)/2E_k^\tau$.

It should be remarked that p-pairing has always lower T_c compared to d-pairing, because $4\Phi_p$ is comparable to or smaller than $\Phi_d/3$ and the coupling g_t $(=g_2-g_3)$ for p-pairing is smaller (could even be negative) than g_s $(=g_2+g_1)$ for d-pairing. Therefore, since g_1 is of order T_K for the Kondo lattice system, the triplet pairing is expected to be still unfavourble even if the paramagnon mechanism works to encourage the triplet pairings.

Among the singlet pairings, whether the highest T_c occurs for s-pairing or d-pairing, depends on the details of the parameters characterizing the Hamiltonian (11), especially on the values of $\tilde{V}$ and g_3. If both $\tilde{V}$ and g_3 were zero, the highest T_c would occur for d-pairing. However, if $\tilde{V}$ and g_3 take their typical values, e.g. $\tilde{V}=10^{-2}\varepsilon_F$ and $g_3 \sim 10^{-1}T_K$, s-pairing has the highest T_c. Figure 3 shows the results of numerical calculation for the latter case.

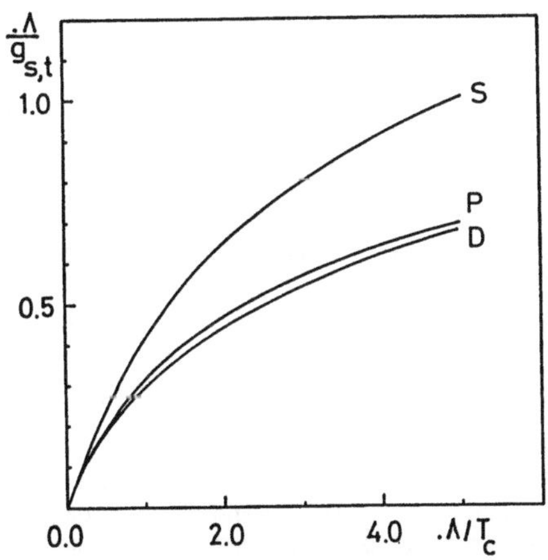

Fig.3

Curves S, D and P give transition temperature $1/T_c$ satisfying (22), (23) and (24), respectively, for $1/g_s$ and $1/g_t$. Parameters are chosen as $\Lambda/\varepsilon_F=10^{-3}$, $\tilde{V}/\varepsilon_F=10^{-2}$, $(\tilde{\Gamma}_{\uparrow\downarrow}-g_0)/\Lambda=\pi$ and $g_3=0.\Lambda$

The above results are considered to be applicable to the Kondo lattice system where the KVC mechanism works. For the system which shows not so marked Kondo-like behavior, there exists a possibility that the triplet pairing potential dominates the singlet one. Cooper pairing in such systems can be treated similarly to above, if g_s and g_t are regarded merely as parameters characterizing the model.

3.3 Structure of the Excitation Gap

In general it is a complicated problem to determine the type of low energy excitations, i.e., whether excitation spectrum has a gap or not, and whether the zeroes of the gap on the Fermi surfact are on points or lines. Here we discuss some special cases. In the Ginzburg-Landau region, the structure of the quartic terms of the free energy determine the mixing of the different kinds of pairings, e.g., 1, γ_k, η_k, and ζ_k in (21).

Example 1 (simple cubic crystal)
If the highest T_c occurs for s-pair $\Delta_k^s \equiv \Delta_1 + \Delta_\gamma \gamma_k$, then the d-components, η_k and ζ_k, do not admix together.[12] There is a possibility that the surface in the k-space where $\Delta_k^s = \Delta_1 + \Delta_\gamma \gamma_k = 0$, i.e., $\gamma_k = -\Delta_1/\Delta_\gamma$, crosses the Fermi surface $\varepsilon_k^\tau = 0$, i.e., $\tilde{\varepsilon}_k \xi_k = \tilde{V}^2$. If so, the excitation gap vanishes along lines.[13] Such a possibility was first pointed out by OHKAWA and FUKUYAMA [14] for the phenomenological model. Whether the two surfaces really cross together or not, of course, depends on the details of the parameters of (11).

If the highest T_c occurs for d-pair with, η_k and ζ_k, the s-components can admix at lower temperatures. As far as the admixtures are neglected, the structure of d-pair is always given by $\Delta_k^d = \Delta(\eta_k \pm i\zeta_k)$ which has nodes in the (1,1,1) and its equivalent directions; so zeroes of the gap on the Fermi surface are on points.[12]

Example 2 (face centered cubic crystal, $d\varepsilon$-pair)

$$\Delta_k^{d\varepsilon} = \Delta S_k^z (S_k^x \pm i S_k^y) \; . \tag{25}$$

Example 3 (body-centered cubic crystal, triplet f-pair)
The triplet pair giving the highest T_c is non-degenerate and is f-pair:

$$\Delta_k^f = \Delta S_k^x S_k^y S_k^z \; . \tag{26}$$

For the pairs like (25, 26), the zeroes of excitation gap are located on the lines on the Fermi surface.

4 Spin-Orbit Coupling and Crystal Field; their Practical Irrelevancy to Cooper Pair Formation

It is said that, in heavy fermion systems, the spin-orbit coupling and the crystal field should be taken into account for discussing Cooper pair formation, especially in the case of triplet (odd parity) pairings. On the basis of group theoretical discussions, ANDERSON [15], VOLOVIK-GORKOV [16], and UEDA-RICE [17] have studied taxonomy concerning the possible type of triplet pairings. Their fundamental assumtion is essentially the transformation properties of d-vector under the lattice rotation $\hat{A}$ of point group of the system:

$$\hat{A} \cdot d_\alpha(k_i) = A_{\alpha\beta} d_\beta (A_{ij} k_j) \; , \tag{27}$$

which means that the spin of Cooper pairs can not rotate independently of the lattice. The restriction (27) excludes the possibility that, for O_h, D_{6h}, and D_{4h} symmetries, excitation gap vanishes on lines on the Fermi surface [16], which disagrees with experiments.[18~20]

However, there is a possibility that remnant spin degrees of freedom, which are associated with the Kramers doublet and independent of the lattice rotation, could survive in spite of strong intra-atomic spin-orbit coupling and crystal field. In this section we present one of such non-trivial examples.

First of all, let us assume that heavy fermion quasiparticles (irrespective of its origin, Kondo lattice, Brinkman-Rice liquid, etc.) form narrow bands well separated each other, i.e., the bandwidth Λ and the splitting Δ between different bands satisfy the condition $\Lambda \ll \Delta$. One particle Hamiltonian describing such bands can be expressed by the tight-binding model as follows:

$$H_0 = -\frac{1}{2}\sum_{ij}\sum_{nm}(t_{ij}^{nm}c_{in}^{+}c_{jm}+\mathrm{h.c.}) + \sum_{i}\sum_{n}\varepsilon_n c_{in}^{+}c_{in} \;, \tag{28}$$

where c_{in}^{+} is a Fermi operator creating a quasiparticle of the state n at the lattice point i. Notations of this section are not to be confused with those of the preceding sections. The quantum number n ($n=\pm1, \pm2,\cdots$) indicates a Kramers doublet of the "atomic orbital"; so it includes all the effects of intra-atomic spin-orbit coupling and crystal field. Let us assume $\varepsilon_{\pm1}<\varepsilon_{\pm2}<\cdots$. The origin of the splitting may be the "atomic orbital" splitting due to the intra-atomic spin-orbit coupling and the crystal field. In terms of $c_{kn}\equiv 1/\sqrt{N}\cdot\Sigma_i\exp(-ik\cdot R_i)c_{in}$, (28) is written as

$$H_0 = \sum_{k}\sum_{nm}(t_k^{nm}+\varepsilon_n\delta_{nm})c_{kn}^{+}c_{km} \;, \tag{29}$$

where $t_k^{nm}\equiv -1/2\cdot\Sigma_{(i,j)}(t_{ij}^{nm}+t_{ji}^{mn})\exp[ik\cdot(R_i-R_j)]$. The Hamiltonian (29) is diagonalized for each wavenumber k as follows

$$H_0 = \sum_{k}\sum_{\zeta}\tilde{\varepsilon}_{k\zeta}a_{k\zeta}^{+}a_{k\zeta} \;, \tag{30}$$

where $a_{k\zeta}$ is related to c_{km} by the unitary transformation

$$a_{k\zeta} = \sum_{m} B_k^{\zeta m}c_{km} \;. \tag{31}$$

If the crystal has inversion symmetry (otherwise odd- and even-parity pairings can not be decoupled), the band $\tilde{\varepsilon}_{k\zeta}$ is doubly degenerate for each k, i.e., $\tilde{\varepsilon}_{k\zeta}=\tilde{\varepsilon}_{k,-\zeta}$.[21] From this fact, the form of t_k^{nm} is restricted to some extent. Let us assume that the Fermi level is located in the band with $\zeta=\pm1$. In the limit of large splitting $\Lambda\ll\Delta=\varepsilon_2-\varepsilon_1$, the 2×2 matrix $(t_k^{nm}+\varepsilon_n\delta_{nm})$, for $n,m=\pm1$, is diagonalized independently of other bands, i.e., those for $n=\pm2,\cdots$. So, the eigenvalue of $(t_k^{nm}+\varepsilon_n\delta_{nm})$ is necessary to be degenerate, which is possible only if $t_k^{nm}\propto\delta_{nm}$. Therefore, to zeroth order of Λ/Δ, $t_k^{nm}=t_k\delta_{nm}$ and $B_k^{\zeta m}=\delta_{\zeta m}$. The effect of other bands is of $O(\Lambda/\Delta)^2$ or higher: For $n,m=\pm1$,

$$t_k^{nm} = t_k[\delta_{nm}+O(\Lambda/\Delta)^2] \;, \tag{32}$$

$$B_k^{\zeta m} = \delta_{\zeta m} + O(\Lambda/\Delta)^2 \;. \tag{33}$$

In other words, to $O(\Lambda/\Delta)$, double degeneracy of the bands $\tilde{\varepsilon}_{k\pm1}$ can be described by quasi-spin (spin 1/2) corresponding to the Kramers doublet of the "atomic orbital" on the lattice point.

Next, let us consider to what extent the quasi-spin ζ of the band is conserved for various kinds of interactions. Even if the interaction conserves the quasi-spin n of "atomic orbital", ζ-nonconserving interaction arises through the unitary transformation from the terms of $O(\Lambda/\Delta)^2$. Here we assume that the interaction V_{direct} which stems from the Coulomb interaction can be written in the following form:

$$V_{direct} = \frac{1}{2}\sum_{nm}\sum_{kk'q} V_{nm}(k,k')c^+_{-k+q/2,m}c^+_{k+q/2,n}c_{k'+q/2,n}c_{-k'+q/2,m}. \quad (34)$$

Using the inverse relation of (33), (34) is reduced to

$$V_{direct} = \frac{1}{2}\sum_{\zeta_1\cdot\cdot\zeta_4}\sum_{kk'q}\tilde{V}^{k,k'}_{\zeta_1\zeta_2,\zeta_3\zeta_4} a^+_{-k+q/2,\zeta_1}a^+_{k+q/2,\zeta_2}a_{k'+q,\zeta_3}a_{-k'+q/2,\zeta_4}, \quad (35)$$

where

$$\tilde{V}^{k,k'}_{\zeta_1\zeta_2,\zeta_3\zeta_4} \equiv \sum_{nm} B^{\zeta_1 m}_{-k+q/2}B^{\zeta_2 n}_{k+q/2}V_{nm}(k,k')[B^{\zeta_3 n}_{k'+q/2}B^{\zeta_4 m}_{-k'+q/2}]^* . \quad (36)$$

From (33), (35) and (36) one can see that ζ-nonconserving interaction $V_{noncons}$ appears from the term of $O(\Lambda/\Delta)^2$, i.e., $|V_{noncons}|\sim(\Lambda/\Delta)^2|V_{direct}|$. Since $|V_{direct}|$ is in general of the order of the bandwidth Λ, $|V_{noncons}|$ is estimated as follows:

$$|V_{noncons}| \sim (\Lambda/\Delta)^2\Lambda . \quad (37)$$

The indirect interaction $V_{indirect}$, discussed in section 2.3, arises through the coupling between heavy fermion and lattice vibrations which are accompanied by only volume change but not by violation of inversion symmetry. So, the relation analogous to (37) holds also for $V_{indirect}$. On the other hand, the lattice vibrations which violate the inversion symmetry give the off-diagonal elements of transfer $t^{n,-n}_k$ (n=±1) even to zeroth order of Λ/Δ. Such heavy fermion-phonon coupling may be expressed in the form

$$\Lambda\sum_{k,q}\frac{u_q}{a}c^+_{k+q,n}c_{k,-n} , \quad (38)$$

where u_q and a denote lattice displacement and lattice constant, respectively. Order of magnitude of ζ-nonconserving indirect interaction $V'_{indirect}$, which is obtained by eliminating the phonon degrees of freedom, is estimated as

$$|V'_{indirect}| \sim (\Lambda/Ms^2)\Lambda \sim 10^{-3}\Lambda . \quad (39)$$

where it has been assumed as in section 2.3 that $\Lambda\sim10K$ and $Ms^2\sim10^4K$.

Both the quasi-spin-orbit interaction V_{so} due to the relative motion of two quasiparticles and the dipole interaction V_{dipole} between quasi-spins violate the conservation of ζ. These are estimated as follows:

$$|V_{so}| \sim \frac{m}{m^*} \frac{h^2}{2m^2c^2} \frac{\Lambda}{a^2} \sim \frac{m}{m^*} \quad 3\times10^{-5}\Lambda \ , \tag{40}$$

and

$$|V^{n.n.}_{dipole}| \sim \frac{\mu_B^2}{a^3} \sim 2\times10^{-2}K \sim 2\times10^{-3}\Lambda \ , \tag{41}$$

where μ_B denotes the Bohr magneton, m/m* is mass ratio of free electron and heavy fermion, and the lattice constant a is chosen as a ~5Å.

As far as the band, where the Fermi level is located is far separated from other bands for each wavenumber k, i.e., $\Lambda \ll \Delta$, the problem of Cooper pair formation can be discussed among the bands associated with the Fermi level. The magnitudes of ζ-nonconserving interactions, $V_{noncons}$, (37), $V'_{indirect}$, (39), V_{so}, (40), and V_{dipole}, (41), are much smaller than the transition temperature $T_c \sim 10^{-1}\Lambda$ as well as the ζ-conserving interaction $|V_{cons}| \sim \Lambda$. Therefore, except for the very narrow region near T_c, it is allowed to discuss Cooper pair formation on the basis of the Hamiltonian which conserves the quasi-spin ζ. In other words, for d-vector of the quasi-spin space, such a restriction as (27) is unnecessary in practice. So, the discussions of section 3 is applicable even to the case where the intra-atomic spin-orbit coupling and the crystal field are strong: A non-trivial example of odd parity pairing in which excitation gap vanishes along lines on the Fermi surface is given by (26).

Acknowledgements

I would like to acknowledge T. Matsuura, H. Jichu, Y. Kuroda and Y. Nagaoka for the collaborations and discussions on which the present paper is based.

References

1 W. Assumus, M. Herrmann, U. Rauchschwalbe, S. Riegel, W. Lieke, H. Spille, S. Horn, G. Weber, F. Steglich and G. Cordier: Phys. Rev. Lett. 52, 469 (1984); and references therein

2 H.R. Ott, H. Rudigier, T.M. Rice, K. Ueda, Z. Fisk and J.R. Smith: Phys. Rev. Lett. 52, 1915 (1984)

3 G.R. Stewart, Z. Fisk, J.O. Willis and J.L. Smith: Phys. Rev. Lett. 52, 679 (1984)

4 M.L. Cohen and P.W. Anderson: Superconductivity in d- and f-Band Metals, ed. D.H. Douglass (New York, 1972), p.17

5 C.M. Varma: Bull. Am. Phys. Soc. 29, 404 (1984)

6 P.W. Anderson: Phys. Rev. B30, 1549 (1984)

7 T. Matsuura, K. Miyake, H. Jichu and Y. Kuroda: Prog. Theor. Phys. 72, 402 (1984)

8 K. Miyake, T. Matsuura and H. Jichu: Prog. Theor. Phys. 72, 652 (1984)

9 H. Jichu, T. Matsuura and Y. Kuroda: Prog. Theor. Phys. 72, 366 (1984); Proc. ICCF5 1985, J. Magn. Magn. Mater. in press

10 H. Razafimandimby, P. Fulde and J. Keller: Z. Phys. B54, 111 (1984)

11 K. Yamada: Prog. Theor. Phys. 53, 970 (1975); K. Yosida and K. Yamada: ibid. 53, 1286 (1975); see also A. Yoshimori: ibid. 55, 67 (1976)

12 K. Miyake, T. Matsuura, H. Jichu and Y. Nagaoka: Prog. Theor. Phys. 72, 1063 (1984)

13 T. Matsuura, K. Miyake, H. Jichu, Y. Kuroda and Y. Nagaoka: Proc. ICCF5 1985, J. Magn. Magn. Mater. in press

14 F.J. Ohkawa and H. Fukuyama: J. Phys. Soc. Jpn. 53, 4344 (1984)
15 P.W. Anderson: Phys. Rev. B30, 4000 (1984)
16 G.E. Volovik and L.P. Gorkov: JETP Lett. 39, 550 (1984); preprint 1984
17 K. Ueda and T.M. Rice: preprint 1984
18 D.J. Bishop, C.M. Varma, B. Batlogg, E. Bucher, Z. Fisk and J.L. Smith: Phys. Rev. Lett. 53, 1009 (1984)
19 D.E. MacLaughlin, C. Tien, W.G. Clark, M.D. Lan, Z. Fisk, J.L. Smith and H.R. Ott: Phys. Rev. Lett. 53, 1833 (1984)
20 Y. Kitaoka, K. Ueda, T. Kohara, K. Asayama, Y. Onuki and T. Komatsubara: Proc. ICCF5, J. Magn. Magn. Mater. in press
21 J.M. Ziman: Principles of the Theory of Solids, 2nd ed. (Cambridge Univ. Press, 1972), §3.11

Heavy Electron Superconductors — Some Consequences of the p-Wave Pairing

K. Ueda
Department of Applied Physics, The University of Tokyo
Tokyo 113, Japan

T.M. Rice
Theoretische Physik, Eidgenössische Technische Hochschule, Hönggerberg
CH-8093 Zürich, Switzerland

Two aspects of p-wave pairing in metals are discussed. First the group-theoretical classifications of p-wave states in metals with strong spin-orbit scattering and cubic, hexagonal tetragonal symmetry is reviewed. In the second part the depairing effect associated with normal impurities is shown to be dependent on the form of the p-wave pairing states. States with lines of zeros have no critical concentration for the onset of a finite density of states at the Fermi energy while states with no zeros or only point zeros have critical concentrations.

1. Introduction

In this lecture we will treat two aspects of the theory of p-wave superconducting states in metals. This work has been stimulated by the extensive experimental effort on the superconducting properties of heavy electron metals. While there are many properties of these superconducting states which differ from conventional superconductors, there is at present no consensus on the key questions. In these circumstances it is clearly important to develop the relevant theory so that meaningful comparisons can be made. We will discuss here two aspects of the theory namely the role of crystalline symmetry and strong spin-orbit scattering on the form of possible p-wave states and secondly the effect of normal impurity scattering on p-wave states.

It was quickly recognized that the symmetry classifications of p-wave states appropriate to an isotropic fluid such as ^{3}He would be modified by crystalline symmetry and the spin-orbit part of the ionic potentials. The first works on the problem were by ANDERSON[1] and VOLOVIK and GOR'KOV[2]. Anderson considered the form of the order parameter near the transition temperature which if we make an analogy to crystalline distortions is the equivalent of a soft mode analysis. Volovik and Gor'kov used a factor group method to obtain the most symmetric possible p-states. Both authors found no cases where the energy gap vanished along lines of zeros (such states are generally called polar states) but found the possible states had point zeros or no zeros at all. States with analogous properties are found in ^{3}He namely ABM and BW states. A general argument to show that lines of zeros are not permitted for odd-parity superconductors has been given by BLOUNT[3]. Elsewhere we [4,5] have extended Anderson's analysis to discuss the effect of the higher terms in a Ginzburg-Landau expansion and determined the possible stable p-wave states in the case of cubic, hexagonal and tetragonal crystalline symmetries. Also VOLOVIK and GOR'KOV[6] have made a complete analysis of the possible odd and even parity states in these cases also. In general there is complete agreement on the possible p-wave states between the various authors and, in particular, all agree that lines of zeros for p-wave states are not allowed.

The depairing effect of normal impurities on p-wave states was recognized in the early work of BALIAN and WERTHAMER[7]. In the present work we will present a detailed analysis of the problem for these anisotropic states and examine how the condition for gaplessness (or a finite density of states at the Fermi level) depends on the nature of the p-wave state. Simultaneously a similar analysis has been made by GOR'KOV and KALUJIN[8]. Again there are no disagreement on the results and on the essential difference between states with point zeros of the gapfunction and any possible state with lines of zeros.

2. Symmetry properties

The crystal structures of heavy electron superconductors all have inversion symmetry; e.g. in UBe_{13} the point group is O_h, in UPt_3, D_{6h} and in $CeCu_2Si_2$, D_{4h}. By combining inversion with time reversal we can define doubly degenerate single particle eigenstates at each point of the Brillouin zone. In the presence of spin-orbit coupling, although the spin of an electronic eigenstate is no longer a good quantum number, we can introduce a pseudo-spin label $\{\alpha,\beta\}$ for the double degenerate states. The order parameter for the superconductivity is defined in the pseudo-spin space

$$\Delta_{\gamma_1\gamma_2}(\vec{k}) = \begin{bmatrix} \Delta_{\alpha\alpha}(\vec{k}), & \Delta_{\alpha\beta}(\vec{k}) \\ \Delta_{\beta\alpha}(\vec{k}), & \Delta_{\beta\beta}(\vec{k}) \end{bmatrix} \tag{1}$$

We are interested in p-wave pairing, which is the simplest case of odd-parity pairing. For odd-parity pairing it is convenient to use the vector notation

$$\Delta_{\gamma_1\gamma_2}(\vec{k}) = i\vec{d}(\vec{k})\cdot(\vec{\tau}\tau_y)_{\gamma_1\gamma_2} \tag{2}$$

where $\vec{\tau}=(\tau_x,\tau_y,\tau_z)$ are Pauli matrices in the pseudo-spin space. For p-wave pairing we can expand the order parameter in a basis set $\{\psi_x(\vec{k}),\psi_y(\vec{k}),\psi_z(\vec{k})\}$ which transforms like a vector

$$\vec{d}(\vec{k}) = \sum_{\ell,m} d_\ell^m\, \hat{\ell}\, \psi_m(\vec{k}) \qquad (\ell,m=x,y.z) \tag{3}$$

where the unit vectors $\{\hat{x},\hat{y},\hat{z}\}$ are basis functions for the spin part.

The transition temperature for the superconducting state is given by

$$T_c = 1.14\, \varepsilon_c \exp(-1/\omega) \tag{4}$$

with the cutoff energy ε_c and the eigenvalue ω of the integral equation

$$-\sum_{\gamma_3\gamma_4} N_0 \langle V_{\gamma_2\gamma_1:\gamma_3\gamma_4}(\vec{k},\vec{k}')\Delta_{\gamma_3\gamma_4}(\vec{k}')\rangle_{\vec{k}'} = \omega\Delta_{\gamma_1\gamma_2}(\vec{k}) \tag{5}$$

where N_0 is the density of states at the Fermi energy and $\langle\cdots\rangle_{\vec{k}}$ denotes the average over the Fermi surface.

Within the subspace where $\{\hat{x},\hat{y},\hat{z}\}$ and $\{\psi_x,\psi_y,\psi_z\}$ are the basis functions for the spin and the spatial part, respectively, we can introduce spin operators s_ℓ by

$$S_\ell\hat{m} = i\varepsilon_{\ell mn}\hat{n} \tag{6}$$

and angular momentum operators L_ℓ by

$$\langle L_\ell(\vec{k},\vec{k}')\psi_m(\vec{k}')\rangle_{\vec{k}'} = i\varepsilon_{\ell mn}\psi_n(\vec{k}) \tag{7}$$

The pairing interaction can be parametrized by these operators. The isotropic part is given by

$$V_{iso} = V_0 + V_1\vec{S}\cdot\vec{L} + V_2(\vec{S}\cdot\vec{L})^2 \tag{8}$$

and the cubic anisotropy term by

$$V_c = V_3\sum_{x,y,z} S_x^2L_x^2 \tag{9}$$

and the uniaxial anisotropy term by

$$V_a = V_4L_z^2 + V_5S_z^2 + V_6S_z^2L_z^2. \tag{10}$$

For a cubic crystal (O_h group, e.g., UBe_{13}) the pairing interaction has a form of

$$V = V_{iso} + V_c \tag{11}$$

and for a hexagonal crystal (D_{6h}, e.g., UPt_3),

$$V = V_{iso} + V_a \tag{12}$$

and for a tetragonal crystal (D_{4h}, e.g., $CeCu_2Si_2$),

$$V = V_{iso} + V_c + V_a \tag{13}$$

In a cubic crystal the basis set $\{\psi_x(\vec{k}),\psi_y(\vec{k}),\psi_z(\vec{k})\}$ as well as $\{\hat{x},\hat{y},\hat{z}\}$ belong to T_1 representations. Thus the order parameter is a product representation which is decomposed by the spin-orbit coupling as

$$T_1\times T_1 = A_1 + E + T_1 + T_2. \tag{14}$$

The eigenfunctions for these irreducible representations are tabulated in [1], [4] and [6]. Generally we can expect that each irreducible representation has a different T_c and that with the highest T_c is the physical one. For two-(E) or three-dimensional (T_1,T_2) representations several phases are possible. We can determine the forms of the order parameter by considering a

Ginzburg-Landau expansion of the free energy. For details we refer the reader to [4,6] and here we will only summarize some important results.

Within the weak coupling theory all stable states have real order parameters except for overall phase factor. These real states are even under time reversal symmetry. However, in general, with other values of the Ginzburg-Landau coefficients, states with complex order parameters are also possible for the E, and T_1 and T_2 representations. In these complex states time reversal symmetry is broken. This broken symmetry could possibly lead to the existence of a Josephson current between a complex p-wave state and a usual s-wave superconductor (which is simply odd under time reversal). As regards the zeros of the gap function which determine the low temperature behavior of physical quantities, there are certain stable states where the gap vanishes at symmetry points but there is never a line of zeros in accordance with a general argument given by BLOUNT[3].

In hexagonal crystals the three-dimensional basis set $\{\psi_i(k)\}$ already splits into A_2+E_1 representations and a similar splitting occurs for the spin part. In the presence of the spin-orbit coupling the 9-dimensional space of the order parameter, Eqn.(3), is decomposed as

$$(A_2+E_1)\times(A_2+E_1) = 2A_1 + A_2 + 2E_1 + E_2. \tag{15}$$

The basis functions for these irreducible representations and all stable states are tabulated in [5,6]. For two-dimensional representations (E_1,E_2), complex states are possible. Except for the A_1 state and the complex E_2 states, all states have points where the gap vanishes. An interesting case is the complex E_2, where electrons with only one pseudo-spin component condense into Cooper pairs. The complex E_2 states have angular momentum $\pm 2\hbar$ around the c-axis. Although the gap has zeros for this pseudo-spin component, the low-temperature thermodynamic behavior is dominated by the other component since it has no gap at all. The low temperature properties would be those of a normal Fermi liquid.

In tetragonal crystals symmetry is lower than cubic and hexagonal ones. An important difference from the hexagonal case is that two-dimensional E_2 representation further splits into one-dimensional representations,

$$(A_2+E)\times(A_2+E) = 2A_1 + A_2 + 2E + B_1 + B_2. \tag{16}$$

All possible states are obtained in [5,6]. For all states except for the two A_1 states the gap has points of zeros.

3. Effect of normal impurities on p-wave pairing

It is well known that normal impurities in s-wave superconductors do not change the thermodynamic properties (Anderson's theorem[9]). On the other hand, it is well known that normal impurities in p-wave superconductors are depairing just like paramagnetic impurities in usual s-wave superconductors [7].

In this section we will discuss the depairing effect due to the normal impurities in the framework of Abrikosov-Gor'kov theory[10,11]. For simplicity we neglect the spin-orbit coupling in this section. The anomalous self-energy due to a pairing interaction V is

$$\Delta_{\sigma\sigma'}(\vec{k}) = -T\sum_{\vec{k}'\omega_n} V(\vec{k},\vec{k}')F_{\sigma\sigma'}(\vec{k}',i\omega_n) \tag{17}$$

where F is the anomalous Green's function. Impurity scattering gives new contributions to the self-energy. One is of the normal type

$$\Sigma^{(1)}(\vec{k},i\omega_n) = n_i\sum_{\vec{k}'} |u(\vec{k},\vec{k}')|^2 G(\vec{k}',i\omega_n) \tag{18}$$

where n_i is the concentration of impurities and $u(\vec{k},\vec{k}')$ is their scattering potential. There is also contribution of the anomalous type

$$\Sigma^{(2)}_{\sigma\sigma'}(\vec{k},i\omega_n) = n_i\sum_{\vec{k}'} |u(\vec{k},\vec{k}')|^2 F_{\sigma\sigma'}(\vec{k}',i\omega_n). \tag{19}$$

In the case of normal impurities in s-wave superconductors, it is the cancellation between the two contributions which leads to Anderson's theorem. If we assume s-wave impurity scattering $(u(\vec{k},\vec{k}')=u)$ for simplicity, then since $F_{\sigma\sigma'}(\vec{k},i\omega_n)$ is an odd function of $\vec{k}$ in a p-wave state, the r.h.s. of Eqn.(19) vanishes. Generally, any lack of cancellation between the two contributions $\Sigma^{(1)}$ and $\Sigma^{(2)}$ leads to depairing effecs. The Dyson equations for the Green's functions are

$$[i\omega_n-\xi_{\vec{k}}-\Sigma(i\omega_n)]G(\vec{k},i\omega_n)+\sum_{\sigma'}\Delta_{\sigma\sigma'}(\vec{k})F^+_{\sigma'\sigma}(\vec{k},i\omega_n) = 1 \tag{20}$$

$$[i\omega_n+\xi_{\vec{k}}+\Sigma(-i\omega_n)]F^+_{\sigma\sigma'}(\vec{k},i\omega_n) + \Delta^+_{\sigma\sigma'}(\vec{k})G(\vec{k},i\omega_n) = 0 \tag{21}$$

where $\xi_{\vec{k}}$ is the energy of $\vec{k}$ state measured from the chemical potential and $\Sigma(i\omega_n)$ stands for $\Sigma^{(1)}(\vec{k},i\omega_n)$ which is independent of $\vec{k}$ as a consequence of the s-wave scattering approximation.

For simplicity we will consider only the Balian-Werthamer (BW) state, the Anderson-Brinkman-Morel (ABM) state and a possible polar state as the representatives of the states where the gap does not vanish, vanishes on points and on lines, respectively. For these states, which belong to the class of unitary states, the Green's functions have the form

$$G(\vec{k},i\omega_n) = (i\tilde{\omega}_n+\xi_{\vec{k}})/[(i\tilde{\omega}_n)^2-\xi^2_{\vec{k}}-\Delta(\vec{k})\Delta^+(\vec{k})] \tag{22}$$

$$F^+_{\sigma\sigma'}(\vec{k},i\omega_n) = -\Delta^+_{\sigma\sigma'}(k)/[(i\tilde{\omega}_n)^2-\xi^2_{\vec{k}}-\Delta(\vec{k})\Delta^+(\vec{k})] \tag{23}$$

with $\tilde{\omega}_n=\omega_n+i\Sigma(i\omega_n)$. By substituting Eqns.(22) and (23) into Eqns.(17) and (18) and integrating over the momentum, we obtain the following selfconsistent equations for $\Delta_{\sigma\sigma'}(\vec{k})$ and $\Sigma(i\omega_n)$.

(a) BW state: $\Delta_{\sigma\sigma'}(\vec{k}) = \Delta\begin{bmatrix} -\hat{k}_x+i\hat{k}_y, & \hat{k}_z \\ \hat{k}_z & , \hat{k}_z+i\hat{k}_y \end{bmatrix}$

$$1 = \pi N_0 \overline{V} T \sum_n \frac{\Delta}{\sqrt{\tilde{\omega}_n^2+\Delta^2}} \tag{17a}$$

$$\tilde{\omega}_n = \omega_n + \Gamma \frac{\tilde{\omega}_n}{\sqrt{\tilde{\omega}_n^2+\Delta^2}} \tag{18a}$$

(b) ABM state: $\Delta_{\sigma\sigma'}(\vec{k}) = \Delta(\hat{k}_x+i\hat{k}_y)\begin{bmatrix} 1 & ; & 0 \\ 0 & , & 1 \end{bmatrix}$

$$1 = \pi N_0 \overline{V} T \frac{3}{2} \sum_n \frac{1}{2\Delta}\{\frac{|\tilde{\omega}_n|}{\Delta} - (1 - \frac{\tilde{\omega}_n^2}{\Delta^2})\frac{i}{2} \log \frac{i|\tilde{\omega}_n|-\Delta}{i|\tilde{\omega}_n|+\Delta}\} \tag{17b}$$

$$\tilde{\omega}_n = \omega_n - i\,\Gamma \frac{\tilde{\omega}_n}{2\Delta} \log \frac{i|\tilde{\omega}_n|-\Delta}{i|\tilde{\omega}_n|+\Delta} \tag{18b}$$

(c) polar state: $\Delta_{\sigma\sigma'}(\vec{k}) = \Delta\hat{k}_z\begin{bmatrix} 1 & , & 0 \\ 0 & , & 1 \end{bmatrix}$

$$1 = \pi N_0 \overline{V} T \frac{3}{2} \sum_n \frac{1}{\Delta}[\sqrt{1+(\frac{\tilde{\omega}_n}{\Delta})^2} - (\frac{\tilde{\omega}_n}{\Delta})^2 \log \frac{\Delta+\sqrt{\Delta^2+\tilde{\omega}_n^2}}{|\tilde{\omega}_n|}] \tag{17c}$$

$$\tilde{\omega}_n = \omega_n + \Gamma \frac{\tilde{\omega}_n}{\Delta} \log \frac{\Delta+\sqrt{\Delta^2+\tilde{\omega}_n^2}}{|\tilde{\omega}_n|} \tag{18c}$$

In the above expressions $\overline{V}$ is the strength of the pairing interaction defined by $V(\hat{k},\hat{k}')=3\overline{V}\hat{k}\cdot\hat{k}'$, N_0 is the density of states at the Fermi energy in the normal state and $\Gamma=\pi n_i N_0 |u|^2$ is the strength of the impurity scattering.

The density of states of single particle excitations is given by an analytic continuation

$$N(\omega) = N_0 \frac{1}{\Gamma} \mathrm{Im}[i\tilde{\omega}_n]_{i\omega_n=\omega+i\delta} \tag{24}$$

where δ is a positive infinitesimal. In Fig.1a we show the density of states for the BW state. The excitation gap parameter is given by $\omega_g=\Delta[1-(\Gamma/\Delta)^{2/3}]^{3/2}$ and thus when $\Gamma/\Delta\geq 1$ the system is in a gapless regime in the same manner as the usual paramagnetic impurity effect in an s-wave superconductor.

The density of states for the ABM state is shown in Fig.1b. The density of states at low energies is given by

$$N(\omega) = \begin{cases} N_0(\frac{\omega}{\Delta})^2[1 - \frac{\pi}{2}(\frac{\Gamma}{\Delta})]^{-3} & (\frac{\Gamma}{\Delta} < \frac{2}{\pi}) \\ N_0 \frac{\Delta}{\Gamma} \cot \frac{\Delta}{\Gamma} & (\frac{\Gamma}{\Delta} \geq \frac{2}{\pi}) \end{cases} \tag{25}$$

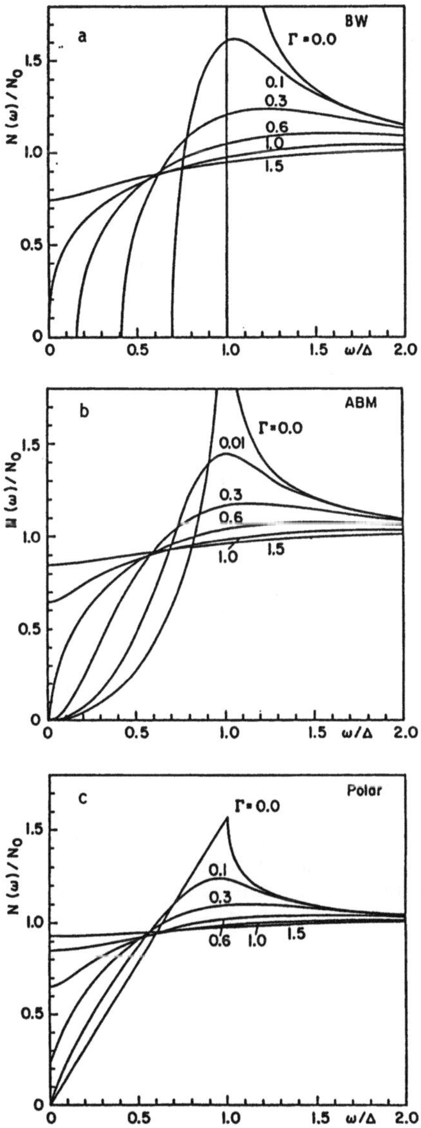

Fig.1 Density of states of single-particle excitations in a) the BW state, b) the ABM state and c) the polar state. Γ is in unit of Δ

For weak scattering impurities increase the coefficient of the ω^2-term but there is finite density of states at $\omega=0$ only when $\Gamma/\Delta \geq 2/\pi$.

The most remarkable result is obtained for the polar state (Fig.1c) where

$$N(0) = N_0\Delta\left[\Gamma\sinh\frac{\Gamma}{\Delta}\right]^{-1}$$

and zero energy excitations exist even for an infinitesimal concentration of normal impurities.

Equations (17a), (17b), (17c) give, of course, the same transition-temperature which is determined by

$$\log\frac{T_{c0}}{T_c} = \psi\left(\frac{1}{2} + \frac{\Gamma}{2\pi T_c}\right) - \psi\left(\frac{1}{2}\right), \tag{26}$$

where T_{c0} is the transition-temperature for the pure system and ψ is the digamma function. This has precisely the same form as for the paramagnetic impurities in an s-wave superconductor.

It is straightforward to calculate temperature-dependence of the gap function in the Ginzburg-Landau region. Then we obtain the jump of the specific heat as

$$\frac{\Delta C(T_c)}{C_n(T_c)} = \frac{12[1 - \psi'(\frac{1}{2} + \rho_c)\rho_c]^2}{-a\psi^{(2)}(\frac{1}{2} + \rho_c) - \frac{1}{6}\rho_c\psi^{(3)}(\frac{1}{2} + \rho_c)} \tag{27}$$

where $C_n(T_c)$ is the specific heat at T_c in the normal state, $\rho_c = \Gamma/2\pi T_c$ and $\psi^{(n)}$ is a polygamma function. The coefficient a is dependent on the states under consideration; a=1/2 for the BW state, 3/5 for the ABM state, 9/10 for the polar state. In Fig.2 we plot the jump of the specific heat as a function of scattering strength. In Figs.3 numerical results for the specific heat below T_c are plotted for the pure case (Fig.3a) and a dirty case (Fig.3b). For illustrative purposes we show the normal state specific heat extrapolated into the superconducting phase by the dashed lines. A T^3-law is also shown by the dot-dashed lines. It is clearly seen that in the polar-type state with normal impurities, the specific heat below T_c is a T-linear one with a smaller coefficient than in the normal state. In the ABM state the T^3-law remains for these parameters but the coefficient is increased as is seen from a comparison of Figs.3a and 3b. The changes in the excitation spectrum should naturally be reflected in the low-temperature behavior of the other quantities; e.g., T_1 and ultrasonic attenuation. We will report on them elsewhere.

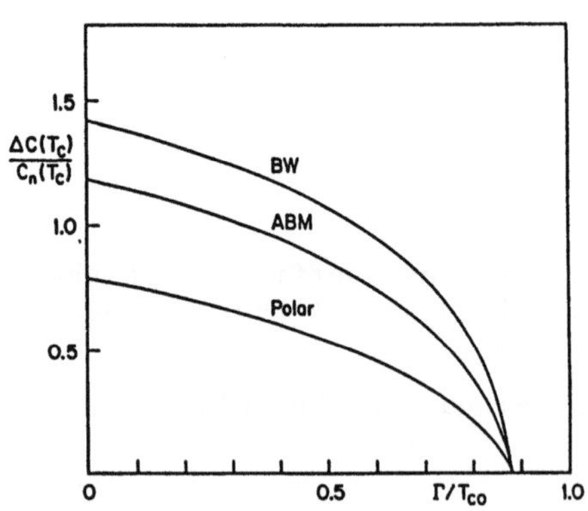

Fig.2 Jump of the specific heat at the transition temperature as a function of scattering strength

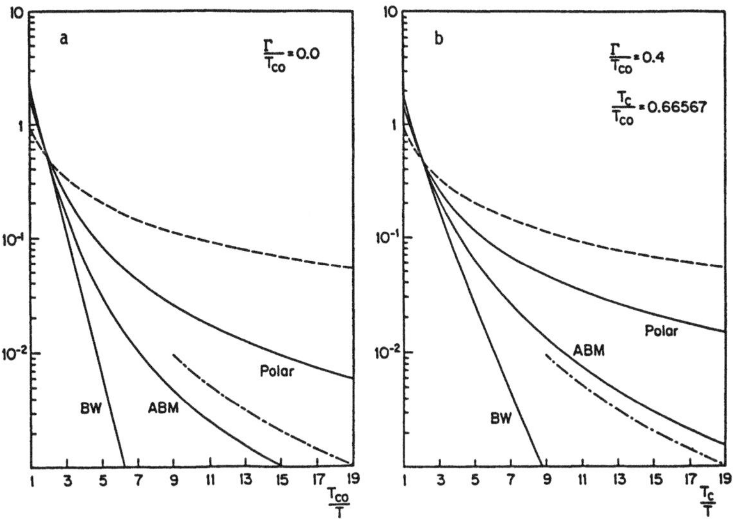

Fig.3 Specific heat in the superconducting state for a) the pure case and b) a dirty case

4. Conclusions

In this work we treated two aspects of the theory of p-wave superconducting states in metals. In the first part we summarized the results of the group theoretical classification of p-wave states in crystals with a strong spin-orbit component of the potential and with cubic, hexagonal and tetragonal symmetry. Perhaps the most important result to come from this analysis is the absence of any polar states — or states with lines of zeros in the set of allowed states. However, several authors have claimed such polar states in UPt_3 and UBe_{13} based on ultrasonic attenuation or nuclear magnetic resonance measurements[12,13]. Such claims clearly raise serious theoretical problems in view of the group theoretical results. Very recently the theoretical analysis behind the interpretation of the ultrasonic attenuation data was criticized by RODRIGUEZ[14] who argues that a hydrodynamic description is appropriate and is consistent with a set of point zeros rather than a line of zeros. More theoretical analysis of this point is required.

Since much of the argument hinges on the behavior of the various response functions at low temperature, a detailed analysis of the depairing role of impurity scattering is required. The conclusions from the analysis presented for the density of state have shown that the most serious effect would be on any polar state since in this case the low-temperature behavior is modified in any concentration of impurities. By contrast a state with point zeros has a critical concentration before an essential modification of the power-laws sets in. Although the analysis was carried through only for the simplest form of p-wave state, the results depend on the generic form of the density of states and therefore should be applicable with only slight modification to any state with the same generic form.

1 P.W. Anderson: Phys. Rev. B30, 4000 (1984).
2 G.E. Volovik and L.P. Gor'kov: J.E.T.P. Lett. 39, 559 (1984).
3 E.I. Blount: to be published.
4 K. Ueda and T.M. Rice: Phys. Rev. B31, (1985).
5 K. Ueda and T.M. Rice: Advances in Solid State Physics XXV.
6 G.E. Volovik and L.P. Gor'kov: preprint.
7 R. Balian and N.R. Werthamer: Phys. Rev. 131, 1553 (1963).
8 L.P. Gor'kov and P.A. Kalujin: preprint.
9 P.W. Anderson: J. Phys. Chem. Solids 11, 26 (1959).
10 A.A. Abrikosov and L.P. Gor'kov: J.E.T.P. 12,1243 (1961).
11 K. Maki: Superconductivity, ed. by R.D. Parks, (Marcel Dekker, 1969) p.1035.
12 D.J. Bishop, C.M. Varma, B. Batlogg, E. Bucher, Z. Fisk and J.L. Smith: Phys. Rev. Lett. 53, 1009 (1984).
13 D.E. MacLaughlin, Cheng Tien, W.G. Clark, M.D. Lan, Z. Fisk, J.L. Smith and H.R. Ott: Phys. Rev. Lett. 53, 1833 (1984).
14 J.P. Rodriguez: Preprint.

Phenemenological and Some Microscopic Aspects of Heavy Fermi-Liquids and Their Superconductivity

C.M. Varma

AT & T Bell Laboratories, Murray Hill, NJ 07974, USA

There appears to be three regimes [1] in the physics of the rare-earth and actinide metals and compounds. The *local moment regime* which is the most frequent occurs when the ionization level of the f-orbitals is well below the chemical potential. The other two regimes may well be called the *confined moment regime* since magnetic moments which are free at high temperatures (Curie-Weiss susceptibility) so not show themselves in any property at low temperatures (and low frequencies). Instead a Fermi-liquid behavior (Pauli susceptibility and linear specific heat) obtains. If the f-ionization potential pins the chemical potential we have the *valence fluctuation regime*, while if it is below, but not too far below, the so-called *heavy Fermion behavior* obtains. Some of the valence fluctuation materials are semiconductors, but when they are not, there is little qualitative distinction in their low temperature properties from the heavy Fermi-liquids. The effective mass is of $0(10-10^2)$ in the former and $0(10^2-10^3)$ in the latter. Although phenemenologically the valence fluctuators are similar to the heavy Fermions, microscopically they are different. The average f-occupation is far from integral, so that real quantum-mechanical f-charge fluctuations are essential, while in the latter case such fluctuations may be ignored. A. Zawadowski and I have recently completed calculations of a two component interacting electron gas model suitable for valence fluctuators. One of our conclusions is that the singularities of the problem are primarily driven by inter-component spin interactions and that inter-component charge (valence) fluctuations do not determine the low temperature state. This puts the valence fluctuators in the same universality class as the heavy Fermions and gives a basis on which to understand the phenemenological similarity.

In the rest of these notes, I will only discuss the heavy Fermions and refer the ready to Ref. (2) for the calculation mentioned above.

1. Heavy Fermion Superconductivity

Following the discovery of superconductivity in the heavy Fermion metals [3], the suggestion [4] was made that it could not be the conventional form of superconductivity. The argument simply is that the conventional phonon-induced superconductivity occurs only because the electron-electron Coulomb repulsion is spread out over the electronic bandwidth and is ineffective compared to the phonon induced electron-electron attraction, which is of similar magnitude, but is confined to about the Debye frequency. In heavy Fermion metals the cut-off for the Coulomb repulsion is in fact smaller than the phonon frequency. A Cooper pair with a node at the origin of the relative co-ordinate overcomes the strong short-range part of the Coulomb repulsion, but this means pairing must be in a finite angular momentum state. One may further invoke spin-fluctuations or Kondo effect induced pairing to suggest that $S = 1$ and odd angular momentum based pairing is favored. This runs into some phenemenological problems which I will discuss later.

The suggestion of unconventional superconductivity appears to be borne out in all three heavy Fermion superconductors that have been extensively investigated. In UPt_3, ultrasonic attenuation [5], thermal conductivity and specific heat [6] favor a polar like state, with a superconductivity gap which is zero on a line (or lines) on the Fermi-surface and a corresponding quasi-particle density of states which is proportional to E for E small compared to the maximum superconductivity gap Δ. In UBe_{13} nuclear relaxation rate [7] T_1^{-1}, and thermal conductivity [6] again favor a polar state. In $CeCu_2Si_2$ T_1^{-1} measurements [8] suggest a polar state also. Point contact tunneling measurement [9] in $CeCu_2Si_2$ have been used to rule out a triplet superconducting state. For $S = 1$ state with perfectly smooth and nonmagnetic barriers, there should be no Josephson effect with ordinary superconductors. The Josephson effect observed is not completely convincing to me. The Josephson current does not have the correct temperature dependence. A proximity effect induced tunneling could give the observed behavior.

Recently in UBe_{13}, (a cubic crystal) a peak in the ultrasonic attenuation has been observed [10] just below T_c. This is characteristic of absorption into orbital collective modes. Such a peak is not observed in UPt_3, a hexagonal crystal. The difference probably has to do with the fact that the pinning of orbital (and spin-orbital) collective modes, which raises their frequency and increases their damping by putting them in a region of higher quasi-particle density of states, is a higher order effect in cubic crystals than in hexagonal crystals.

At this point, why the polar state should be favored in all the three superconductors mentioned is a great mystery. By Group theory [11] such a state cannot exist in the $S = 1$ manifold for a general interaction. If the pair-interaction has negligible spin-orbit interaction contribution, a polar $S = 1$ state is allowed. The experiments mentioned above are all consistent with either a $S = 1$, odd parity state or a $S = 0$ even parity but not $L = 0$ state. There is one recent experiment in UBe_{13} which is not consistent with a $S = 0$ state. This is the magnetic susceptibility measurement through magnetic form measurement [12] by neutron-scattering. Within the experimental error, $\lesssim 5\%$ there is no change in the susceptibility in going from the normal to the superconducting state. I do not think this is possible with any singlet pairing. We are therefore left with the mystery.

A quantity of importance for transport experiments and, indeed, for the very existence of anisotropic superconducting states, is the quasi-particle lifetime τ due to elastic scattering from impurities. It is easy to see that for a quasi-particle of energy E, $\tau(E) \sim E^{-1}$ in the polar state (and $\sim E^{-2}$ in the axial state). Quasi-particles become much longer lived in such superconducting states than in the normal metallic state.

2. Phenemenological Aspects

The thermodynamic and the recent transport experiments in the heavy Fermion solids can be used to put severe constraints on the nature of the many body effects responsible for this phenomena. They can be used to formulate phenemenological theory in terms of which different experiments may be correlated and specify results that a microscopic theory must obtain.

The experiments I will focus on are the specific hear C_v, the susceptibility χ, the ultrasonic attenuation α_s, and the local magnetic moment relaxation rate T_1^{-1}, in the normal state of UPt_3 or UBe_{13}. Similar behavior is expected in the other heavy

Fermions. A clue to the nature of the phenomena is provided by the following two contrasting facts: (1) The thermodynamic experiments [3] viz. the specific heat and the susceptibility yield an enhancement of $0(10^2-10^3)$ compared to a free Fermi-gas of similar electronic density (2) BATLOGG, BISHOP et al. [13] have made the remarkable observation that the ultrasonic attenuation in UPt_3 in the normal phase is of similar magnitude to that of an ordinary metal, say Sn or Pb. Also S. SCHULTZ finds [14] that the relaxation rate of Er moments in UBe_{13} is the same order as that of Er moments in $ThBe_{13}$ which has a susceptibility a few hundred times smaller.

In the usual Fermi-liquid, the specific heat enhancement C_v/C_{vo} is related at low temperatures to the enhancement of the quasi-particle density of states at the Fermi-surface or equivalently to that of an average inverse Fermi-velocity or to that of an average effective mass

$$C_v/C_{vo} = m^*/m \, . \tag{1}$$

We shall take Eq. (1) to define m^*/m.

Ultrasonic attenuation can be shown to be proportional to τ_s^*/m^*, where τ_s^* is the lifetime for sound wave to decay into particle-hole pairs. Experiments therefore suggest that both τ_s^* and m^* are renormalized such the τ_s^*/m^* remains at its unrenormalized value. Similarly the spin-lattice relaxation rate indicates that τ_m^*/m^* is unrenormalized where τ_m^* is the lifetime for local magnetization to decay to spin-fluctuations of the electron gas.

Following the suggestion [4] that the heavy Fermion superconductors could not be ordinary phonon induced superconductors and are likely to be anisotropic superconductors, as is liquid ^3He, there have been arguments [15] put forth that the physics of the heavy Fermion metals is itself akin to that of liquid ^3He. Rice et al. for example have tried to explain the mass renormalization by a Hubbard-Gutzwiller model which is quite suitable for liquid ^3He, (where the maximum mass enhancement is about 6, before the transition to the solid state). The cancellation in the renormalizations in τ^*/m^* that we have noted above does not occur for ^3He however. Transport properties like ultrasonic attenuation, nuclear relaxation rate etc. are renormalized in the usual Landau Fermi-liquid theory as applied to liquid ^3He or electrons in metals.

The cancellation in τ^*/m^* is reminiscent of renormalizations in electronic properties due to electron-phonon interactions [16]. What is relevant here is not the electron-phonon interactions but only that an enhancement of m^*/m without a renormalization in τ^*/m^* follows if the momentum dependence of the self-energy $\Sigma(k,\omega)$ is negligible compared to the energy dependence. Let the single particle Green's function $G(k,\omega)$ be given by

$$G_\sigma^{-1}(k,\omega) = [(\omega-\epsilon(k)) - \Sigma_\sigma(k,\omega)] \, , \tag{2}$$

where near the Fermi-surface the band structure energy $\epsilon(\underset{\sim}{k}) = \underset{\sim o}{v} \cdot (\underset{\sim}{k} - \underset{\sim F}{k})$. For

$$\partial\Sigma/\partial\omega >> v_o^{-1}\, \partial\Sigma/\alpha k \, , \tag{3}$$

$$G_\sigma^{-1}(k,\omega) = Z^{-1}[\omega - Z^{-1} \underset{\sim o}{v} \cdot (\underset{\sim}{k} - \underset{\sim F}{k})] \tag{4}$$

where Z is the quasi-particle amplitude or wave-function renormalization

$$Z = (1+\partial\Sigma/\partial\omega)^{-1} \, . \tag{5}$$

The specific heat enhancement or m^*/m is then given by

$$m^*/m = Z^{-1} . \qquad (6)$$

In calculating any transport property, like the ultrasonic attenuation by a process as in Fig. 1, the matrix element (vertex) renormalization cancels against the self-energy under the condition (4) for long wavelength perturbations. This follows from a Ward identity as shown by PRANGE and KADANOFF [16]. Physically the cancellation in τ^*/m^* can be understood from the condition (4), which amounts to saying that the self-energy rides with the chemical potential, so that in any perturbation where the local chemical potential is altered and relaxes towards equilibrium, the renormalization is not felt.

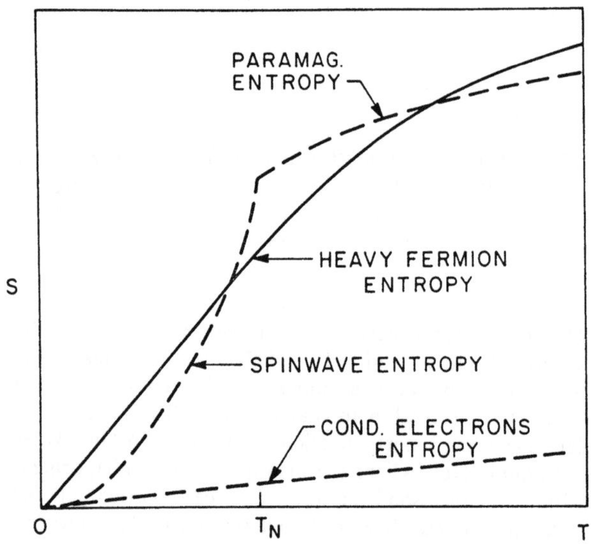

Fig. 1 Schematic diagram illustrating conditions under which a heavy Fermion mass is mandated.

There is one important difference between the physics of electron-phonon renormalizations and the phenomena in the heavy Fermions. Magnetic susceptibility is not renormalized by electron-phonon interactions. This is due to the fact the Landau interaction function $\phi(\underset{\sim}{k}\sigma,\underset{\sim}{k}'\sigma')$ in this case is diagonal in spin-indices. We obviously need, beside the condition (4), an interaction function with off-diagonal spin components. This means that the Bosons that are exchanged by the conduction electrons must carry spin.

Just as in the case of electron-phonon mass enhancement [16], most transport experiments: Ultrasonic attenuation, thermal conductivity, resistivity, spin-lattice relaxation, de Haas van Alphen effect, etc. are unaffected by the renormalizations, while some like cyclotron resonance frequency are affected. Photoemission measures the spectral function of the single-hole Green's function. This means that the narrow feature near the Fermi-level in such experiments will have a spectral weight proportional to Z. I present in Table I, the renormalization of various properties for

Table I

Property	Usual Landau Theory	Heavy Fermi-Liquids
$dn/d\mu$	$1 + F_0^S$	Unrenormalized
m^*/m	$1 + F_1^S$	$1 + F_0^S$
χ	$(1 + F_1^S)/(1 + F_0^a)$	$(1 + F_0^S)/(1 + F_0^a)$
Resistivity, thermal conductivity, ultrasonic Attn, Spin–lattice rd rate de Haas van Alphen	Renormalized	unrenormalized
Cyclotron resonance	$1 + F_1^S$	$1 + F_0^S$.

heavy Fermions and contrast it with the usual Landau theory. $F_\ell^{s,a}$ is the spin ^3He symmetric or antisymmetric part of the ℓ-th partial wave of the Landau scattering function.

Recent inelastic neutron scattering [17] from UPt_3 is consistent with the form of the self-energy deduced here. $\chi(q,\omega)$, within the experimental uncertainty, is observed not to depend on q and as a function of ω at low ω is that of a Fermi-liquid characterized by $(m^*/m)\ 1+F_0^a)^{-1}$. $\chi(q,\omega)$ satisfies sum rules implying that the total magnetic response, which at high temperatures is that of free spins, appears at low temperatures in the form of a Fermi-liquid.

I will now give a simple argument for the conditions in which a very heavy Fermion mass of $0(10^2-10^3)$ is mandated. Consider the contributions to the entropy of an ordinary rare earth metal or compound. There is a small linear contribution γT from the Fermions corresponding to a mass of 0(1), and a contribution from the local moments which at room temperature is nearly temperature independent and two to three orders of magnitude larger than the γT contribution. The spin contribution to the entropy decreases as temperature is decreased, and, as illustrated in Fig. 1, has a change in slope at the magnetic ordering temperature T_N which is typically of 0(10K). Below T_N, one gets the spin-wave contribution to the entropy. The heavy Fermions are described by the same Hamiltonian as the typical rare-earth solid albeit with somewhat different parameters. Suppose we demand that there be no magnetic phase transition and that the low energy excitations in these solids be purely Fermionic, i.e. there is a gap of $\sim T_F$ which is the same order of magnitude as T_N for spin fluctuation excitations, than a heavy Fermion mass of $0(10^2-10^3)$ is necessary, since total entropy in the present problem is the same as for the ordinary rare-earths and the high temperature entropies must be identical.

The discussion above may be summarized by the statement that heavy Fermions arise from ordinary conduction electrons by exchange of spin-fluctuations with total spectral weight similar to that carried by the local moments and that if these spin fluctuations must have at low temperatures, a negligible spectral weight below $0(T_F)$, the specific heat of the Fermions sill be $0(T/T_F)$. One really observes only the spin-fluctuation entropy but in a Fermionic form, i.e. $\sim T$. The effective induced interaction between the Fermions is highly retarded in time but local in space. This ensures the disappearances of the renormalization in most transport experiments.

The isolated impurity Kondo effect satisfies the conditions listed above. On phenemenological grounds, the heavy Fermions cannot be regarded as a collection of independent Kondo impurities however. The resistivity in the heavy Fermions at low temperatures [2] decreases with decreasing temperature, unlike the Kondo effect and has a high temperature logarithmic form with $T_K >> T_F$. This suggests that there is a characteristic coherence temperature in the problem, below which the local spin-fluctuations phase themselves so as not to scatter electrons backward. This is also implied by a recent microscopic theory which we will soon discuss.

It is amusing to note the phenemenological connection of the Heavy Fermi problem to that of quark confinement. Asymptotic freedom in our case means existence of free spins at high energies. The transition from this regime without any magnetic symmetry breaking to a regime where only Fermionic excitation exist, corresponds to the low energy regime of a color-singlet state of the quarks with only the baryons free. Magnetic fluctuations or bosons appear massive.

3. Perturbation Theory for Two Kondo Impurities

I will now describe some perturbative calculations and conclusions by E. ABRAHAMS and myself [18] on the two impurity Kondo problem and the light they shed on the lattice problem.

Perturbation methods can be used to answer some interesting questions:

(1) Is the Kondo singularity due to a given impurity affected by the scattering of conduction electrons from another impurity? In Fig. (2a) such a process is depicted. When the whole series including all the self-energy corrections of the intermediate electron lines is summed, the answer is the same as Fig. (2b), i.e. there is not effect of the other impurities. This result was obtained by ABRIKOSOV [19].

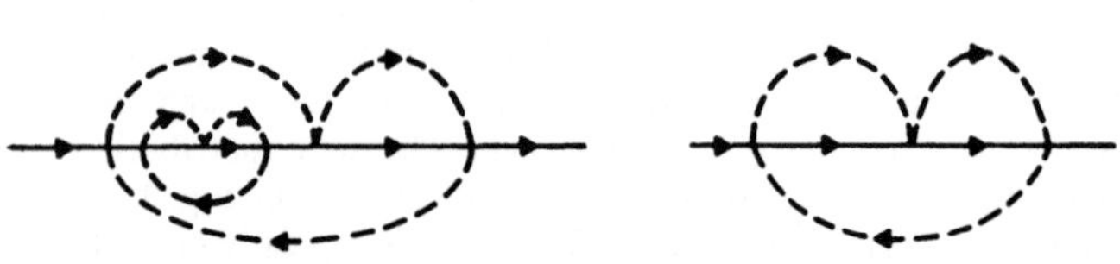

Fig. 2 Lowest-order graph for the effect on the Kondo singularity at one impurity due to the same at another impurity.

(2) Is the RKKY interaction affected by Kondo scattering? One might imagine that as a singlet state begins to form at a magnetic impurity, its interaction with other impurities is reduced. Figure (3a) can be calculated to investigate this possibility. There is a logarithmic divergence if the intermediate particle-hole line is cut, but it is eliminated by the extra integration is Fig. (3) so that it is nondivergent. The same result is obtained for higher order processes as well. The conclusion is that the RKKY interaction is unaffected by Kondo scattering.

One might wonder if the mass renormalization of the conduction electrons affects the RKKY interactions. Figure (3b) investigates this possibility. Not only is it non-divergent, but if vertex corrections like in Fig. (3a) are included, the answer is the same as with neglect of Kondo scattering. This is one more instance of the cancellation of vertex and self-energy discussed earlier.

Fig. 3 Low-order graphs for the effect on RKKY interaction of the Kondo singularity.

(3) Is the Kondo effect affected by RKKY interaction of a magnetic impurity with others? Some processes to investigate this question are depicted in Fig. (4). The answer in every case is that there is no singular correction to the Kondo effect.

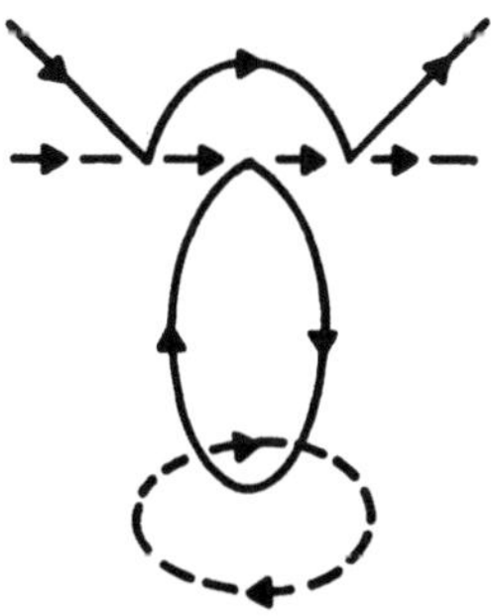

Fig. 4 Low-order graph for the effect on the Kondo singularity at an impurity due to RKKY interaction with another.

So far all the questions asked relate to the mutual interference of the Kondo effect and the interaction among the magnetic moments. The next question to ask is about the interaction itself; is it modified by higher order processes. To third order in J, the answer is negative; to this order only interference processes such as Fig. (3) arise. To fourth order in J, various interference diagrams, Figs. (5), arise; These are nonsingular. Singular connections to the RKKY interactions are obtained if we consider just the iteration of the lowest order process, Fig. (6a), to get Figs. (6b) and (6c). These two graphs give identical contributions because of the anticommutative nature of the spin operators. As in the Kondo effect, these singular contributions cancel exactly for potential scattering.

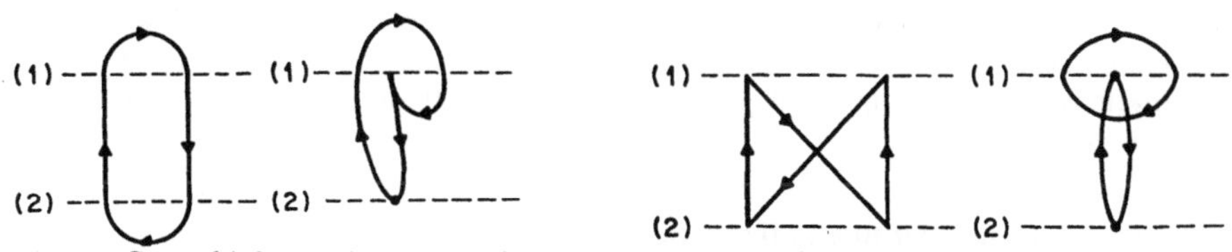

Fig. 5 Some higher-order graphs for coupling between impurities.

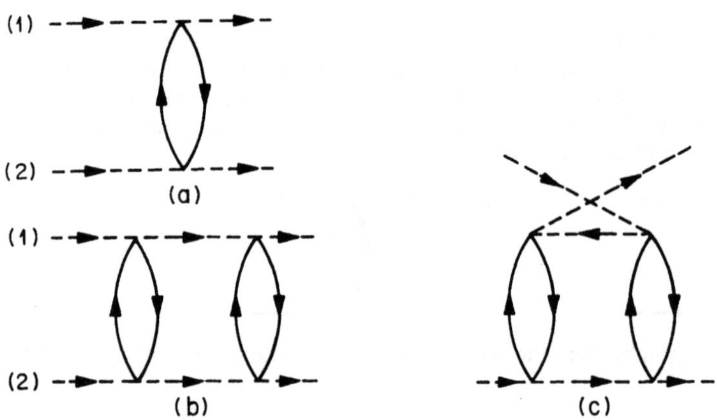

Fig. 6 (a): lowest - order RKKY coupling, (b) and (c) graphs giving singular contributions to the coupling of two impurities.

The similarity of Figs. (6) to the diagrams giving the Kondo singularity suggests that we write the singular contribution to the effective RKKY interaction from Fig. 6 in a form similar to the Kondo problem. Thus, if we denote the frequency RKKY coupling by $K(\omega,R)$, we obtain, for large $k_F R$, the result

$$K(\omega,R) = K_0 \cos 2k_F R \left[1 + \frac{2}{\pi\tilde{D}} K_0 \sin^2 k_F R \, \ell n \, \frac{D}{\omega} \right] \tag{7}$$

where K_0 is the lowest order RKKY coupling, i.e. Fig. (6a), $\tilde{D}$ is an effective bandwidth for RKKY. Thus $\tilde{D} = v_F/R$ and $\tilde{\rho}$ is the corresponding density of states, $\tilde{\rho} = \rho D/\tilde{D}$. The two terms of Eq. (7) have the same sign so that irrespective of the sign of $\cos(2k_F R)$, the perturbative results we have obtained show a tendency toward strong coupling just as in the Kondo problem but now with a characteristic energy scale $T_i(R)$, where

$$T_i(R) \sim \tilde{D} e^{-\pi \tilde{D}/K_0} . \tag{8}$$

Not only are the fourth-order RKKY corrections similar to the second order Kondo ones, but for every singular diagram of the Kondo problem, one can draw an RKKY diagram for K which has a similar infrared divergence. We have not succeeded in finding any singular diagram which mixes the RKKY coupling K with the Kondo coupling J. Therefore, we suggest that K and J scale in a similar manner and from the known behavior of J, we guess the properties of K and the physics of the two-moment problem.

The conjecture is then that a $T \rightarrow 0$, we have, as usual, $J \rightarrow \infty$ and now also $K \rightarrow \infty$, with $J/K \rightarrow$ const. We also conjecture that below T_{int}, J/K is a slow function of temperature (or frequency). These proposals can be tested by numerical calculations for the two-moment problem analogous to those of WILSON [20] for the Kondo problem.

The effective fixed-point Hamiltonian (for $J/t \rightarrow \infty, K/t \rightarrow \infty$) for the two-moment problem can then be written in the form

$$\frac{1}{2} J[(\vec{S}_1+\vec{S}_2) \cdot (c_1^+\vec{\sigma}c_1 + c_2^+\vec{\sigma}c_2) + (\vec{S}_1-\vec{S}_2) \cdot (c_1^+\vec{\sigma}c_1 - c_2^+\vec{\sigma}c_2)]$$
$$+ K(R)\vec{S}_1 \cdot \vec{S}_2 \tag{9}$$

If K/J at the fixed point were zero, then the two moments would each have an independent Kondo effect and $\vec{S}_1$ and $\vec{S}_2$ would have an arbitrary phase relation between them. However, if $K/J \rightarrow$ const. as we suppose, then for $K(R) \gtrless 0$, there is a specific phase relationship between the moments $\vec{S}_1$ and $\vec{S}_2$ with a net singlet state formed with the appropriate combination of conduction electron orbitals for $J \geq |K(R)|$. In this picture, at low temperature there is a complete phase coherence, driven by K, between the local moments. Conduction electrons scatter, just as in Nozières' Fermi liquid picture [21], from the virtual excitations out of the singlet ground state to the magnetic excited states.

As the temperature is lowered through T_K, which we assume to be greater than $T_i(R)$, the individual moments begin to form Kondo singlets with the appropriate local combination of electron orbitals. The RKKY correlations enter at T_i where the Kondo effects at the two impurities begin to acquire the phase coherence we have described. As $T \rightarrow 0$, the system becomes locked into the coherent singlet state. The eigenstates of the problem are those of the Hamiltonian of Eq. (9) with $J \rightarrow \infty$ and $|K| \rightarrow \infty$.

4. Extrapolation to the Lattice

We are tempted to extrapolate these ideas qualitatively to the case of a concentrated moment system — when there is a moment at every site as in Eq. (1). A pairwise approximation to the fixed-point Hamiltonian at T = 0 would be

$$J\Sigma_i \vec{S}_i \cdot c_i^+\vec{\sigma}c_i + \frac{1}{2} \Sigma_{ij} K(R_{ij}) \vec{S}_i \cdot \vec{S}_j \tag{9}$$

where $J \rightarrow \infty$ and $K(R) \rightarrow \infty$ for $R < R_c$. Interactions involving more than two moments are surely also singular but probably will not affect the qualitative conclusions from Eq. (9). The K(R) in general impose a phase coherence between the $\vec{S}_i$ at sites throughout the lattice. For $J >> |K(R)|$, a net singlet for the whole lattice again results just as for the two impurity case, and the excited states (spin fluctuations) are separated by a gap from the ground-state. The rest of the conduction electrons virtually scatter off these excitations. This point of view is quite different from the usual Kondo lattice picture in which each moment is supposed to become an independent Kondo singlet, and RKKY is neglected completely.

Finally, we remark that the resistivity behavior found in the heavy Fermion metals can be roughly accounted for by our picture. As the temperature is lowered through T_k, the resistivity increases as in the single-moment Kondo effect. Somewhat below T_k, the resistivity then begins to decrease again. We identify the crossover temperature between the two behaviors with $T_i(R)$ below which the spin fluctuations at the different sites which are responsible for the extra Kondo resistance begin to acquire coherence. The results mentioned here are meaningful for $T_K \lesssim K_o$; otherwise magnetic order renders moot the quantum-mechanical renormalization effects considered above. For $T_K \cong K_o$ and $D \sim 2eV$, one gets Ti (nearest neighbor) about an order of magnitude below T_K. This is about the right ratio for the resistivity results, but then T_K itself comes out too high. To make useful comparison with the experimental results, the calculations discussed above have to be extended to take orbital degeneracy into account. This is being done.

5. Acknowledgements

I have benefitted enormously through discussions with E. Abrahams, G. Aeppli, B. Batlogg, D. Bishop, Z. Fisk, H. Ott and F. Steglich. Most of the work presented here is a collaboration with some of these, as noted in the text.

1. C. M. Varma, Comments in Solid State Physics (to be published).
2. C. M. Varma and A. Zawadowski, Phys. Rev. (submitted).
3. F. Steglich, et al., Phys. Rev. Lett. *43,* 1892 (1979); H. R. Ott, et al., Phys. Rev. Lett. *50,* 1595 (1983); G. R. Stewart et al., Phys. Rev. Lett. *52,* 679 (1984).

3a. For a review, See G. R. Stewart, Rev. Mod. Phys. *56,* 755 (1984).

4. C. M. Varma, Bull. Am. Phys. Sol. ???.
5. D. Bishop, C. M. Varma, B. Batlogg, E. Bucher, Z. Fisk and J. L. S.
6. D. Jaccard, et al., J. App. Phys. *57,* 3082 (1985).
7. D. E. MacLaughlin, et al., Phys. Rev. Lett. *53,* 1833 (1984).
8. Y. Kitaoka, et al. (to be published).
9. F. Steglich et al., J. App. Phys. *57,* 3053 (1985).
10. B. Golding, et al. (private communication, to be published).
11. G. E. Volovik and L. P. Gorkov, Zh. Eksp. Teor. Fiz. *45*, 1493 (1963); P. W. Anderson, Phys. Rev. *B30,* 4000 (1984); E. I. Blount, Phys. Rev. (to be published).
12. C. Stassis (private communication).
13. B. Batlogg, J. App. Phys. *57,* 3060 (1985).
14. S. Schultz (private Comm., to be published).
15. T. M. Rice, et al., Phys. Rev. *31,* 591 (1985); O. Valls and Z. Tesanovic, Phys. Rev. Lett., *53,* 1497 (1984); P. W. Anderson, Phys. Rev. *B30,* 1549 (1985).
16. R. E. Prange and L. P. Kadanoff, Phys. Rev. *A134,* 566 (1964).
17. G. Aeppli, E. Bucher and G. Shirane, Phys. Rev. Lett. (to be published).
18. E. Abrahams and C. M. Varma (unpublished).
19. A. Abrikosov, Physics *2*, 5 (1965).
20. K. Wilson, Rev. Mod. Phys. *47,* 773 (1975).
21. P. Nozieres, J. Low Temp. Phys. *17,* 31 (1974).

Index of Contributors

MIX
Papier aus verantwortungsvollen Quellen
Paper from responsible sources
FSC® C105338

If you have any concerns about our products,
you can contact us on
ProductSafety@springernature.com

In case Publisher is established outside the EU,
the EU authorized representative is:
Springer Nature Customer Service Center GmbH
Europaplatz 3, 69115 Heidelberg, Germany

Printed by Libri Plureos GmbH
in Hamburg, Germany